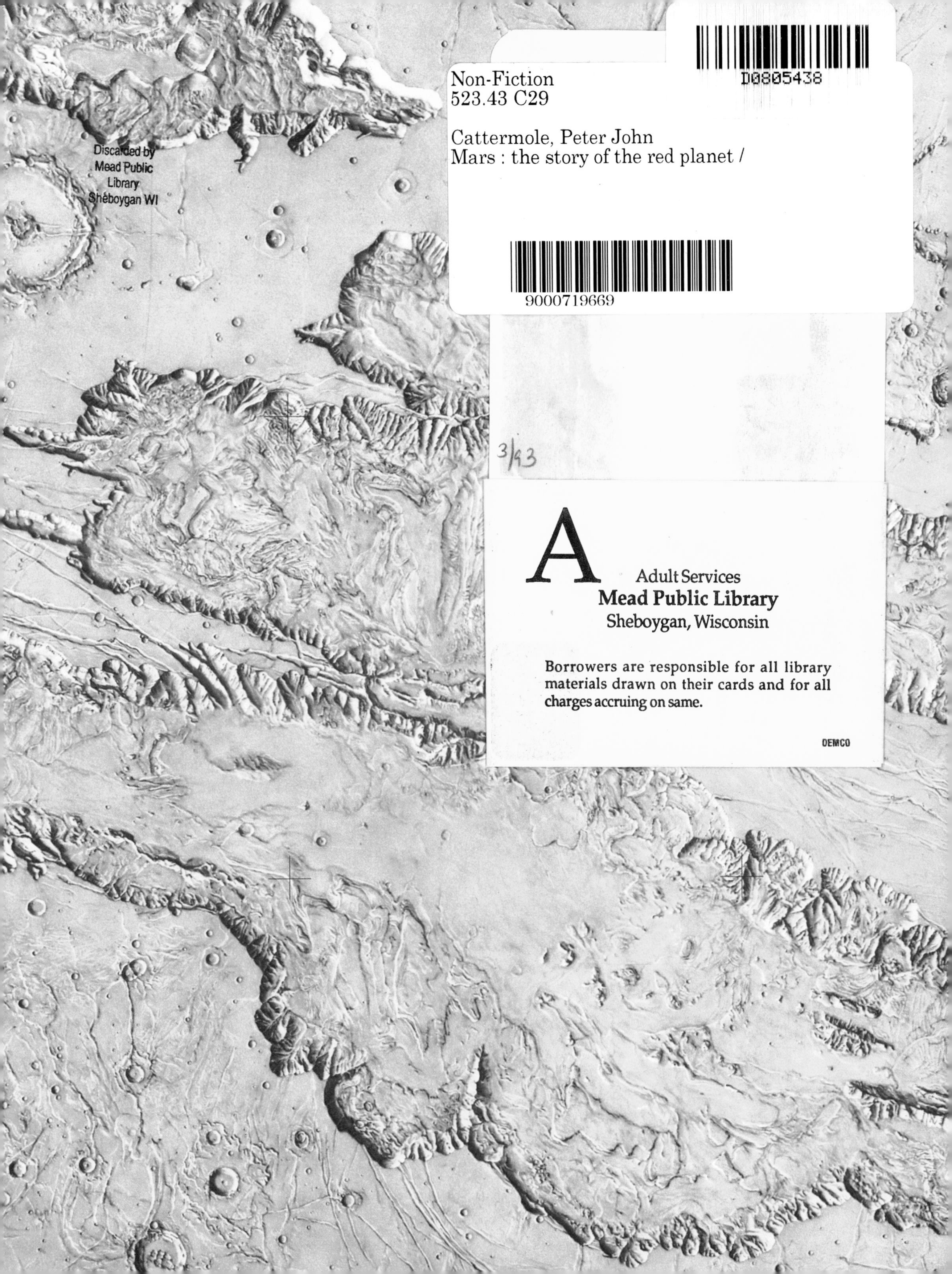

MARS

MARS

The story of the Red Planet

PETER CATTERMOLE PhD, FRAS
Scientific writer, lecturer and consultant.
NASA principal investigator for the
Planetary Geology and Geophysics Program

With a Foreword by Patrick Moore

CHAPMAN & HALL
London · Glasgow · New York · Tokyo · Melbourne · Madras

Published by Chapman & Hall, 2–6 Boundary Row, London SE1 8HN

Chapman & Hall, 2–6 Boundary Row, London SE1 8HN, UK

Blackie Academic & Professional, Wester Cleddens Road, Bishopbriggs, Glasgow G64 2NZ, UK

Chapman & Hall, 29 West 35th Street, New York NY10001, USA

Chapman & Hall Japan, Thomson Publishing Japan, Hirakawacho Nemoto Building, 6F, 1-7-11 Hirakawa-cho, Chiyoda-ku, Tokyo 102, Japan

Chapman & Hall Australia, Thomas Nelson Australia, 102 Dodds Street, South Melbourne, Victoria 3205, Australia

Chapman & Hall India, R. Seshadri, 32 Second Main Road, CIT East, Madras 600 035, India

First edition 1992

Typeset in 10/12pt Palatino by Best-set Typesetter Ltd., Hong Kong
Printed in Singapore

ISBN 0 412 44140 3

A catalogue record for this book is available from the British Library

Library of Congress Cataloging-in-Publication data available

To Ron.
Thanks for all your help.

CONTENTS

Colour plates appear between pages 112 and 113

ACKNOWLEDGEMENTS

The basis for much of this book is the superb spacecraft imagery which has been returned to Earth by NASA spacecraft and the various excellent maps produced by the United States Geological Survey (USGS). During my lengthy involvement with Mars research I have had unfailing support from the various Directors of NASA's National Space Science Data Centre at Greenbelt Maryland, and from Larry Soderblom, Jody Swann and Alfred McEwen at the USGS Branch of Astrogeology at Flagstaff; I tender them my sincere thanks. I also make formal acknowledgement of NASA for permission to use imagery they have kindly provided.

Naturally, no one person can have a detailed grasp of all Martian matters and, needless to say, I have had to lean heavily on the work and opinions of colleagues from many institutions. I would like to take this opportunity to thank them all here, and encourage them to continue the quest for answers to the many remaining questions which their vaulable research has prompted. I would like to make specific mention of Ron Greeley of Arizona State University and those other of his colleagues and staff who during my various visits have made me welcome and given me help. I would also like to thank John Guest of University College, London, for his unfailing help and guidance and also to his colleagues at Observatory Annexe, Mill Hill. Others I would like to mention by name are Jo Boyce, Hank Moore, Peter Mouginis-Mark, Ken Tanaka and Lionel Wilson.

Finally, I would like to thank Paul Doherty for his excellent illustrations.

PREFACE

As I write this short preface, the red orb of Mars is high in the eastern sky, and is brighter than it has been for many years. Last night my telescope again revealed the strange polar hood which is a feature of the planet at this time in its cycle. Because of its current prominence in the night sky, it is a very appropriate time to bring together and reappraise what we know of Mars and look forward to the next wave of planetary exploration.

The initial notion of writing a book about Mars is an exciting one; the practicalities involved in working through and completing the project are, however, more than a trifle exacting. The first problem I encountered was the sheer vastness of the library of information about Mars which now exists. The second was the natural extension of the first, that is, how best to analyse it and reach widely acceptable interpretations. I have tried to write the story of Mars in a logical and unbiased way, however, we all have our individual prejudices, and I would be less than truthful if I did not admit to personal bias here and there. With this in mind, I apologise to any authors who may feel either misinterpreted or less than adequately acknowledged.

The project is now completed and has been superbly prepared by Chapman & Hall. I hope the results of my labours will provide anyone interested in planetary science with a stimulating survey of the Red Planet, and provoke them into fresh thinking about this fascinating world.

Peter Cattermole

FOREWORD

Mars, the Red Planet, is of special interest to us. Indeed, it may justly lay claim to being the most fascinating world in the entire Solar System. Though we no longer believe in the brilliant-brained, canal-building Martians, it is true that our neighbouring world is much more like the Earth than any other planet, and in so far as manned missions are concerned it must surely be our next target.

We have learned a great deal about it during the past few years and there have been many books about it, but most of these tend to be either purely observational or else too technical for any but the qualified reader to appreciate. Peter Cattermole's book is different. He is an eminent geologist – a specialist in volcanology – who also happens to be an astronomer, and he has undertaken important researches with NASA in connection with Mars, particularly with respect to the giant volcanoes which rise from the surface. He is, therefore, exceptionally well qualified to write a book in which astronomy and geology come together; this is what he has achieved. He gives a comprehensive account of Mars in all its varied aspects with emphasis upon the nature and evolution of the surface; moreover he has done so in a way which will be calculated to hold the interest of the general reader as well as the specialist. In reading his book, a basic knowledge of geology is certainly a help but it is not essential. I have been observing the planet for more than half a century, often with some of the world's most powerful planetary telescopes, but even so I found that I learned a great deal from what Peter Cattermole has written here.

I do not believe that there is any other book which covers so much ground in quite this way. If there is, I can only say that I have not come across it.

I commend this book to readers of all kinds and I am confident that it will have the wide circulation which it so richly deserves.

Patrick Moore
Selsey
21 January 1992

INTRODUCTION

Two Soviet spacecraft were launched towards Mars in July 1988. This initiated the Phobos mission, in which scientists from fourteen countries and the European Space Agency (ESA) were involved. It heralded a new and very welcome era in international scientific collaboration. Although contact was lost with Phobos 1 soon after it was launched, in March 1989 Phobos 2 obtained a variety of data about the solar wind and magnetosphere in the vicinity of Mars, about Mars itself and about its larger moon Phobos which had been the principal objective of its mission. These included nine TV pictures of Phobos and the first new spacecraft images of the mother planet obtained since the US spacecraft Viking 2 ceased transmitting data during August 1980, after sending information back to Earth almost continually for five years. The new batch of imagery was obtained in multispectral mode – a new departure – and the returned thermal infrared pictures gave scientists their first real chance of studying the heat radiation of Mars, as opposed to the reflected solar radiation.

Currently, an international team of scientists and technologists is well into the development of two further Soviet probes (Mars 94, Mars 96) whose goal is also the Red Planet. These are planned for launch during 1994 and 1996 and, if successful, will set down the first mobile landing vehicles (rovers) and put balloon-carried payloads above the Martian surface. Furthermore, the US Mars Observer mission (a geoscience/climatology orbiter) is also at an advanced stage of planning and is due for launch in 1992. It is clear, therefore, that we have a new wave of planetary exploration on our hands and that Mars is one of its prime targets.

Why Mars? Well, there are several reasons. First, along with Earth's Moon, it must rank as the most suitable potential site for a manned base – something which has received considerable press attention during the past couple of years – and this makes it a focus of great interest for those planning future missions. Only as recently as 1987, in her report to NASA entitled Leadership and America's Future in Space, did US astronaut Sally Ride propose the establishment of a staffed base on Mars by the year 2010. Then again, the Russian scientist, Roald Sagdeyev, has pressed for a joint US-Russian mission to be launched by the year 2007.

Second, the quarter of a century or so that has been spent studying the data returned from various planetary probes, beginning with Mariner 4 in 1965, reveals that because of its particular position within the Sun's family, Mars has developed in a way that is transitional between the less highly oxidized, more refractory worlds of the inner solar system and the more highly oxidized, volatile-rich worlds of its outer regions. It therefore occupies a key position in our understanding of how planets evolved.

Third, its surface features are recognized as having developed in response to familiar geological processes which have been controlled very differently from those of the Earth. In consequence, it provides a natural laboratory for studying both exogenic and endogenic phenomena which have been constrained by an entirely different set of boundary conditions. The study of Mars therefore should bring new perspectives to our studies of comparative planetology and enhance our knowledge of planet Earth.

I have endeavoured to set out what we have learned about Mars since Galileo first turned a telescope in its direction, nearly four hundred years ago; however, because there are numerous excellent books which have very adequately covered the history of telescopic observation, the emphasis is on more recent discoveries, particularly those made since the successful US probe, Mariner 9, first revealed the immense canyons and huge volcanic shields of our sister world. We should, however, never forget what we owe to our predecessors, not only those who achieved a degree of notoriety, but also those less well known and often amateur observers whose tireless and painstaking telescopic observations built the springboard from which modern exploration took off. For those who wish to delve into the history of telescopic observations, the Bibliography at the end of the book lists some of the more important books that were published prior to the space age.

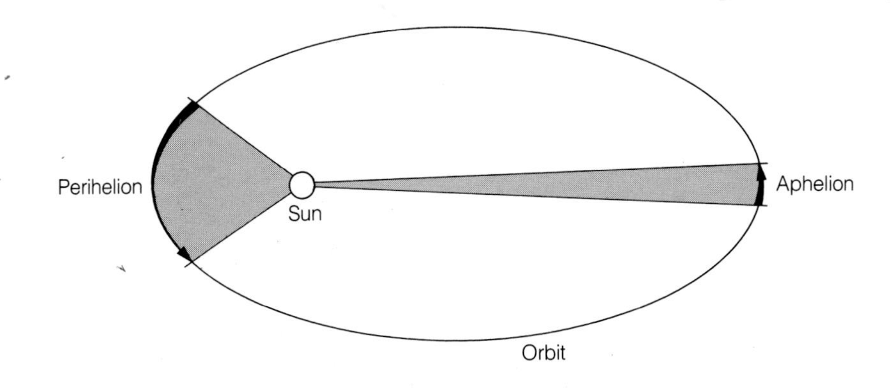

Fig. 1.1 Kepler's second law. The radius vector defines an equal area whether Mars is near perihelion or aphelion; however Mars moves quicker at perihelion than at aphelion

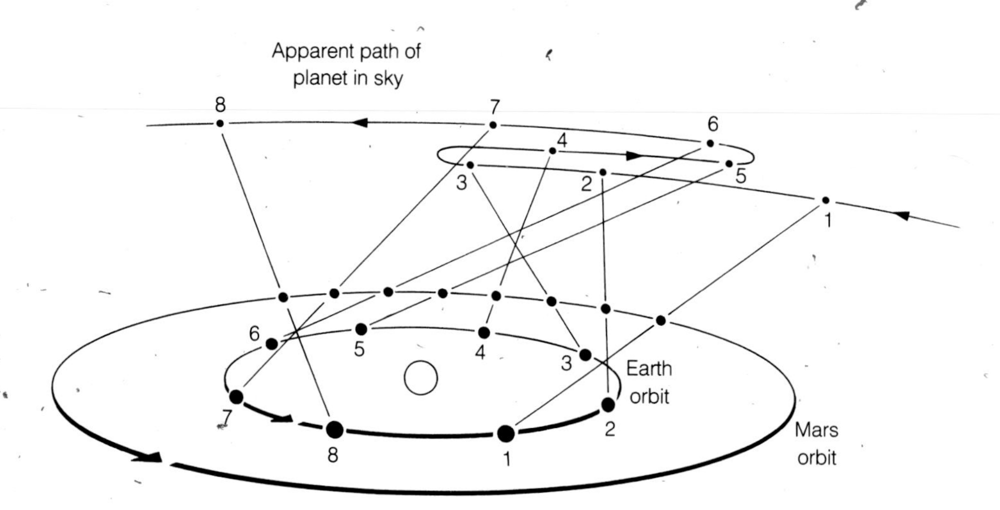

Fig. 1.2 The apparent motion of Mars in the sky. When the two planets are close to opposition, Mars appears to move in a retrograde manner.

MARS IN THE SOLAR SYSTEM

1

1.1 THE ORBIT AND MOVEMENTS OF MARS

Mars has occupied a very special place in the history of scientific thinking since the late sixteenth century. It was then that the imperious Danish astronomer, Tycho Brahe, made careful observations of the movements of Mars which subsequently enabled the brilliant German mathematician, Johannes Kepler, to formulate his three fundamental laws of planetary motion. The first two of these were published in 1609 after the latter realised, having calculated the distance of Mars from the Sun at different points in its circumsolar journey, that its orbit was not, as hitherto had been supposed, a circle but an ellipse, one focus of which was occupied by the Sun. He appreciated, too, that this was true of the other planets, a conclusion which led him to formulate the first of his three laws which stated simply that planets move around the Sun in elliptical orbits, with the Sun at one focus. As it happens, the Earth has an orbit which departs little from a circle (eccentricity 0.017) but Mars follows an orbit which is much more eccentric. Thus its distance from the Sun is 207 million km when at its closest (perihelion), while at its furthest point (aphelion) it is 249 million km distant. This gives the eccentricity a value of 0.093.

Kepler's second law also derived from studies of Mars. He had observed that the Red Planet travels more quickly near perihelion than aphelion and his second law states that a straight line joining the Sun and each planet – known as the radius vector – sweeps out an equal area in equal time (Fig. 1.1). Thus it is that when a planet is near perihelion it travels more quickly than when near aphelion, and that those planets nearer to the Sun travel more quickly than those at greater distances. Mars, for instance, travels at a mean rate of about 24 km s^{-1}, compared with 30 km s^{-1} for the Earth and 13 km s^{-1} for Jupiter.

The third law, published nine years later, states that the square of a planet's orbital period is proportional to the cube of its mean distance from the Sun. Taken together, laws two and three explain why Mars (and other planets beyond the orbit of Earth) shows occasional retrograde motion against the star background. Figure 1.2 shows the apparent motion of Mars in the sky, together with the relative positions of Mars and the Earth in their respective orbits. It will be seen that between points 3 and 6 the Earth is chasing Mars and eventually overtakes it, during which period Mars appears to move backwards amongst the constellations.

Every 780 days Mars, the Earth and the Sun become aligned; Mars is then said to be at opposition (Fig. 1.3), and the planet is then well-placed for observation from the Earth. Because the orbits of Mars and the Earth are

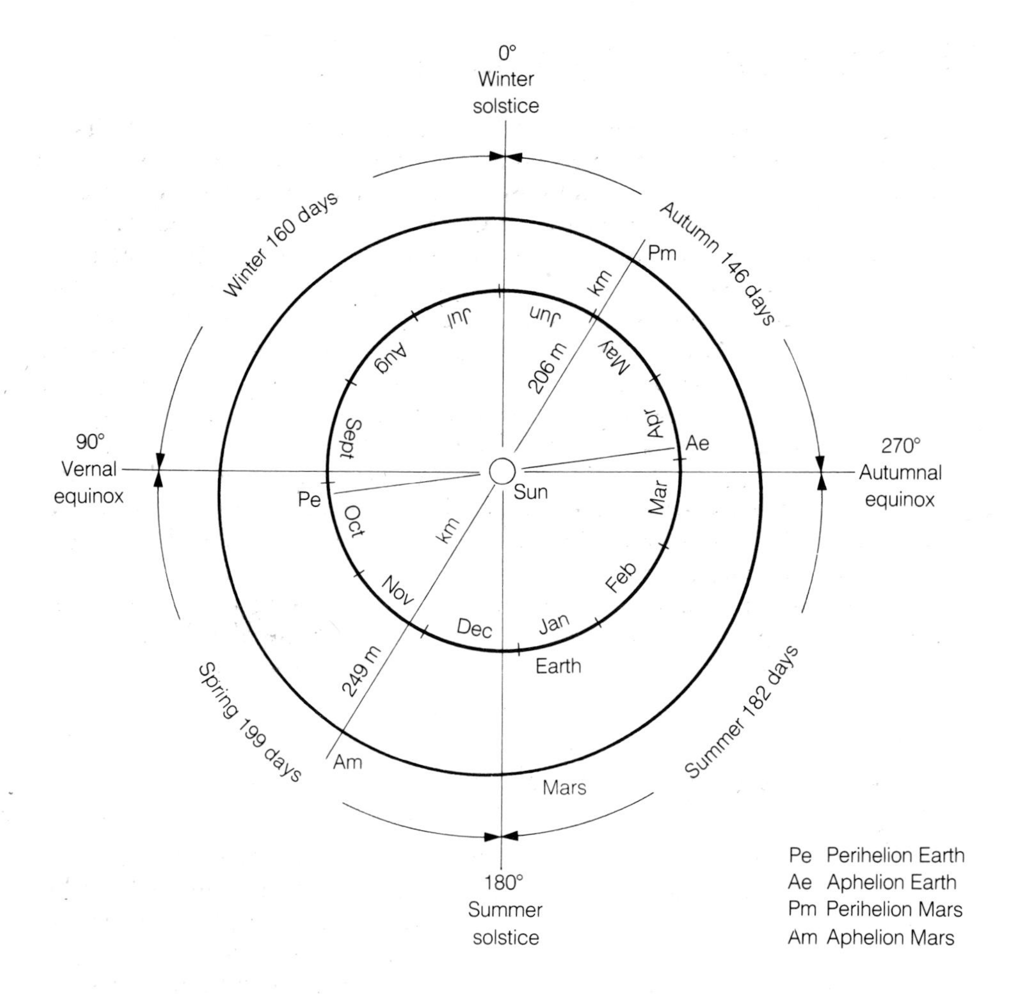

Fig. 1.3 *The movements of Mars and the Earth around the Sun.*

elliptical, the distance between the two bodies is not the same at each opposition and in consequence some oppositions are more favourable to Earth-based observers than others. Thus, if Mars is near perihelion, the distance may be as little as 55.7×10^6 km; if it is at aphelion, the distance increases to 101.3×10^6 km. The interval between successive oppositions is known as the planet's synodic period. This is longer than the Martian year, or sidereal period, which is 687 days.

Because Mars is significantly further from the Sun than Earth it receives proportionately less benefit from insolation and, as might be anticipated, its mean surface temperature is much lower. Because of precessional effects, the northern and southern hemispheres have different temperature regimes. The lowest temperatures currently are experienced by the south pole during its winter, where Viking IRTM (infra-red thermal mapper) data showed that temperatures may plummet as low as −133 °C (the frost point of carbon dioxide). During the same mission, a temperature of −63 °C was recorded at the north pole during its summer, while the highest temperatures occur in southern mid-latitudes during summer, where mid-day temperatures of +23 °C were recorded. During perihelic dust-storm activity, temperature values and regimes may be modified significantly.

1.2 DIMENSIONS AND MASS

The equatorial diameter of Mars is 6788 km, which is approximately half that of the Earth. Its small size, coupled with the fact that even at its closest it can never approach nearer than 140 times the distance of the Moon, conspires to render it a difficult telescopic object; it subtends an angle roughly equivalent to a largish lunar impact crater. It is this fact which renders it such a tantalizing telescopic subject; its disk is just large enough for details to be seen, but never very clearly. Most of the more revealing series of Earth-based observations of the planet have been made at August-September oppositions, which is when the Earth passes the perihelion of Mars' orbit; unfortunately, these are separated from one another by roughly 15 years. Even then there are difficulties for northern observers, for Mars has a far southerly declination at such times!

The mass of Mars is 6.44×10^{20} kg, roughly one tenth that of the Earth. It is also considerably less dense, the average density being $3.906 \times 10^3 \text{ kg m}^{-3}$, or about 70% that of the Earth. This implies that any dense metallic core that exists must be substantially smaller than Earth's. The consequently lower escape velocity of 5.1 km s^{-1} means that it has only managed to hold on to a very tenuous atmosphere which exerts a surface pressure of only 8.1 mbar, a mere one-hundredth that of the Earth's. This is composed predominantly of carbon dioxide.

1.3 THE MARTIAN SEASONS

The length of the Martian day is a trifle longer than the terrestrial one (24 h 37 min 22 s), while the rotational axis is inclined at 25° to the plane of the orbit, just a little more than Earth's (23.5°). Consequently, Mars experiences seasons much like our own, although longer; however, because the eccentricity of Mars' orbit is considerably greater than Earth's, there are significant differences in the lengths of the respective seasons. As Kepler's second law predicts, the time taken for Mars to traverse any section of its orbit varies, the time spent close to perihelion being much shorter than that near aphelion. Because the southern hemisphere is tilted towards the Sun near perihelion, spring and summer seasons in that hemisphere are 52 and 25 terrestrial days shorter, respectively, than autumn and winter (Table 1.1). The virtual circularity of Earth's orbit means that the eccentricity effect gives only a three-day difference.

Table 1.1 The length of Martian and terrestrial seasons compared

Season		Duration of seasons		
		Mars		*Earth*
Northern hemisphere	*Southern hemisphere*	*Martian days*	*Terrestrial days*	*Terrestrial days*
Spring	Autumn	194	199	92.9
Summer	Winter	178	183	93.6
Autumn	Spring	143	147	89.7
Winter	Summer	154	158	89.1
		669	687	365.3

The eccentricity also has an effect upon the maximum summer temperatures experienced by the different hemispheres; for instance, southern summers are currently shorter and hotter than northern ones, maximum temperatures in the south being about 30 °C higher. This is because Mars is 20% closer to the Sun at perihelion than at aphelion, and receives 45% more incoming solar radiation.

There are also longer-term changes in the seasons which are a manifestation of very slow variations in the various orbital and rotational parameters. These come under the general heading of precession, a term which describes the slow conical motion of a rotational axis, such as that experienced by a spinning top. The rotational axis of Mars precesses over a period of 175 000 years, during which time there is a slow rotation of the line of intersection of the equator with the orbital plane (Fig. 1.4). The axis of Mars' orbit also precesses and this cause a line joining aphelion and perihelion (termed the line of apsides) to rotate. This cycle is completed every 72 000 years and causes a gradual shift in the position of perihelion in space. The combined effect of these two phenomena is to produce a 51 000 year long cycle of climatic change (Leighton and Murray, 1966). As we have seen, at present the southern hemisphere is tilted towards the Sun at perihelion; however, in about 25 000 years the northern hemisphere will be in a similar position, and will experience short, hot summers. The cycle will then repeat itself.

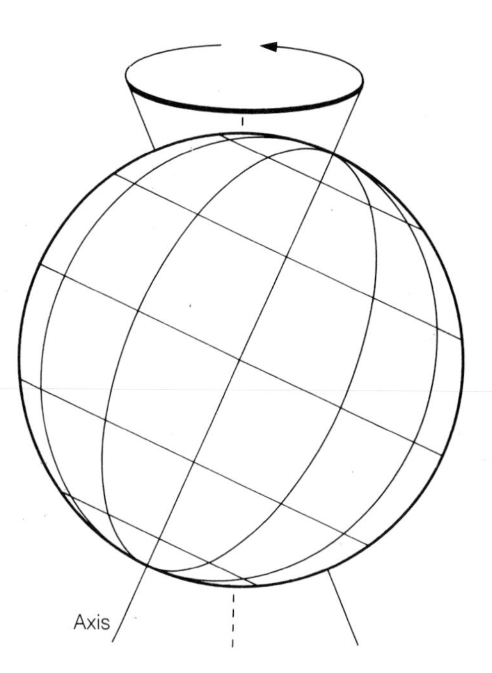

Fig. 1.4 *Precession of the axis of Mars' orbit.*

There are also some longer-term climatic changes which occur because of variations in orbital eccentricity. As we have seen, today the eccentricity has a value is 0.093, but this ranges from 0.004 to 0.141 over a shorter period of 95 000 years and also over a longer period of about 2 million years (Ward, 1974). Obliquity also changes; at present it is 25°, but over a period of 1.2 million years it varies between 14.9° and 35.5° (Ward, 1974). This has an effect on the amount of solar radiation which hits different latitudes; thus when obliquity is high, relatively more radiation hits the polar regions, with the result that polar temperatures rise while equatorial temperatures fall. At low obliquities the reverse applies. As we shall see later, this has an effect on the volume of volatiles which can be stored in the polar caps.

1.4 THE PATTERN OF DISCOVERY

Galileo is held to have been the first person to observe Mars through a telescope, which he did in 1610. He could make little out on its disk and was able to confirm its phases, as he had done previously for Venus. His inadequate telescope, coupled with the small size of the planet's disk (it subtends an angle of only 25 seconds of arc at opposition and a mere 3.5 seconds of arc at superior conjunction) were the reasons for this. A few years later, in 1638, drawings made by the Neapolitan observer Francesco Fontana, do show phase effects but little else. It remained for Christiaan Huygens to produce the first really useful drawing of Mars, which he did on November 28th 1659; this shows a dark triangular-shaped region at the centre of the disk which may well have represented Syrtis Major. It was christened the 'Hourglass Sea' and, as was customary at this point in history, believed to be an area of open water. Huygens also was able to show that the planet rotated on its axis in a period little different from Earth's day.

By 1666, the Italian, Giovanni Cassini – who subsequently became the first director of the Paris Observatory – had discerned the white polar caps; Huygens had done so by 1672. The Englishman, Robert Hooke, also made observations at this time. During the next quarter of a decade, the periodic advance and retreat of the polar caps in response to the changing seasons was noted by several observers. The next major step forward was made in 1777, when William Herschel entered the scene and made an important series of observations which he published in 1781 and 1784. He determined, first, that the planet's obliquity was 25° and rightly deduced that Mars must have seasons like those of the Earth; second, by tracing the movement of features across the disk, he estimated the length of the day to be 24 h 39 min 22 s, only 13 seconds less than the currently accepted figure. His perspicacity also led him to speculate that the icy polar caps must be relatively thin, and also that transient brightenings he observed at various points on the disk, were clouds.

The particularly favourable opposition of 1830 presented the German pair of Wilhelm Beer and Johann von Mädler, using Beer's excellent Fraunhofer 95 cm refractor, with an unprecedented opportunity to draw the outlines of the light and dark regions in a way that, for the first time, strongly resembled modern telescopic charts. In 1837, these two observers noted a dark band surrounding the north polar cap which they suggested might represent a wet region adjacent to the melting ice; a similar observation was made by Webb in 1856, for the southern cap. This 'wave of darkening', as it was called by Gerard de Vaucouleurs, subsequently has been noted by many; it was, for instance, particularly noticeable at the favourable 1956 opposition. That it has something to do with the transference of moisture from the shrinking snowfields to the surrounding region was a popular view held right up until the point where high-resolution imagery was obtained by Mariner 9 and Viking Orbiters. We now know the effect has a more complex origin.

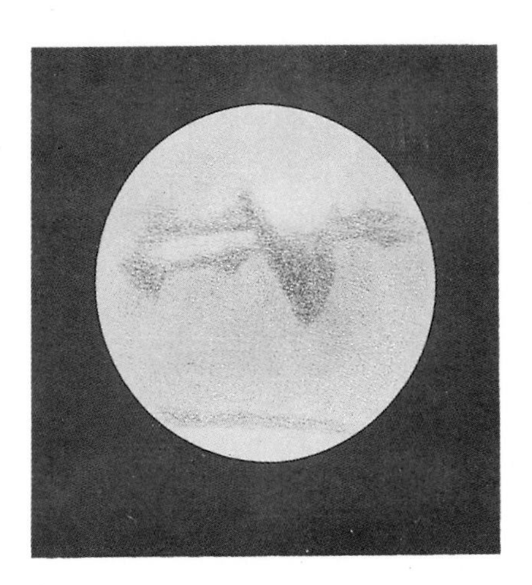

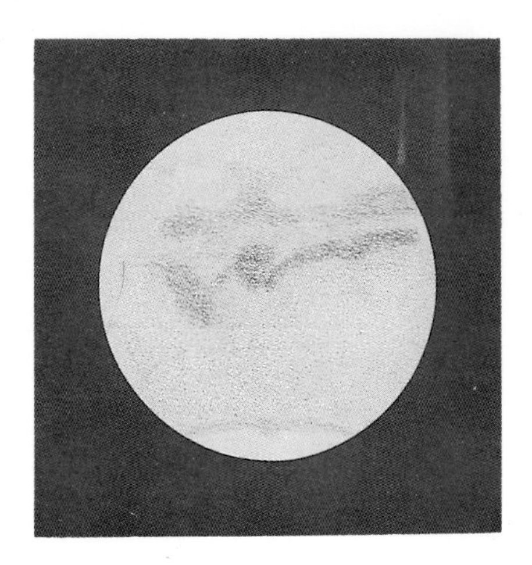

Fig. 1.5 *The telescopic appearance of Mars (left) November 5th 1990; (right) November 6th 1990. Drawings by Patrick Moore.*

Since the mid-nineteenth century Mars has, of all the planets, been the most widely observed by amateur and professional astronomers alike (Fig. 1.5). The particularly excellent opposition of 1864 spawned a fine series of observations by the Englishman, Willam Rutter Dawes. A skilful draftsman, Dawes' drawings were superior not only to those of Beer and Mädler but also to the later ones by Schiaparelli and Lowell. Important maps of the period were published by the English astronomer, Richard Proctor, and the Frenchman, Camille Flammarion; the most important, however, was undoubtedly that of the Milan-based observer, Giovanni Schiaparelli, which was published in 1878. This was influential in that it introduced a new system of nomenclature wherein permanent features were given classical or biblical names, a policy that continued until the era of spacecraft demanded a revised method. It also showed a large number of linear markings – the famous Martian canals.

The term 'canal' appears to have been first introduced during the 1860s by Cardinal Pietro Angelo Secchi, to describe faint linear features he had observed on several occasions. Schiaparelli used the same term in 1877, for markings just visible during spells of particularly stable seeing while using a 21 cm Merz refractor. During the more favourable opposition two years later, he was amazed to observe that several of the canals appeared double, an appearance which became known as gemination (Fig. 1.6). Although Schiaparelli's observations attracted relatively little attention at first, as more reports of the features came in, people began to take more notice. By the end of the century, the canals had become the focus of attention on Mars.

Schiaparelli himself, while not in any way doubting the authenticity of his observations, clearly had conceptual problems; he believed them to be due to some kind of geological process, yet could not conceive what that might be, yet while Percival Lowell later was prepared to see them as the handiwork of intelligent Martians (Lowell, 1906, 1909, 1910), at no point did the Italian concede to this notion, although he wrote 'I am very careful not to combat this suggestion, which contains nothing impossible.' Lowell himself, who in 1894 founded the famous Lowell Observatory in Flagstaff, Arizona, was assuredly the champion of Martian canals; he devoted a large part of his life to their observation and his drawings showed about five hundred of them.

Despite the immense momentum imparted by the work of Schiaparelli and Lowell to the 'canal movement' at and just before the turn of the century,

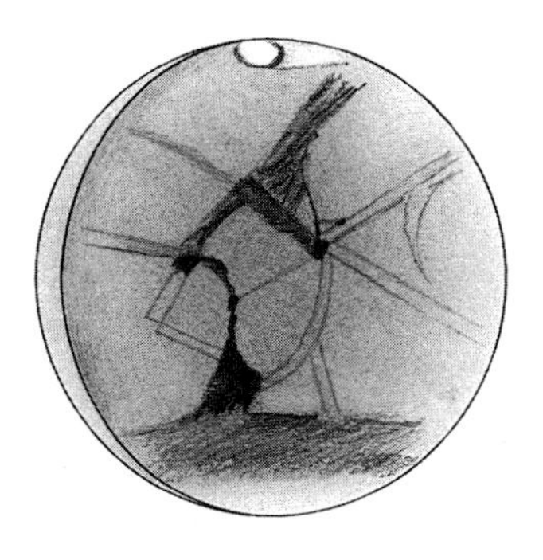
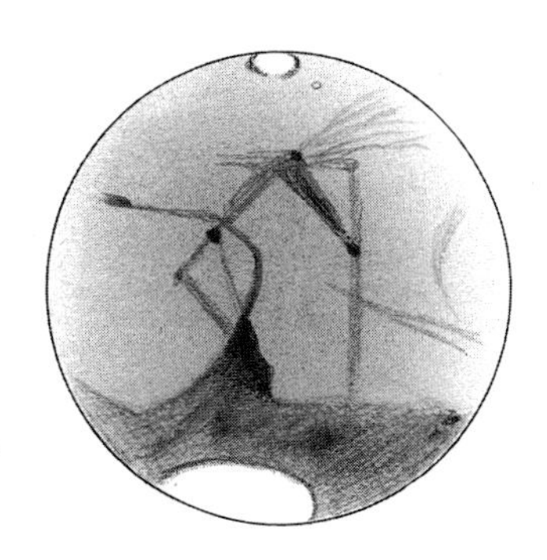

Fig. 1.6 Gemination of a Martian canal according to Schiaparelli.

considerable scepticism was expressed by some astronomers. The Greek astronomer, Eugenios Antoniadi, who, incidentally, spent most of his life in France, was particularly scathing in his criticism. Rightly, he pointed to the fact that, against all laws of perspective, Schiaparelli's canals were too straight to be realistic (Lowell avoided this criticism by projecting his drawings on to globes). Furthermore, because he was using much improved telescopes, Antoniadi's observations, over a twenty-year period, showed not continuous canals, but lines of discontinuous blotches and streaks. He, like Evans and Maunder (1903), saw the canals as an artifice, a trick played when either poor seeing or too small an aperture caused discontinuous features to be joined up by the eye . A most stimulating account of the controversy can be found in the excellent recent book by William Sheehan (1988).

Once spacecraft imagery was returned to Earth, the matter of the canals was resolved once and for all. Whereas most of the irregular features drawn by Earth-based observers do correspond with albedo markings recorded on Mariner and Viking images, of the canals there is no sign; they have to be seen as illusory, as Antoniadi and others had, unpopularly, suggested. Intelligent Martians must, rather sadly, be relegated to the annals of fiction.

1.5 ALBEDO MARKINGS

The light and dark albedo markings which have been depicted on telescopic drawings and photographs for a century or more have retained their general outline and position for at least this long. The fine maps of Schiaparelli and Antoniadi show these clearly and, in 1877, the former introduced a new scheme of annotation which appended both biblical and mythical names; many of these names (or parts of them) persist to the present. In an effort to standardize and clarify nomenclature a committee of the International Astronomical Union (IAU) published a list of 128 'permanent' albedo features together with an accompanying chart (Fig. 1.7).

Once detailed orbital mapping of the planet began, in the early 1970s, it became clear that few of the albedo markings showed any correlation with topographic features. There were exceptions; for instance, one light area corresponds to the impact basin Hellas, while a prominent dark marking coincides with the equatorial canyon system, Coprates Chasma; however, generally, there was little correspondence. The modern view sees the darker regions as being those relatively free of wind-blown dust-grade sediment and the lighter

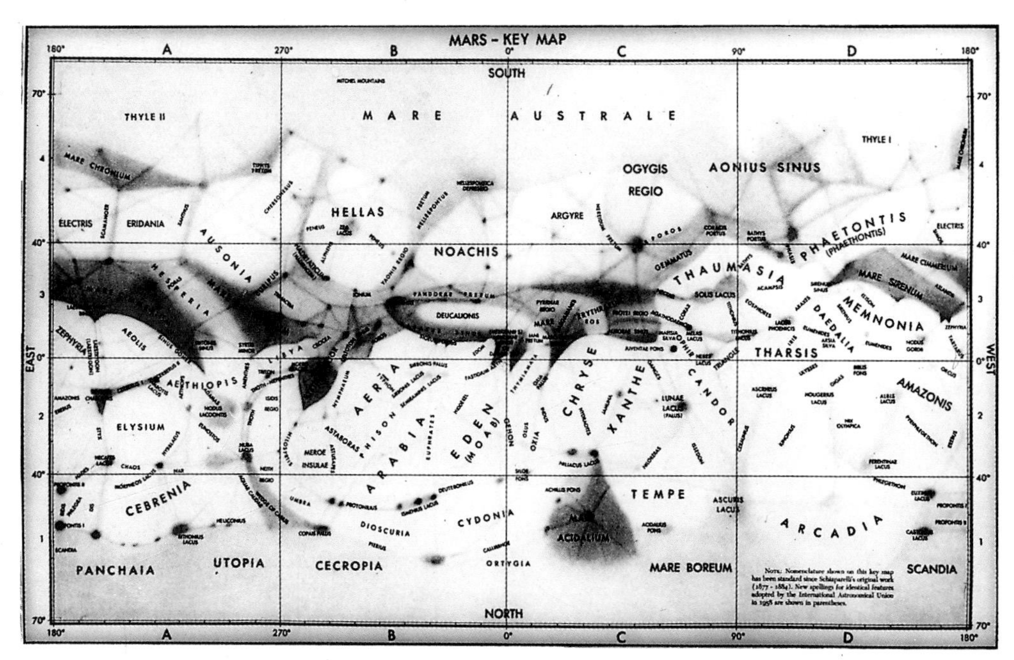

Fig. 1.7 Chart of Mars, Lowell Observatory 1938, showing revised nomenclature of albedo markings.

areas as having a significant mantle of such material. On this basis it is not difficult to see how the ephemeral changes seen in the shape of areas like Solis Lacus and Syrtis Major are related to perihelic dust-storm activity, when all or nearly all of the characteristic albedo markings may temporarily disappear (as Mariner 9 was to discover, to its cost!). That the general shape and appearance of the planet's telescopic markings has remained fairly constant over the last century, appears to imply that the broad wind circulation pattern must have remained constant over the period.

While dust storms may periodically obliterate the familiar markings, other changes are related to the formation of clouds in the polar regions and to the seasonal melting of the ice caps. A 'hood' of polar clouds gradually extends over the north pole during its autumn and may eventually extend to within 40° of the equator; it remains hovering there through autumn and winter, obscuring the growing ice cap below. During mid-spring, however, the hood disappears and the northern cap may be seen shrinking at a rate of up to 20 km per day.

Other clouds are due either to condensates of carbon dioxide or water, or to dust. The former are known as white or blue clouds, and tend to focus on elevated regions such as Tharsis, Alba and Elysium; yellow clouds do not necessarily show any correlation with elevation and are generated by dust storms.

More recently, exploration of Mars has proceeded apace. Since July 1965, several spacecraft have either flown past, orbited or landed on the planet. Our knowledge of its atmospheric structure, weather, climate, magnetic field and geology have largely derived from this modern period. The immense amount of data returned, particularly by the Mariner 9 and Viking spacecraft, is still being studied by a large number of scientists. It is these missions which have provided the basis for planning future Mars exploration and it is their data which can be used in unravelling the fascinating history of this cold, red world.

SPACECRAFT EXPLORATION OF THE RED PLANET

2

2.1 EARLY MARS MISSIONS

The very first spacecraft to reach the vicinity of Mars was the Soviet probe Mars 1, a fly-by mission which arrived on November 1 1962 but sent back no useful data. It had been launched a mere eighteen months after Yuri Gagarin's historic flight into space. Two years later, the first successful Mars mission was launched. On November 28th 1964, the US spacecraft Mariner 4 began its long journey to the Red Planet, carrying on board a variety of experiments to study the plasma and magnetic fields in the inner solar system. During July 1965 Mariner 4 reached Mars whereupon its magnetometer recorded for the first time that Mars had a weak magnetic field whose strength was only 0.03% of that of the Earth's. Onboard, too, was a radio occultation experiment that allowed NASA scientists to place a figure of 5 mbar on the surface atmospheric pressure of Mars.

The probe also carried a single TV camera and this obtained 22 images along a strip of ground running from south of Thaumasia to western Amazonis. Unfortunately the image contrast was disappointingly low, while the images themselves indicated a high incidence of large impact craters, which gave little encouragement to those scientists who had hoped to see a greater diversity of landforms than had been found on Earth's Moon. Surely after all this effort, Mars was not to be just another dead Moon-like world?

Almost as if they were willing Mars to be more forthcoming, NASA launched two further probes in 1969. Mariners 6 and 7 had three times the instrument payload of their predecessor, and enjoyed a level of sophistication and computer flexibility not bestowed on Mariner 4. Of particular significance was the carrying of both IR and UV spectrometers to make measurements of the Martian atmosphere. These instruments confirmed that CO_2 was the predominant atmospheric gas, with nitrogen being present to the extent of less than one per cent. Solid CO_2 was also confirmed to be present at the poles.

Two TV cameras were mounted on each instrument platform: a wide-angle camera which had a ground resolution of about 2 km but covered an area 12 times larger than the camera aboard Mariner 4, and a narrow-angle camera which, although having a field of view only one-tenth that of the wide-angle instrument, gave a tenfold improvement in resolution. The returned imagery largely reinforced the impression gained from the previous Mariner encounter, that Mars showed extensive cratered surfaces like those of the lunar highlands (Fig. 2.1). However, some images also afforded the first glimpses of blocky or 'chaotic' terrain in the region of Margaritifer Sinus, of sinuous ridges,

Fig. 2.1 *Mariner 6 image of Martian cratered plains, showing large impact craters and smoother intercrater areas. This image was obtained on 31st July 1969. The frame width is 902 km.*

polygonal landforms and sub-parallel grooves in south polar latitudes and of the very smooth floor of the Hellas basin. Contemporary reports of these findings can be found in Leighton *et al.* (1969) and Collins (1971).

Although Mariners 6 and 7 certainly gave hints that a greater diversity of landforms existed on Mars than had been revealed by Mariner 4, the images had neither the resolution nor the width of coverage to whip up any great excitement. However, an analysis of contemporary reports suggests there was an underlying note of scepticism that belied considerable unease with the Mars imagery data set at this time. How truly representative was it of Mars as a whole?

2.2 MARINER 9

The answer to this question was to be provided eventually by Mariner 9, one component of the next pair of US spacecraft, Mariners 8 and 9, whose objectives were the systematic mapping of about 70% of Mars, the collection of information relating to the composition and structure of the atmosphere, temperature profiling and a the measurement of topography. However, this was not until there had been some drama during the launch of the reconnaissance probe, Mariner 8, which, on May 9th 1971, found itself abruptly and unceremoniously dumped in the Atlantic Ocean after the failure of the second-stage rocket, posing NASA scientists with the unwelcome and taxing task of reassigning its companion craft in the space of only three weeks.

NASA, however, rose to the occasion, and three weeks later Mariner 9 was successfully launched. It was to take 167 days to complete its journey, a distance of 400×10^6 km. The main eyes of the probe were two TV cameras: a wide-angle mapping camera of 50 mm focal length and with a potential resolution of 1 km from periapsis altitude (the A-camera), and a narrow-angle camera of 500 mm focal length and potentially 100 m resolution (the B-camera). The mission team finally had settled on a 12-hour orbit inclined at 65° to the equator which gave somewhat higher solar incidence angles than had originally been planned for it, but which was the best compromise after the loss of Mariner 8.

Also on board Mariner 9 was a UV spectrometer, to be used for characterizing the scattering properties of the tenuous Martian atmosphere, an IR radiometer to measure temperature, and an IR spectrometer to determine the composition of the atmosphere and its entrained dust. As the spacecraft went behind Mars and then reappeared each revolution, the observed attenuation of the radio signals would be used to obtain temperature and pressure profiles and to establish surface elevations.

As the time of encounter approached, telescopic observations showed the all-too-familiar development of a perihelic dust storm of global proportions – something which had been predicted earlier by Chick Capen of Lowell Observatory. Surely Mariner had not gone all that way to have Mars clouded out?

At last, on November 10th 1971, when Mariner was 800 000 km away from Mars, the cameras were switched on. They revealed a virtually blank disk! The only recognizable feature was the bright south polar cap. It was, however, for just such an eventuality that the probe's computer had built into it a high degree of flexibility, in order that last-minute changes of plan could be made. Thus on receiving a signal from Mission Control, the spacecraft was able to sit out the dust storm, not wasting valuable energy in triggering camera shutters that would reveal a featureless Mars.

About two weeks after the spacecraft's arrival, the dust began to clear a little; in addition to the bright polar cap, several prominent dark spots became visible in the region known as Tharsis, one of them corresponding with the telescopic feature called Nix Olympica. As the atmosphere cleared, this was found to have a prominent complex depression at its summit. Could this be a volcanic caldera? Three similar spots lay to its south and east, lying along a roughly NE-SW line; these also had summit depressions. Since they all protruded above the dust clouds, it was clear that the spots were high mountains, and in fact we now know the group of three dark spots to be the massive shield volcanoes of Tharsis Montes, while the fourth is Olympus Mons.

Almost one month after Mariner 9's arrival, the decision was made to commence the systematic mapping of the planet's surface. Thereafter, each day the cameras generated two swaths of images 180° apart, successive swaths lying adjacent to the previous day's strip. By this method planetary geologists were provided with a new terrain traverse each day. What surprises were in store . . .

During the first phase of the mapping cycle, when the region between 25° and 65°S was imaged, not only was the whole of the Hellas basin gradually revealed but also a previously unsuspected 900 km diameter basin, Argyre, situated over 2500 km to its west (Argyre, as an albedo feature, was of course well known to observers). Then there were pictures of extensive fracture systems, on a scale hitherto found nowhere else within the solar system, of plains units traversed by lunar-like wrinkle ridges, strange tributary networks

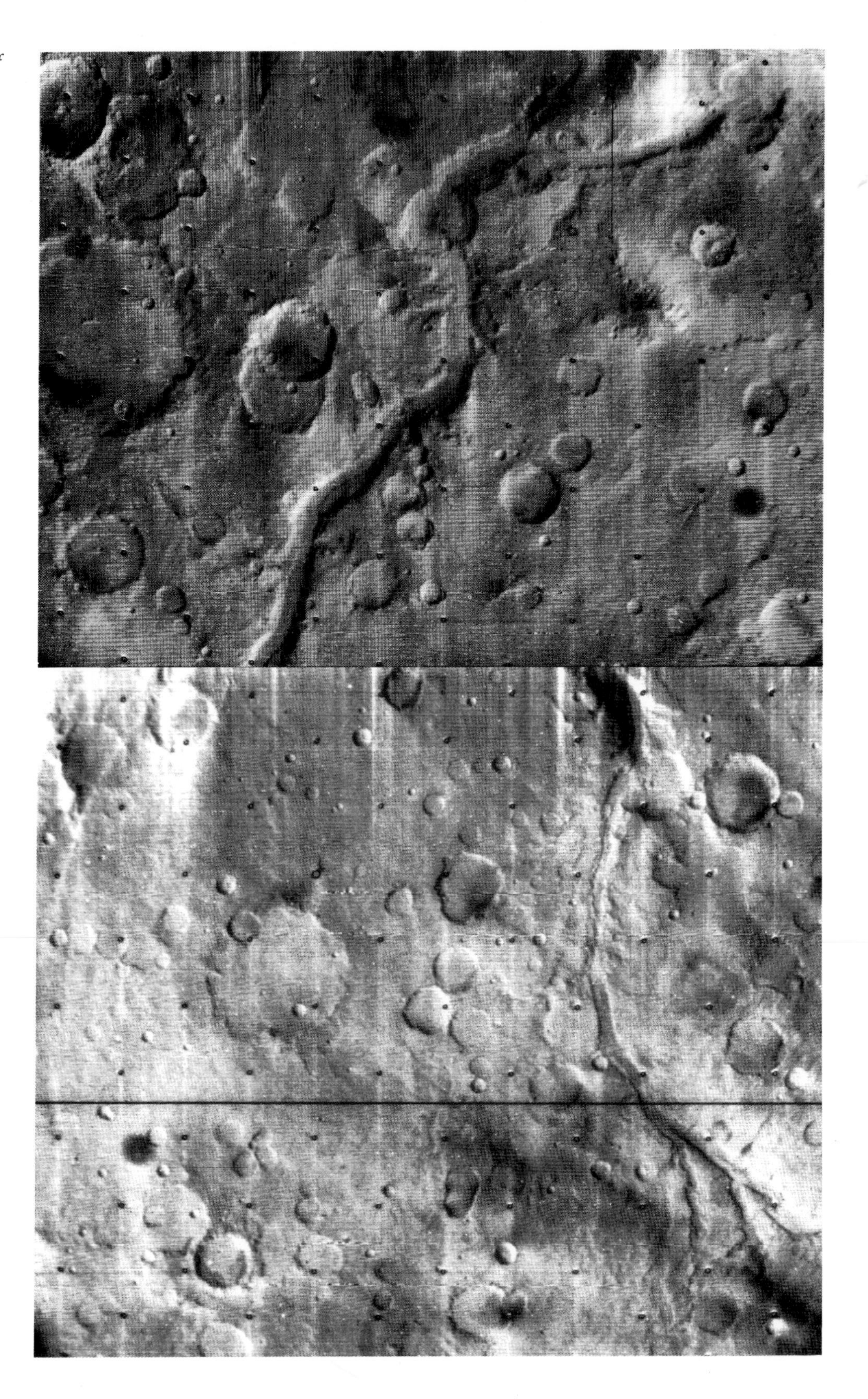

Fig. 2.2 *Mariner 9 mosaic of Martian channel networks. Mariner 9 image 4167–18/24.*

and various flow-like features that appeared to have been carved by running water (Fig. 2.2). At long last Mars was beginning to reveal the secrets investigators had hoped it was hiding.

The second phase saw the mapping extended towards latitude 25°N. This not only revealed the details of the Tharsis Montes shield volcanoes but also an immense canyon system which extended for at least 4000 km along the Martian equator and which was fittingly named Valles Marineris (Mariner Valleys) after the spacecraft which discovered it. The full extent of these equatorial canyons was only revealed as the mapping extended gradually eastwards, and it was also discovered that they ended in vast regions of blocky chaotic terrain of the type first revealed during the Mariner 5/6 encounter.

By the time the mapping had been accomplished, Mariner 9 had completed 698 orbits and had returned to Earth 7329 images. When the control gas supply finally ran out, on October 27th 1972, a signal was sent that switched off the probe and left the scientific community with the enormous but fascinating task of interpreting the new data set and rewriting the history of Mars.

2.3 SOVIET MARS EXPLORATION

The Soviet Mars exploration programme has never achieved the high degree of success typical of their Venus exploits and a degree of bad luck has dogged them right up until the most recent launch, that of Phobos 1 in 1988. Contemporary with the US Mariner 9 mission were Mars 2 and Mars 3, which were launched on May 19th 1971 and May 29th 1971, respectively. Although both spacecraft reached Mars in late 1971, as planned, and a small lander capsule was successfully sent down on to the floor of Hellas, the latter sent back no data at all, while the two orbiters suffered much the same fate as did Mariner 9 when it first reached the planet, their cameras recording a virtually featureless disk. The difference here was that since both probes were preprogrammed, it was not possible for them to sit out the storm while in Mars orbit, hence their singular lack of useful imagery.

The Soviet probes did, however, record the temperature at the Martian surface. It ranged from 13 °C above zero to 93 °C below, depending on latitude and time of day. At the northern polar cap the temperature dropped as low as −110 °C! The Mars 3 radio telescope also recorded that the soil temperature at depths of between 30 and 50 cm remains practically unchanged throughout the day. Presumably this is a reflection of the low thermal conductivity of the Martian soil.

Undeterred, in 1973 the Soviets launched four more spacecraft, two orbiters and two landers. Of these, the two orbiters (Mars 4 and Mars 5) reached Mars on schedule, but the former failed and simply flew past Mars. Mars 5, however, was more successful and completed 20 orbits during which it collected over 70 images similar in quality to those returned by Mariner 9 (Fig. 2.3). It also made measurements of atmospheric water vapour content and recorded Mars' weak magnetic field. Of the remaining probes, Mars 6 descended to the surface and is presumed to have made a soft-landing, but regrettably the controllers lost contact just 0.3 s before it did so. During its descent it made useful measurements of the atmosphere. Mars 7 unfortunately missed Mars completely.

A particularly controversial spin-off from one of the engineering experiments associated with Mars 5 was the detection of an unspecified inert gas in the

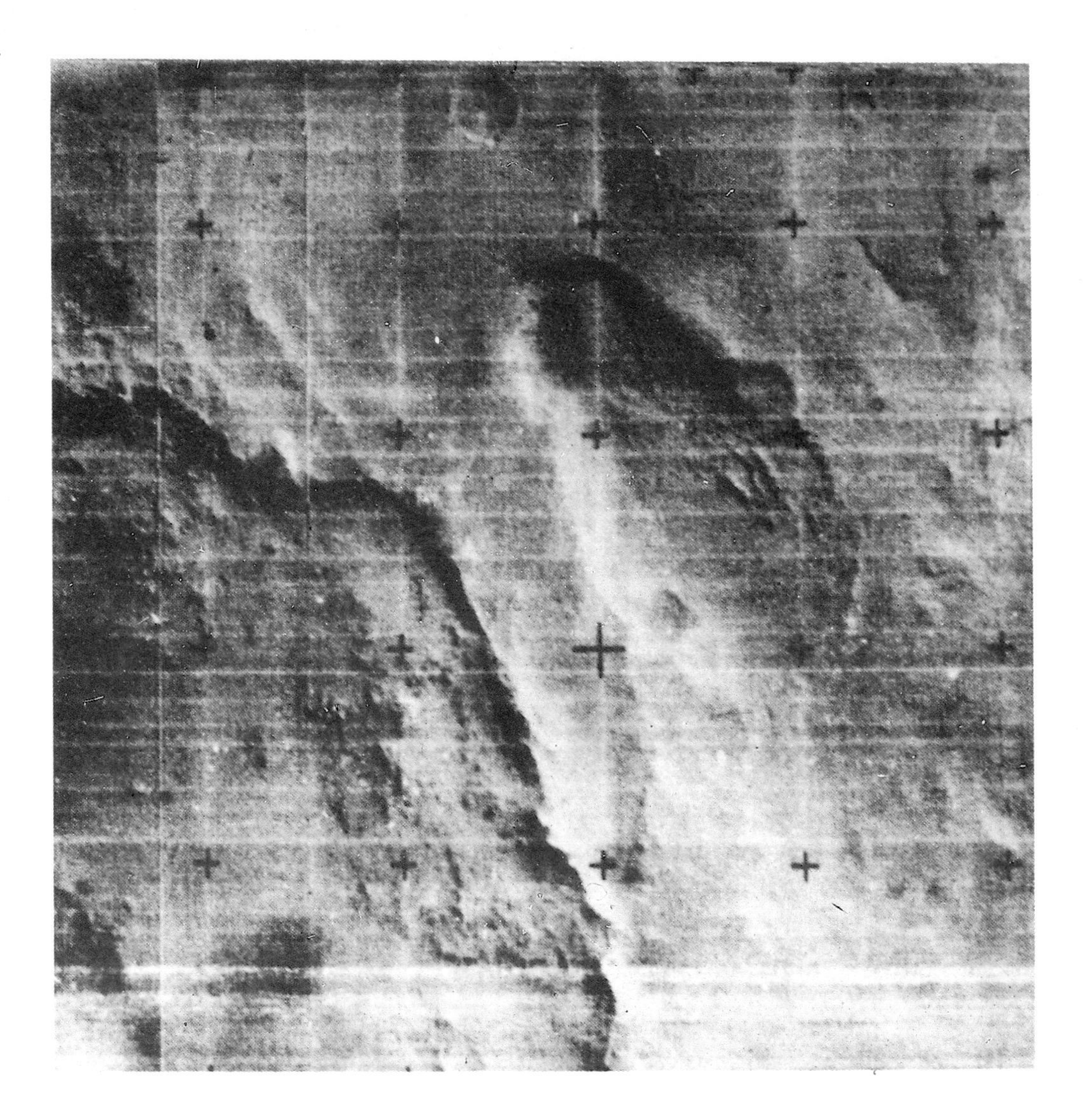

Fig. 2.3 *Mars 5 image of Mars.*

Martian environment. Because the only likely candidate for this was argon (later confirmed to be present to the extent of over 1% in the Martian atmosphere), it seemed more than likely that the present atmosphere of Mars could only be a residue from an originally much denser primary atmosphere. This was a particularly interesting possibility in terms of the past climate of the planet which, from the photogeological evidence for fluvial features, may once have been very different from today.

2.4 THE VIKING MISSIONS

Of all the missions which have been sent to Mars, undoubtedly the Viking missions have been the greatest successes and have contributed the most to our current understanding of the Martian environment. Planning for this major project had already begun well before Mariner 9 was launched and each of Viking 1 and 2 was to consist of an orbiter and a lander. Since one of their prime objectives had been identified as the search for life on Mars, on board

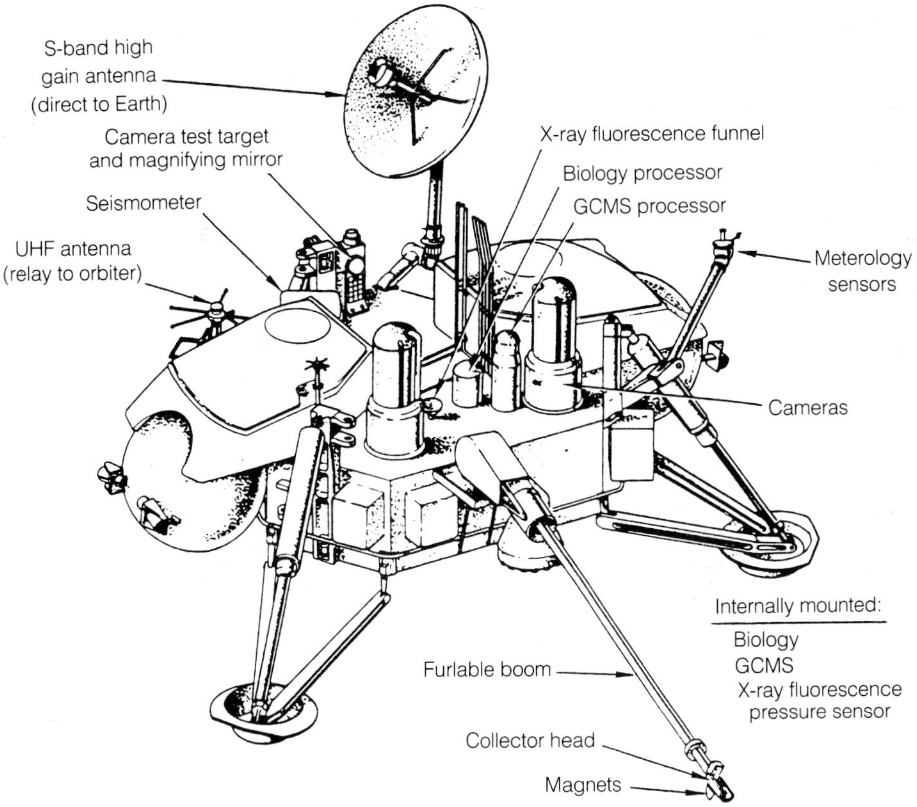

Fig. 2.4 *Instrumentation on board a Viking lander.*

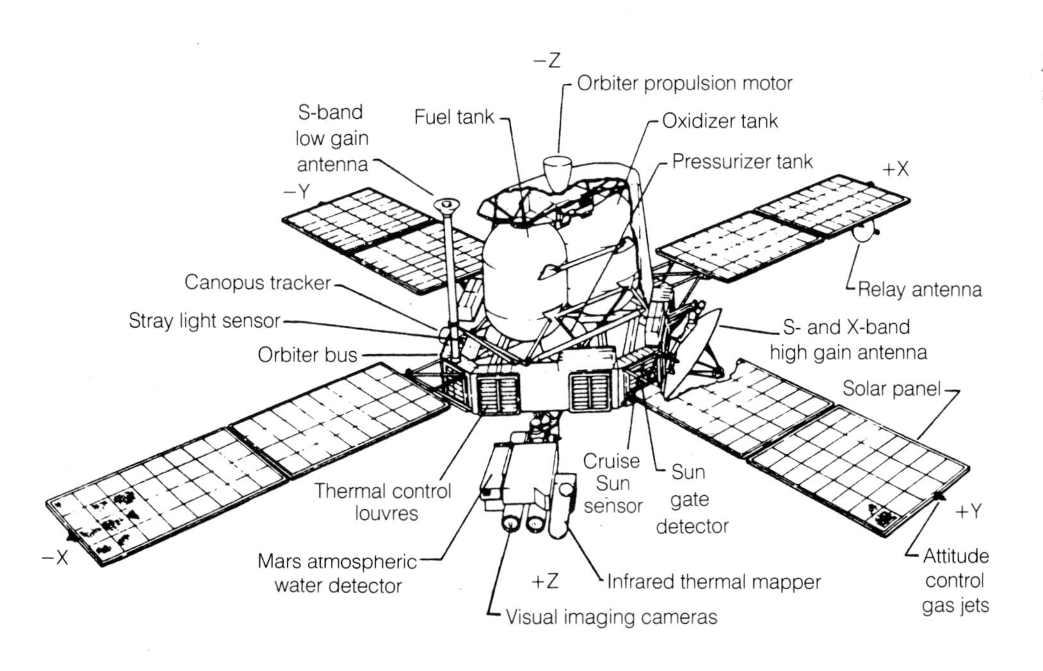

Fig. 2.5 *Viking orbiter instrumentation.*

each lander was to be a gas chromatograph–mass spectrometer, in order that any organic content in the soil could be detected, and a four-component biological experimental package. Additionally, there was to be an X-ray fluorescence analytical device, a seismometer and a set of meteorological instruments (Fig. 2.4).

The Viking orbiters were to have three main packages: a pair of high-resolution slow-scan vidicon cameras, an IR spectrometer to detect levels of atmospheric water vapour (MAWD experiment), and a series of IR radiometers (IRTM experiment) to measure the thermal properties of the Martian atmosphere and surface. In addition, the spacecraft's radio was to be used to measure the planet's gravity field and atmospheric profiles and for topographic profiling (Fig. 2.5). The cameras, in particular, were different from those flown on earlier missions, and were arranged so that they could be shuttered alternately once every 4.48 s, thus providing a continuous swath of images, two frames wide. Each of the pair of cameras was identical, having a 475 mm focal length telescope attached to a 37 mm diameter vidicon; each image subsequently was constructed from around 1.25×10^6 individual picture elements (pixels).

The mission controllers had arranged that each orbiter–lander pair be coupled on arrival at Mars. In this way, the orbiting probe would check out the suitability of selected landing sites; if the original one or ones were found unsuitable, then others could be inspected before the final decision was made to send the lander craft down. This degree of flexibility was considered vital for the Viking mission to be a success. Furthermore the orbiting probe would act as a relay for the lander below, as well as conducting its own experiments.

Viking 1 reached Mars on June 19th 1976 and within three days the orbiter was relaying back to Earth pictures of the first of the proposed landing sites, in Chryse Planitia. The significantly better resolution of these images revealed that the site was somewhat more rugged than had been anticipated, while radar measurements suggested a degree of surface roughness that could best be explained by the presence of numerous boulders. Consequently, a landing was delayed while other sites were reconnoitred.

It took three weeks to finally hit upon the ideal site, also in Chryse Planitia, but further west than the original, at 22.5°N 28.0°W. The Viking 1 lander made its historic descent and landing on July 20th and provided us with the first ever close-up views of the Martian surface (Plate 1). On the reddish boulder-strewn surface were small dunes and in the distance, several small impact craters. The boulders themselves were dark in colour and full of holes, rather like vesicular basalts. Indeed this is what they appear to be.

Just over two weeks later, on August 7th, Viking 2 reached Mars. Eventually, on September 3rd 1976, the Viking 2 lander also landed safely, this time in the region of Utopia Planitia, at 44°N 226°W. Here too, the original landing site was considered unsafe and a search had to be made for an alternative (Fig. 2.6).

Both landers worked exceedingly well and pursued an intensive series of experiments, including the biological search which scientists looked forward to with much anticipation. Disappointingly, the gas chromatograph–mass spectrometer experiments both showed a singular lack of organic molecules in the Martian soil and the general feeling which emerged was that life does not currently exist on Mars, and possibly never did.

The other lander instruments worked well, too, and sent back a detailed record of the changing local weather patterns, of amounts of dust suspended in the air, of diurnal temperature variations and the varying water vapour

Fig. 2.6 *Viking 2 lander site in Utopia.*

content of the Martian atmosphere. The cameras took a large number of panoramic shots of the local region, from which details of the geology were learned. Mechanical arms allowed scoops to dig up samples of the local soil which were analysed on board the spacecraft by the XRF-device and the results sent back to Earth.

Table 2.1 Chemical composition of Martian surface materials

Major elements (wt%)	Viking 1	Viking 2
SiO_2	44.7	42.8
Al_2O_3	5.7	
Fe_2O_3	18.2	20.3
TiO_2	0.9	1.0
MgO	8.3	
CaO	5.6	5.0
K_2O	<0.3	<0.3
SO_3	7.7	6.5
Cl	0.7	0.6
Trace elements (p.p.m)		
Rb	<30	<30
Sr	60 + 30	100 + 40
Y	70 + 30	50 + 30
Zr	<30	30 + 20

Twenty-two samples were scooped up at the two Viking lander sites; several of these were sufficiently large to allow for XRF analysis. Naturally, the materials had to be relatively easily obtainable; consequently, the analysed rocks were loose, friable materials, known as **fines**. The samples from both sites contained abundant Si, together with significant amounts of Mg, Al, Ti, Ca and S (Table 2.1). The relative proportions of the components, however, were unlike any known terrestrial rock, and the analyses are presumed to represent admixtures of different Martian materials. The gas chromatography experiment on board the landers detected about 1 wt% H_2O; this amount of water is probably bound into the samples collected.

The minerals present were largely silicates, with some oxides and carbonates. The grains themselves appear to have been cemented together with varying degrees of effectiveness, by a sulphate-rich cement, to give what is termed a **duricrust**. Between 3% and 7% of the material was magnetic and this suggests the presence of iron-rich minerals like magnetite or maghaemite, but some contribution from Fe/Ni meteoritic material cannot be ruled out. The concensus of opinion suggests that most of the soil is composed of iron-rich clays like nontronite, iron oxides, hydroxides and minor amounts of carbonate.

The orbiters, meanwhile, were mapping the entire surface of Mars at a resolution of 200 m; large regions benefitted from even higher resolutions, down to 10 m in some places. While most images were in black and white, colour imagery was also obtained, as were several sets of stereoscopic frames. The mapping programme continued for almost two Martian years and, in addition to the visual imagery – which revealed not only the permanent geology but also the changing face of dust-storm activity and global weather patterns – the MAWD and IRTM packages collected data on the continually changing amount of water vapour and dust in the atmosphere and on thermal inertia values within the surface layer. The Martian gravity field had also been constrained by noting the effect of the gravity field upon the spacecraft's orbit. Finally, high-resolution pictures of both of Mars' satellites were also obtained.

The Viking 2 orbiter finally ran out of attitude control gas in July 1978, and Viking 1 in August 1980; by that time over 55 000 images had been returned. Most of the images reproduced in this book have been lifted from the superb Viking data set. A full account of the Viking Mission results can be found in *Scientific Results of the Viking Project*, published by the American Geophysical Union in September 1977.

2.5 THE SOVIET PHOBOS MISSION

The most recent Mars launches occurred in the summer of 1988, when the Soviet Phobos 1 and 2 spacecraft started their 10 month long journeys from the Baikonur Cosmodrome. Each craft had a three-fold purpose: to study the activity of the Sun while en route for Mars, to study Mars itself, and, in particular, to study Mars' larger moon Phobos. This was the first time a mission had the prime objective of studying one of the smaller satellites of the solar system.

Unfortunately, early in September 1989, the ground controllers sent an incorrect signal to Phobos 1 and lost contact with it forthwith. Phobos 2, however, reached Mars on January 29th 1989, then went into a highly elliptical orbit above Mars' equator prior to entering a more nearly circular orbit just

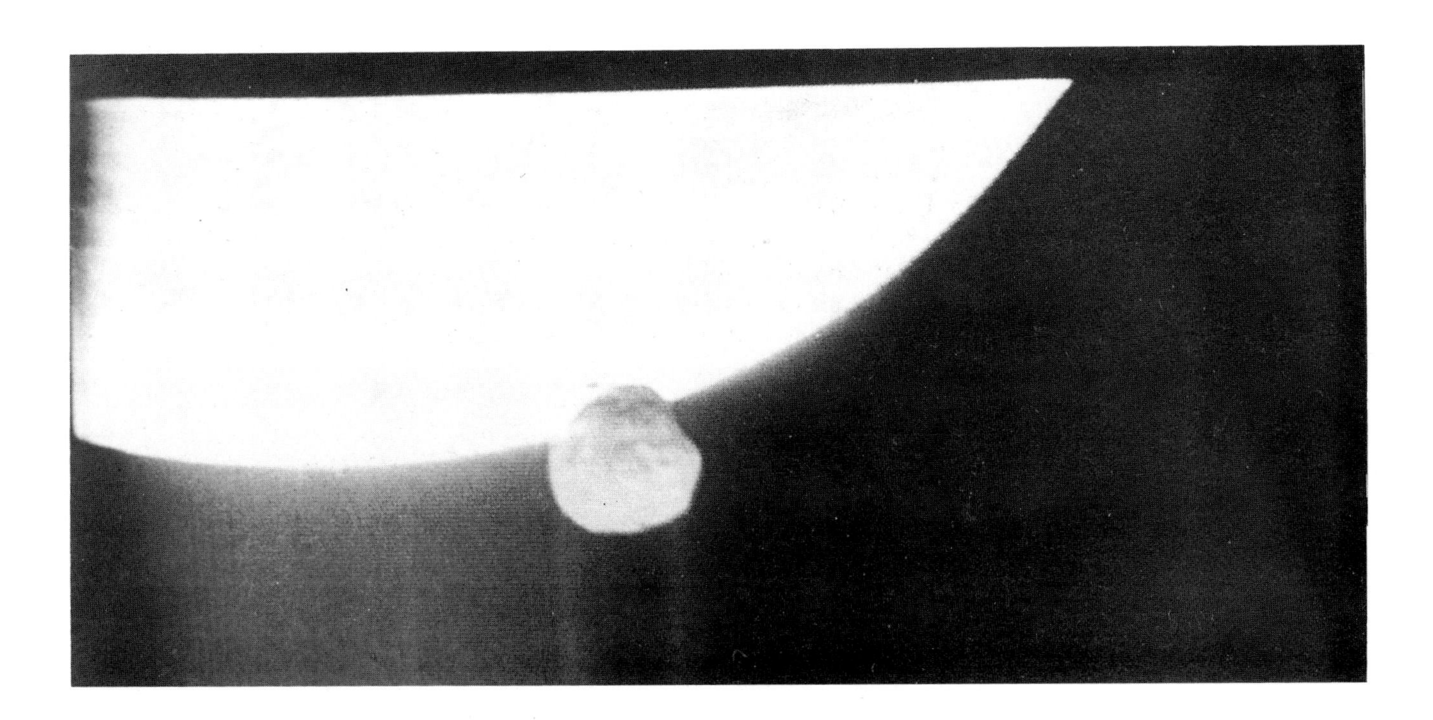

Fig. 2.7 *Soviet Phobos 2 image of the moon Phobos, taken on 28th February 1989 from a distance of 400 km.*

350 km above the orbit of Phobos. While in the first, or 'parking' orbit, Phobos 2 continued to observe Mars itself for three days then, with the probe at a distance of between 860 and 1130 km above the moon, it collected the first nine TV images (Fig. 2.7), which it followed with two further picture-taking sessions, as the orbit was trimmed to a lower periapsis.

On March 21st, Phobos 2 entered a new orbit that took it in as close as 120 km from the moon's surface, during which time it took further TV pictures, and the mission controllers prepared to drop it as low as 35 km above Phobos on the side not facing Mars. Alas, bad luck hit again. On March 27th, with all systems apparently working perfectly, radio contact was lost and never regained. To this day it is not completely clear why this happened.

The mission, although prematurely terminated, was certainly not a failure. Important astronomical observations were made of the Sun, in particular of X-ray emissions from it, and included 140 very high-quality images of the solar corona, as well as plasma measurements. Phobos 2 also completed detailed studies of the Martian magnetosphere which behaves rather differently from Earth's. The very weak magnetic field surrounding Mars was found to be closely interactive with the interplanetary field, creating a natural pathway for the Sun's plasma to reach deep into the Martian magnetosphere. Furthermore it means that the solar wind interacts closely with ions in the Martian atmosphere, and Phobos 2 measured the flow of plasma in the solar wind and the rate at which ions of carbon dioxide and molecular and atomic oxygen were leaving Mars. Surpisingly, it was found that the atmosphere of the planet is escaping at the rate of between 1 and 2 kg every second; this may seem insignificant (and indeed it would be for a planet like Earth) but for Mars, which in any case has such a tenuous mantle of air, it means that it could lose

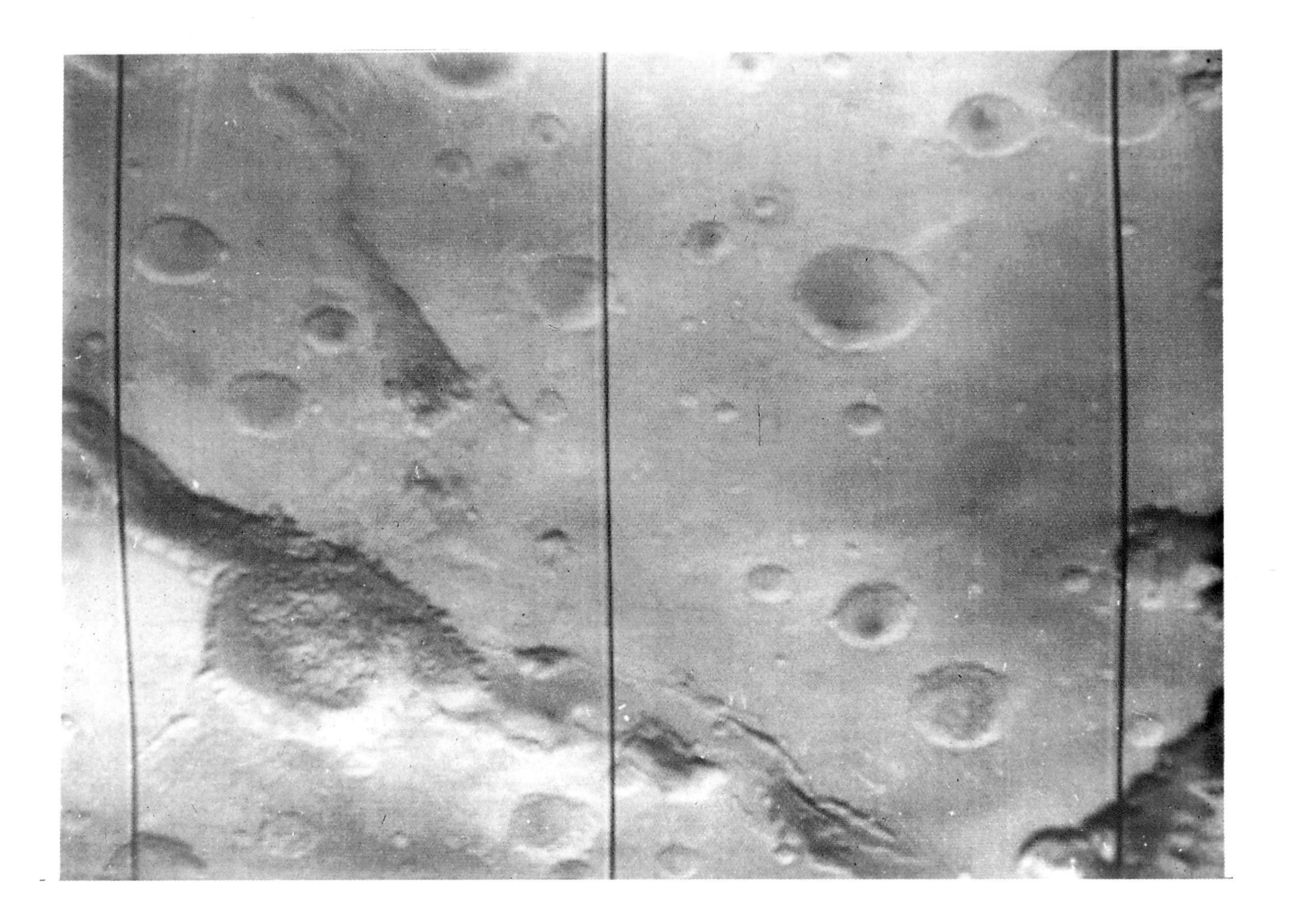

Fig. 2.8 Thermal image of Mars obtained by Phobos 2 spacecraft on 1st March 1989.

its entire inventory of volatiles in much less than the life of the solar system. The extreme weakness of the magnetic field may therefore have played an important role in the gradual loss of water from Mars.

The images which Phobos 2 transmitted of Phobos were extremely interesting and added to the set made by Viking 1 and 2. Perhaps even more interesting were the infra-red images that it returned of Mars itself. The Soviet-made Thermoscan instrument which did this essentially consisted of a very sensitive IR detector cooled by liquid nitrogen. Over a region of the Martian equator roughly 1500 km wide, it obtained exceedingly sharp thermal images with a resolution approaching 2 km (Fig. 2.8). The very high contrast achieved brought out very clearly differences in the degree of soil fragmentation from one area to the next and allowed the Soviet scientists to relate this to regional geomorphological features.

Other instruments recorded the planet's spectrum at 128 different wavelengths and resolved peaks in the absorption spectrum that corresponded to different minerals. The same data should also allow scientists to establish how much volatile material is locked up in these minerals' lattices. Preliminary results suggest a relatively generous distribution of volatile-bearing rocks, presumably sedimentary in origin.

Before leaving this brief discussion of the latest Mars mission, mention must made of the multispectral images collected of Phobos. The various instruments allowed for the collection of data between wavelengths of 0.32 and 3.2 μm, that is, from the ultraviolet into the thermal infrared. The measurements showed that Phobos is very dark indeed, reflecting at most 4% of the incident light, the amount remaining roughly constant for the entire wavelength range. This is a property Phobos shares with certain kinds of carbonaceous chondrite, although the latter contain somewhat less water than the small moon. It is a great pity that the various analytical experiments could not be completed, since these would have provided detailed information on the chemistry of this chunk of primitive asteroidal material and would have enhanced our knowledge of the primitive stuff from which all of the planets were made. However, it did act as a proving ground for technology which may be incorporated in the next batch of Mars missions, due to arrive at Mars in the mid-1990s. This prospect is indeed exciting.

3 THE PRESENT FACE OF MARS

The telescopic albedo markings bear little general resemblance to the major landforms revealed by spacecraft imagery. True there are some correlations; for instance, the whitish spot of Nix Olympica represents the massive volcano, Olympus Mons, while the impact basins of Hellas and Argyre are clearly visible with a telescope as being paler in hue than the surrounding uplands. In consequence, some of the old names from telescopic maps have survived, but the plethora of major landforms only discovered when spacecraft images were returned to Earth, has demanded a largely different nomenclature than would have been familiar to, say, an observer of the early twentieth century (Fig. 3.1).

3.1 THE TOPOGRAPHY OF MARS

Mars is approximately pear-shaped, having an equatorial radius of 3393.4 km and a polar radius of 3375.7 km. Because there are no oceans, a sea level datum is not available against which to measure topography. In view of this, before topographic maps of the planet could be prepared, a reference datum had to be selected by NASA scientists. The datum chosen was that defined by a gravity field described by spherical harmonics of the fourth order and fourth degree (Jordan and Lorell, 1973), combined with a 6.1 mbar atmospheric pressure surface, equivalent to the triple point of water (Christensen, 1975). Global topographic information, derived in the main from radio-occultation and gravity data, has been related to this 6.1 mbar datum to produce a general topographic map (Fig. 3.2, Plate 2). The map illustrates how, in terms of topography, Mars is markedly asymmetric, with the majority of the southern hemisphere lying between 1 and 3 km above datum and most of the northern hemisphere standing below it. The line of dichotomy separating these two elevation zones describes a great circle inclined at approximately 35° to the Martian equator.

In the relatively elevated southern hemisphere, the principal exceptions to the positive topography are the deeper parts of the two impact basins Argyre and Hellas and the region southward of latitude 70°S. The most recent topographic map shows the deepest part of Argyre to descend to a depth of 3 km and Hellas to over 5 km (USGS Map I-2030, 1989). In the northern hemisphere the main exceptions to the subdatum elevations are the heavily cratered terrain between 30° and 270°W and the elevated volcanic provinces of Tharsis and Elysium.

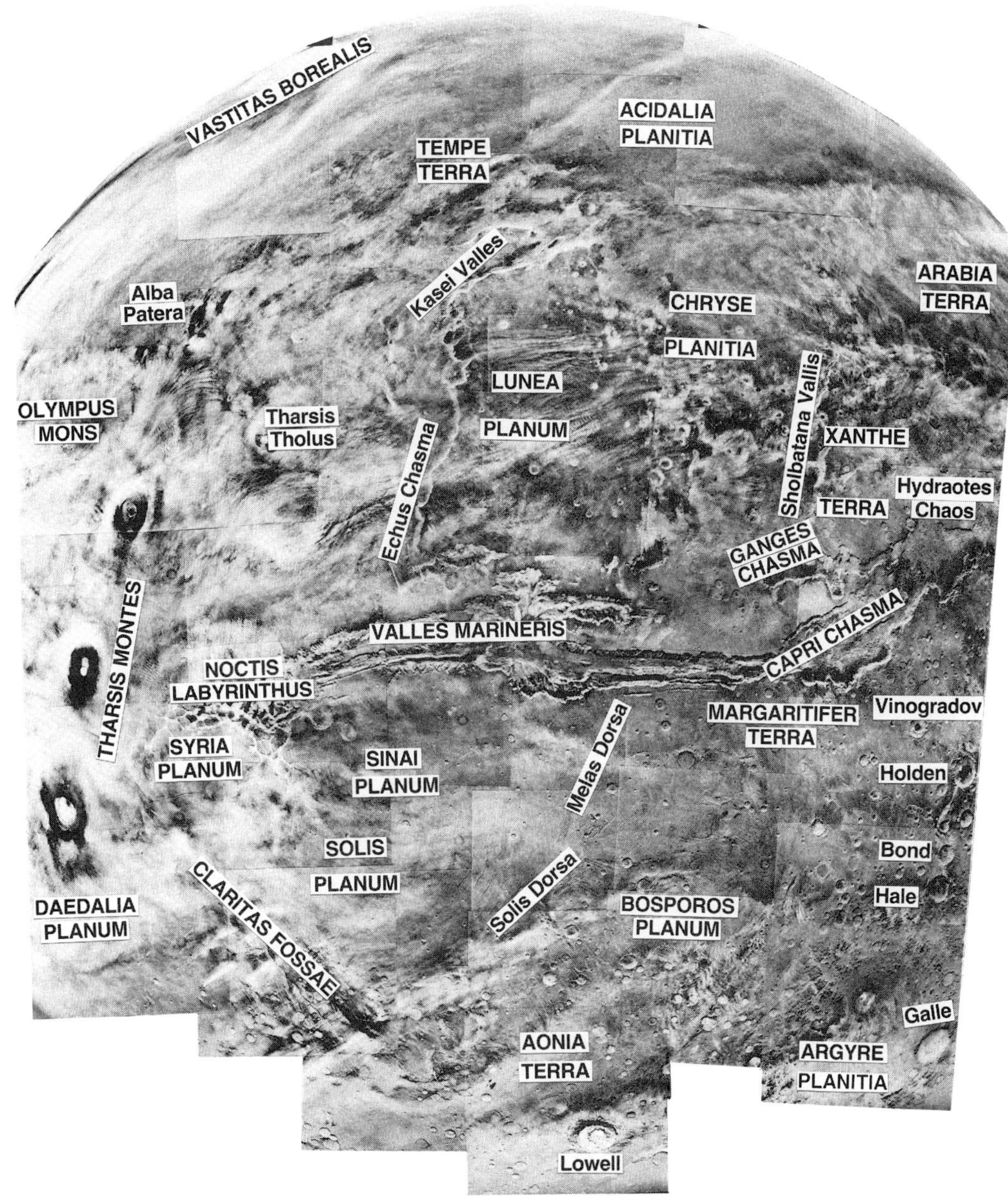

Fig. 3.1 The modern face of the western hemisphere of Mars.

Mention of Tharsis introduces another major topographic feature of Mars – the Tharsis 'Bulge' – a broad crustal upwarp centred at 14°S, 101°W, which rises 10 km above datum and measures roughly 4000 km across. This continent-sized structure straddles the boundary between the two topographic hemispheres. A similar but smaller and lower bulge is centred on 28°N, 212°W in the Elysium province, but the latter falls almost entirely within the lower northern hemisphere.

In addition to the regional variations, there are local elevation differences which are substantially greater than those that characterize the Earth. For instance, some sections of the vast equatorial canyon system, Valles Marineris, descend 7 km below the adjacent plains, while the summits of several shield

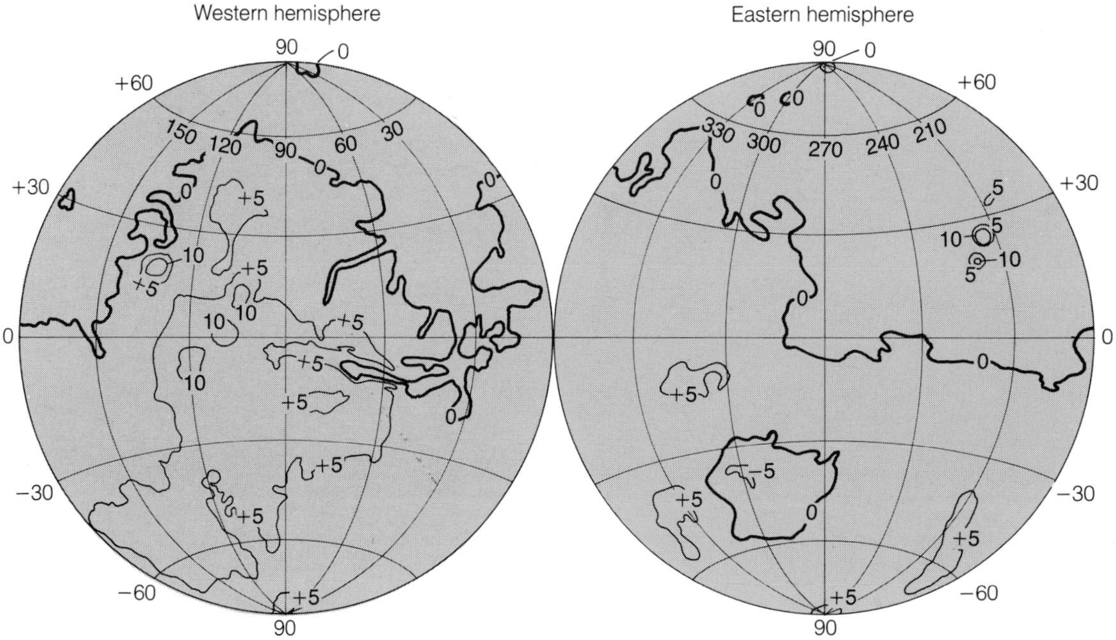

Fig. 3.2 *Topographic map of Mars.*

volcanoes within the Tharsis province rise to over 20 km. This particular characteristic shows up clearly when the planetwide topography distribution of Mars and the Earth are compared (Fig. 3.3). Incidentally, the highest point on Mars is reached at Olympus Mons, whose summit rises to 27 km above datum.

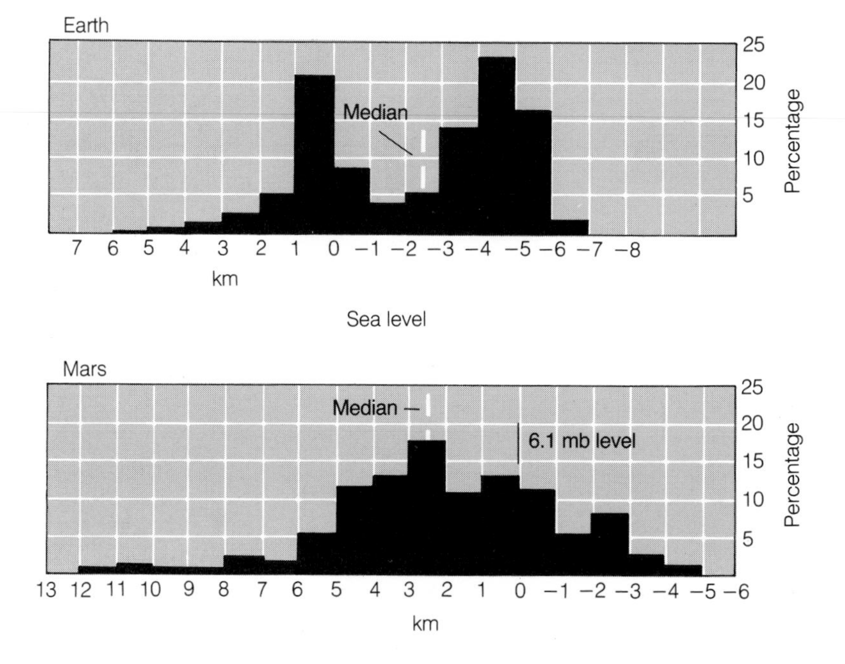

Fig. 3.3 *Comparison of distribution of topography for Mars and the Earth.*

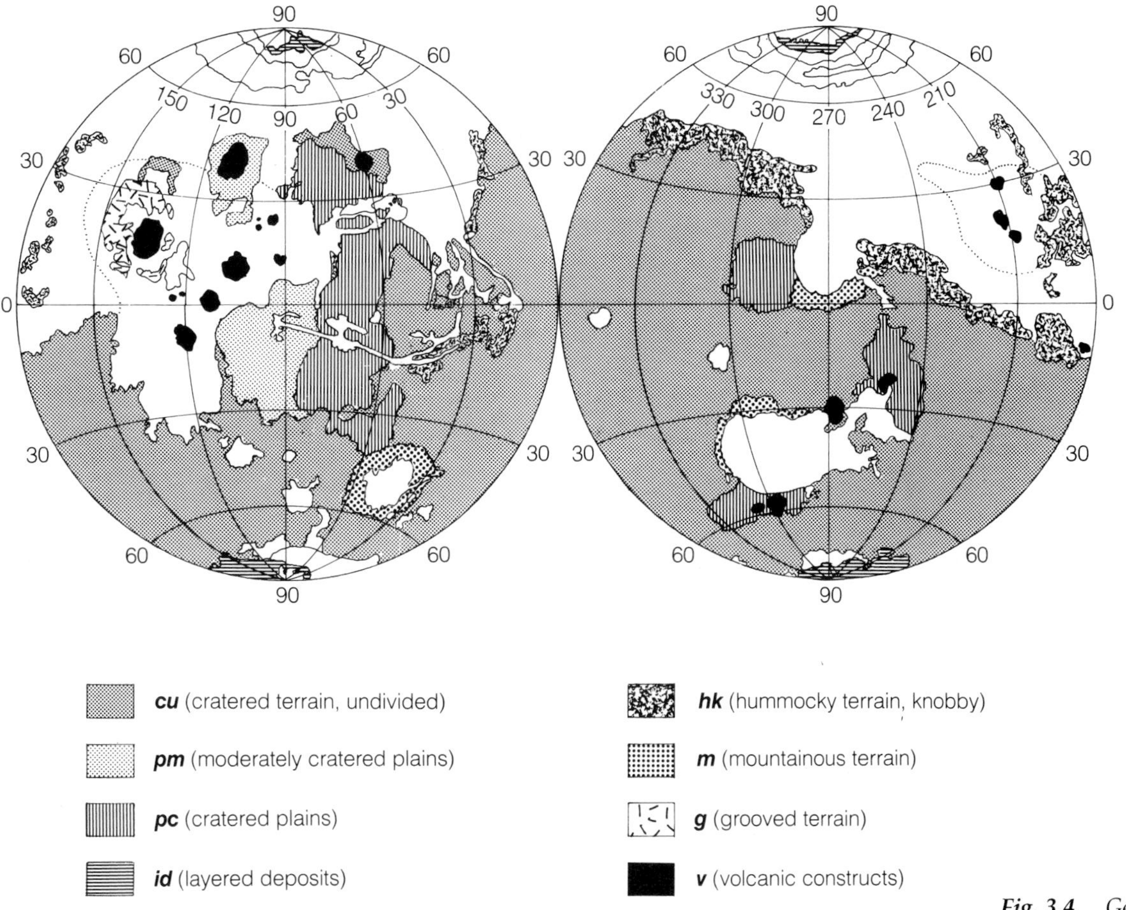

Fig. 3.4 *Generalized physiographic map of Mars.*

3.2 PHYSIOGRAPHIC PROVINCES

Just as the topography of the planet shows a marked asymmetry, so does the distribution of physiographic units. In fact, the physiographic features show a strong correlation with the hemispheric asymmetry in topography (Fig. 3.4). The greater part of the southern hemisphere, for instance, is given over to a landscape carved from an ancient, heavily cratered surface which, at first sight, is similar to the lunar highlands. Closer inspection, however, reveals a major difference, since the Martian cratered plateau is incised by large numbers of valley networks which evidently had their origin in running surface water (Fig. 3.5). This heavily cratered terrain also extends northward across the equator as a broad tongue between longitudes 300° and 10°W, that is, north of the telescopic regions of Sinus Sabaeus and Sinus Meridiani.

Within the heavily cratered regions, the density of craters with a diameter >20 km is high, but there is a relative paucity of smaller craters compared with the lunar highlands. Furthermore, the large impact craters of Mars are relatively shallow, having flat floors and low rims that bear witness to long periods of erosion and weathering. This gives an overall less rugged aspect to the Martian cratered landscape than is typical of the Moon. There are also

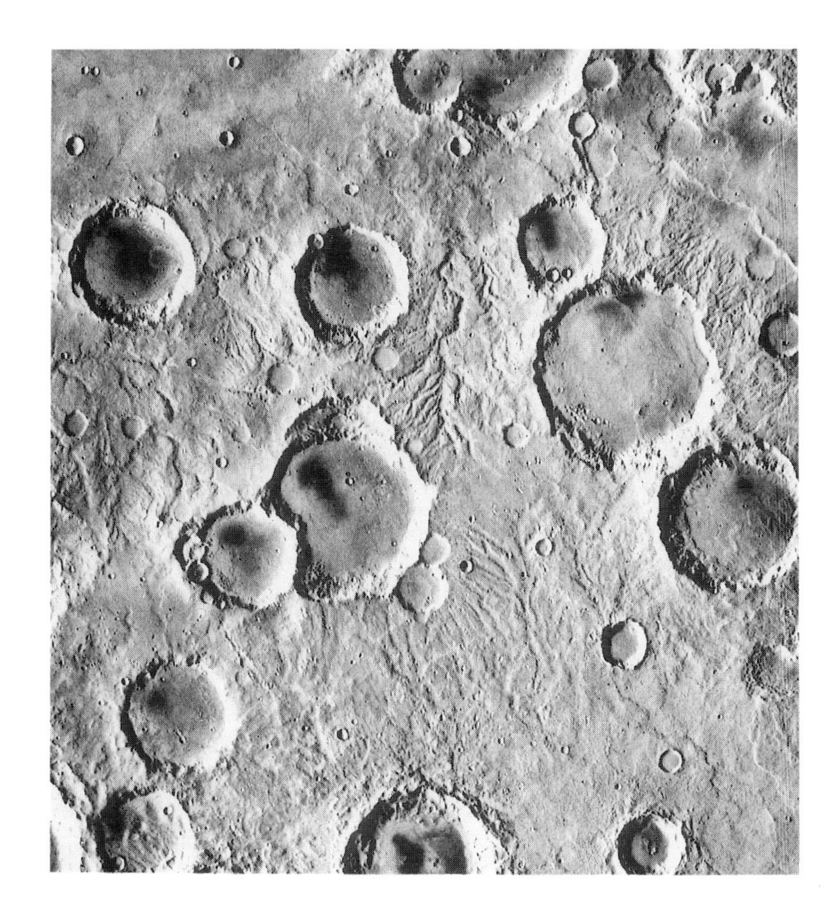

Fig. 3.5 *Channel networks in the cratered uplands of Mars, Terra Cimmeria. Viking orbiter frame 370S54. Frame width 245 km. Centred at 36.95°S, 212.99°W.*

many areas of smoother, less densely cratered plains between the large impact craters. The relative proportion of these intercrater plains compared to that of the more rugged cratered units varies from region to region.

As has been mentioned, many of the intercrater surfaces are cut by short channel networks; others are ridged. Sometimes ridges may be seen crossing the intercrater plains and continuing along the same strike, inside large impact craters. Spudis and Greeley (1978) estimated that the ancient cratered plains cover an area of about $2.9 \times 10^7\,km^2$, of which about 36% is ridged. Quite extensive remnants of such ridged plains are located in a belt extending eastwards from Noachis Terra, in Memnonia and the southern part of Sirenum Terra (Fig. 3.6). Cratering studies suggest that these are some of the oldest plains found on Mars (Scott and Tanaka, 1986). The ridges themselves, in appearance very like lunar wrinkle ridges, have been invoked by many workers to support a volcanic origin (see Cattermole, 1989).

Other ridged plains outcrop in the western hemisphere, within a broad zone about 1000 km wide on the east flank of the Tharsis Bulge where they extend over an area of about $4 \times 10^6\,km^2$. The plains extend from Tempe Terra, through Lunae Planum, then across Valles Marineris into parts of Sinai and Solis Planae. Similar plains occur in the eastern hemisphere, particularly in Hesperia Planum, northeast of Hellas. Upland depressions such as Syrtis Major Planum and the floors of the large impact basins also show a development of similar plains.

Fig. 3.6 *Ridged plains southwest of the Hellas impact basin. Viking orbiter frame 361S13. Frame width 280 km. Centred at 59.25°S, 313.23°W.*

As with the Moon, there are several large impact basins on the Red Planet, the two most obvious being Hellas, with a diameter of 1800 km and Argyre with a diameter of 800 km. The Isidis basin measures 1900 km across but has been substantially modified and is therefore less striking on synoptic images. The larger Martian basins are multi-ringed and of similar dimensions to those on the Moon; the lunar Imbrium basin, for instance, being 1300 km across. In a non-exhaustive search for other impact basins, Chuck Wood lists ten further structures with diameters >200 km which qualify for the name basin, the largest of these being the south polar basin (Wood, 1980).

The polar regions, whose seasonal changes are well known from Earth-based observations, are markedly different from the rest of the planet. Around both poles and extending to about latitudes 80°, are found unique laminated deposits which bear witness to prolonged periods of erosion and deposition on the planet. The layered deposits are exposed in the walls of extensive scarps which girdle the polar ice caps (Fig. 3.7). Because of the very low superimposed impact crater density, these must be amongst the youngest deposits on Mars. Around the north pole the laminated units overlie plains and are girdled by an extensive zone of dunes. The latter are absent from the south pole, where the laminated materials unconformably overlie old cratered terrain and a large polar impact basin with a diameter of about 850 km. The ice caps themselves, naturally, change their extent and shape as Mars' seasonal cycle progresses.

Fig. 3.7 *Laminated deposits near the north pole of Mars. Viking orbiter frame 566B67. Frame width 57 km. Centred at 78.70°N, 349°W.*

The northern hemisphere of Mars is largely given over to plains which are less heavily cratered than the older terrain to the south. The majority of these are believed to have a volcanic origin, although they may have been much modified by surficial processes. Similar plains extend into the southern hemisphere around the Tharsis Bulge and to the south of the equatorial canyons; also they are found both within and around the Hellas impact basin. The impact crater densities characteristic of these plains are within a factor of two or three of those of the Moon's mare plains. Their surfaces exhibit landforms which bear witness to volcanic, aeolian and fluvial activity; furthermore, there is much geomorphological evidence for the presence of sub-surface ice.

Less heavily cratered than the above are the extensive volcanic flow plains that encircle the Tharsis and Elysium volcanoes. These younger deposits show widespread volcanic features, such as flow scarps, lobes, lava channels and wrinkle ridges; many of the individual units are radial about major volcanic centres (Fig. 3.8). The volcanoes themselves are extremely large and impressive on synoptic views of the planet. The most prominent group of shield volcanoes – the Tharsis Montes – are spaced about 700 km apart and aligned northeast/southwest on the northwest flank of the Tharsis Bulge (Plate 3). The even larger structure Olympus Mons (Nix Olympica on telescopic charts) is situated

Fig. 3.8 *Prominent lobate volcanic flow on plains to the east of Olympus Mons. Viking orbiter frames 659A02–04. Frame width 30 km.*

1600 km to the northwest of this tectonic lineament and resides at the edge of the Bulge. It is the youngest Martian shield volcano and also the highest (27 km). All of the volcanoes are immense; the main shield of each of the three Tharsis Montes is over 4000 km across, while Olympus Mons measures over 550 km in diameter. Other, smaller volcanoes also lie close to the northeast/southwest line; most are older than the above. Lastly, on the northern flank of the bulge is the lower-profile volcano, Alba Patera; this immense structure has a diameter of over 2500 km and has associated with it a prominent circumferential fracture belt.

The volcanoes of the Elysium region are three in number and have a somewhat different morphology than their Tharsis counterparts. Effusive and explosive volcanism both appear to have played a part in their development. Lava plains associated with them extend over an area of about 3×10^6 km^2. To the southeast of Elysium there is another, rather isolated major volcano, Apollinaris Patera.

Further volcanoes and associated volcanic deposits are found around the Hellas basin; these are significantly older than either of the Tharsis of Elysium structures. This group comprises what have been termed highland paterae which are the manifestation of a much earlier phase of volcanism which took place at least 3.1×10^9 years ago, well before any of the Tharsis volcanoes had begun to grow. Isolated volcanic structures also occur in Syrtis Major Planum and in Tempe Terra, both regions of relatively ancient cratered or fractured terrain.

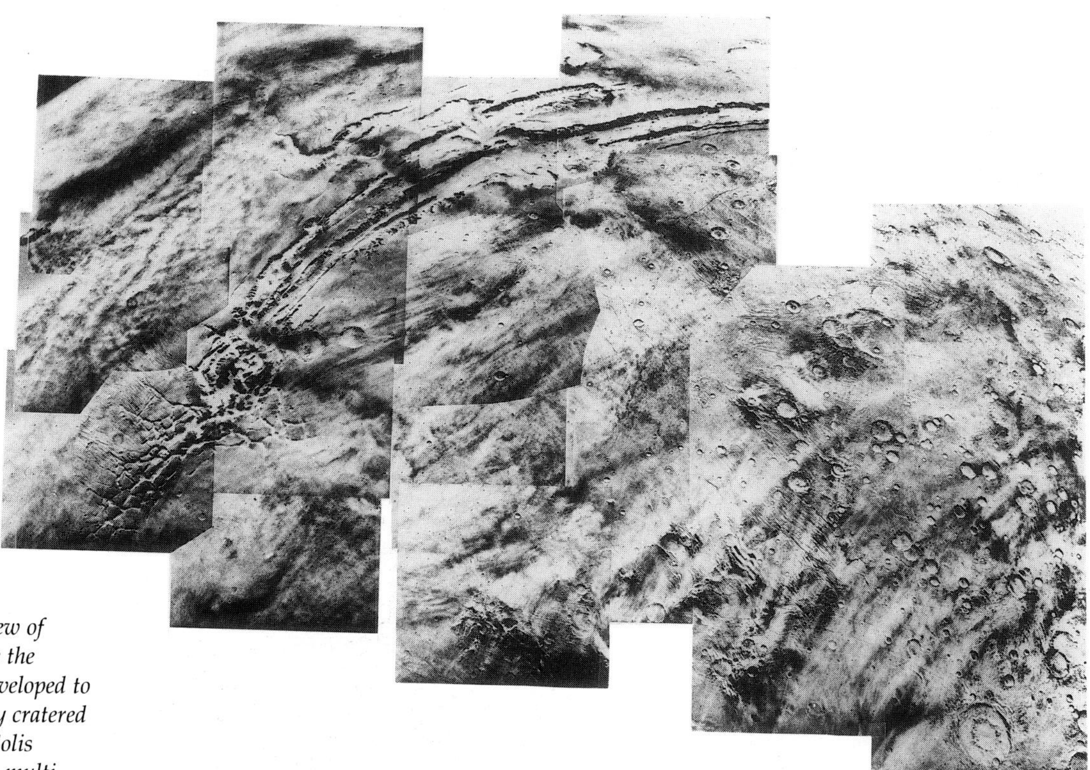

Fig. 3.9 *Synoptic view of Valles Marineris. Note the extensive fracturing developed to the south on the heavily cratered plateaux of Sinai and Solis Planae. The prominent multi-ring structure at bottom right is Lowell. Viking mosaic 211–5048.*

The character of the plains at latitudes higher than 40°N is significantly different from those at lower ones. Little or no direct evidence for a volcanic origin is found here, instead there is a vast blanket of sedimentary debris which gives rise to a smoothed or 'softened' landscape. Impact craters present often show the effects of deep burial, sometimes their interiors being completely infilled with debris. Additionally, their ejecta blankets usually stand out as pedestals, at a slightly higher level than the surrounding plains, as if the ejecta had somehow resisted erosional processes that had lowered the surrounding regions.

Equally as impressive as the giant Tharsis shield volcanoes is the amazing equatorial canyon system which extends eastwards from the crest of the Tharsis Bulge, at 5°S, 100°W, running almost parallel to the equator for a distance of approximately 4000 km. This great landform, known as Valles Marineris, commences in the west as a series of interconnecting box canyons and depressions which comprise Noctis Labyrinthus, changes into a series of sub-parallel and approximately parallel-sided canyons along its central section and finally merges at its eastern end with vast areas of blocky chaotic terrain at around 15°S, 40°W, just west of the telescopic feature known as Margaritifer Sinus. If one could fly over the length of this great series of depressions, one would have flown along almost one quarter of the planet's circumference and would have traversed canyons over 200 km wide and up to 7 km deep (Fig. 3.9). Numerous smaller canyons lie to the north of the main system.

The equatorial canyon network itself is only one part of a much more widespread family of extensional fractures which splays out from the Tharsis

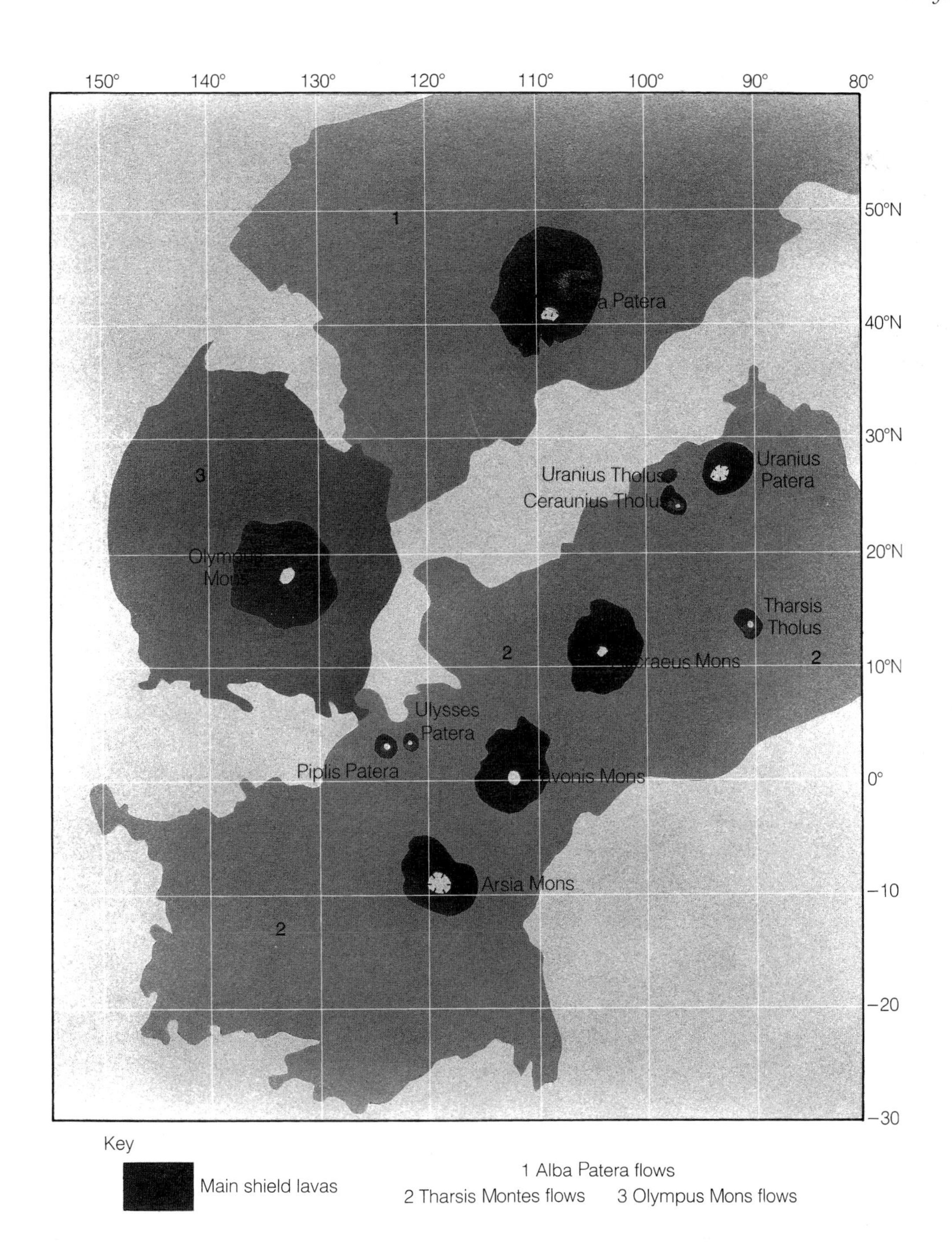

Fig. 3.10 *Fractures associated with the Tharsis Bulge.*

Bulge and extends over almost an entire Martian hemisphere (Fig. 3.10). Fractures are particularly strongly developed to the north and northeast of Tharsis (Tempe and Mareotis Fossae), and also to the south (Claritas Fossae). Families of long curving graben also extend into the telescopic regions of Mare Sirenum and Memnonia (Sirenum and Memnonia Fossae). Less well developed fracturing also is associated with the Elysium province.

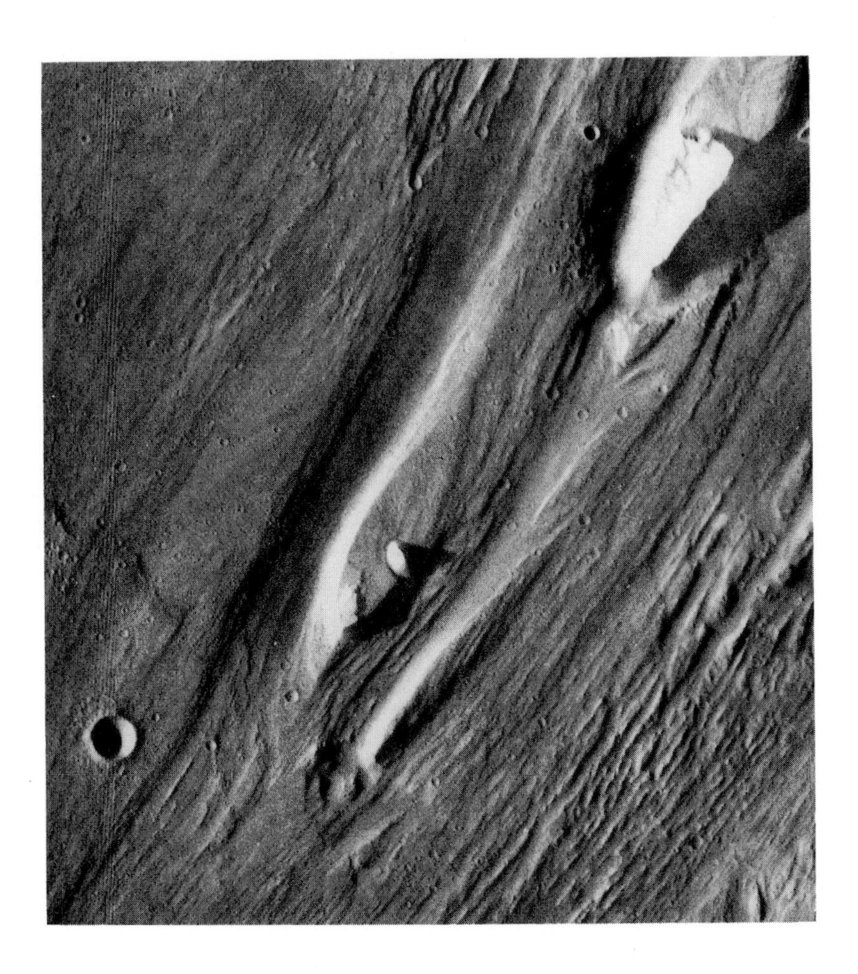

Fig. 3.11 *Scoured channel floor of Kasei Vallis. Viking orbiter frame 665A41. Frame width 38 km. Centred at 27.42°N, 61.95°W.*

The regions of chaotic terrain into which the Valles Marineris runs at its eastern termination, outcrop approximately within the region defined by latitude 5°N and 15°S, and longitude 15° and 40°W. Emerging from the northernmost areas of chaos is a series of major channels which run northwards towards Chryse Planitia, converging there with similar ones that emerge from north of the central section of Valles Marineris, and enter Chryse from the west. These channels exhibit various landforms indicative of an origin in fluvial processes, such as scoured floors and teardrop islands. Their dimensions are typically impressive: individual channels incised into the Chryse plains are at least 25 km wide, while the major channel system (Kasei Vallis), which emanates from valleys that run along the western side of Lunae Planum, entering Chryse from the west, is at least 2000 km long (Fig. 3.11). Other large channels are located in Amazonis, Memnonia, Hellas and Elysium.

A further and peculiarly Martian kind of landscape type is developed along sections of the boundary between the upland southern and lowland northern hemispheres, termed the line of dichotomy; this is termed fretted terrain. Its classic development is found between latitudes 30° and 45°N, and longitudes 280° and 350°W, where it consists of extensive remnants and isolated mesas of the original high plateau between which are lower plains units which take the form of isolated depressions and sinuous flat-floored channels. These fretted channels reach deep into the cratered plateau. Channels sometimes start from

Fig. 3.12 *Fretted channel development, Deuteronilus Mensae. V.O. frame 302S06. Frame width 82 km.*

large impact craters; elsewhere they begin, abruptly, without any obvious connection with specific landscape features. The floors of these channels everywhere contain glacier-like sheets of sedimentary debris. The characteristic surface striae they exhibit represent the results of laminar flow in clastic material that is actively being worn from the cratered hemisphere and being slowly transported, downslope, on to the lower, less heavily cratered plains of the north (Fig. 3.12). This zone is one of active erosion, transportation and deposition.

3.3 THERMAL INERTIA MAPPING

The infrared radiometer aboard each of the Viking orbiters made a series of measurements of the temperatures at the Martian surface as they orbited the planet. As has already been discussed, a wide range in temperature values was recorded. Temperature varies as a function of several factors. It is dependent on the latitude, season, time of day and the properties of the surface materials where the measurements are being made. With respect to the time of day temperature values were found to be at their lowest just prior to dawn, rising quickly during the morning and reaching a maximum just after noon; there-

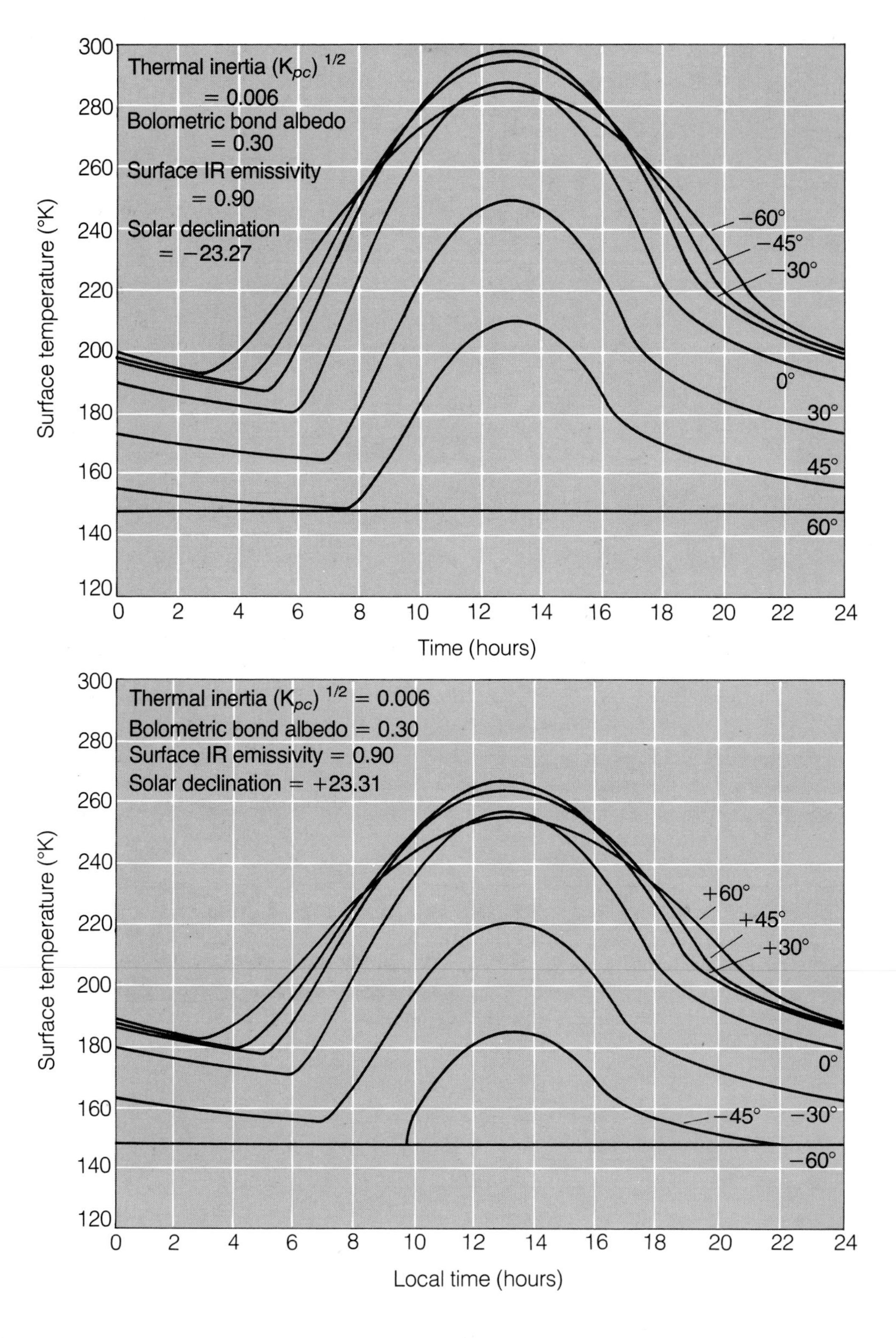

Fig. 3.13 *Diurnal temperature variation as a function of latitude (after Michaux and Newburn, 1972).*

after, temperatures fell rapidly during the afternoon, this fall slowing down during the night and reaching the predawn minimum (Fig. 3.13).

Predawn temperatures are a sensitive indicator of thermal inertia which, along with albedo, is one of the two most important properties of the Martian surface materials. Thermal inertia can be represented by $(K\rho c)^{1/2}$, where K is the thermal conductivity, ρ is the density, and c the specific heat. It is a measure of the responsiveness of materials to changes in temperature; thus if a material has a high thermal inertia, it will respond slowly to temperature changes; if it has a low thermal inertia, the response will be much quicker.

The dominant cause of variations in thermal inertia values on Mars is almost certainly the grain size of the surface layer. It is easy to see why grain size has this effect for, as it gets smaller, the number of interfaces between the constituent grains increases; consequently, because the interfaces conduct heat less efficiently than the grains themselves, the thermal conductivity diminishes. The upshot of this fact is that thermal inertia measurements can tell us where the Martian surface layer is relatively coarse grained and where it is finer.

Hugh Kieffer and colleagues (Kieffer *et al.*, 1977) used the Viking IRTM experiment to produce a global map of thermal inertia values, corrected for latitude and seasonal variations (Plate 4). Values were found to range over one order of magnitude and show only a partial correlation with global physiography. However, the work showed how much of the Tharsis region, together with Elysium and Amazonis, have low thermal inertia, indicative of their being mantled by relatively fine-grained debris. A further region of similar values was found north of the equator, on either side of longitude 330°W. High values, on the other hand, were found throughout the equatorial canyons, the regions of chaotic terrain and the major channel systems, suggesting that such areas have a relatively high incidence of blocks and boulders. Interestingly, a strong negative correlation exists between albedo and thermal inertia. Bright regions have low thermal inertia and presumably are fine-grained, while dark ones have high thermal inertia and so are probably coarser-grained.

3.4 THE STRATIGRAPHY OF MARS

Mapping of the planet, largely by scientists at the US Geological Survey, but also in collaboration with workers from other countries and organizations researching within the NASA framework, have produced a global stratigraphy of Mars. A broad time-stratigraphic classification into three periods – Noachian, Hesperian and Amazonian – was achieved by Scott and Carr who produced the first global geological map in 1978 (Scott and Carr, 1978). Subsequently, after detailed study of Mariner 9 and Viking orbiter imagery, more detailed geological maps have been produced for the western hemisphere (Scott and Tanaka, 1986), for the eastern hemisphere (Guest and Greeley, 1986) and for the polar regions (Tanaka and Scott, 1987). Kenneth Tanaka of the USGS Branch of Astrogeology at Flagstaff, published a detailed stratigraphy in 1986 (Tanaka, 1986), subdividing the three-period time-stratigraphic units into a number of series. This system is used throughout the book. The impact crater density boundaries for each of the series is given in Table 3.1, and the distribution of the rocks of each series is shown in Plate 5.

Table 3.1 Impact crater-density boundaries for Martian series.
(after Tanaka (1986), with permission)

Series	Crater density (N = no.craters >(x) km diam./10^6 km^2)			
	$N(1)$	$N(2)$	$N(5)$	$N(16)$
Upper Amazonian	<160	<40		
Middle Amazonian	160–600	40–150	<25	
Lower Amazonian	600–1600	150–400	25–67	
Upper Hesperian	1600–3000	400–750	67–125	
Lower Hesperian	3000–4800	750–1200	125–200	<25
Upper Noachian			200–400	25–100
Middle Noachian			>400	100–200
Lower Noachian				>200

3.5 SUMMARY

The present face of Mars has developed due to the complex interplay of internal and external forces. The oldest surfaces on the planet are mainly found in the southern hemisphere, where heavily cratered terrain is dominant. Several large impact basins also occur. This region also shows evidence of early fluvial activity and of the emplacement of smoother intercrater plains whose production may partly have been due to volcanism. Volcanism has been responsible for the growth of massive shield volcanoes and for the widespread emplacement of volcanic plains in the northern hemisphere. Volcanic episodes also generated low-profile volcanoes in the southern hemisphere at a much earlier time.

Extensional faulting gave rise to widespread fracturing, particularly around the Tharsis Bulge. A vast canyon system, Valles Marineris, developed along one of the principal equatorial fracture lines, and is associated with large regions of collapsed or chaotic terrain, largely north of the Martian equator. Major channel systems emerge from several of these areas of chaos and must once have flowed towards the Chryse and other northern plains. The active erosion of the southern cratered uplands has produced a belt of fretted channels along the line of dichotomy separating the two topographic/physiographic hemispheres.

Surrounding the poles are considerable thicknesses of layered deposits, exposed in circumpolar canyon networks. At lower latitudes, extensive dunefields have developed around the poles in the northern hemisphere. High-latitude plains appear to be mantled extensively in sedimentary debris.

4 THE ATMOSPHERE AND WEATHER

In many ways the dynamics of the Martian atmosphere are less complicated than those of our own planet. For a start, the atmosphere of Mars is very much more tenuous and therefore responds more slowly to changes in temperature; it has a smaller heat capacity, too. Then again, Mars has no oceans to affect the transport of heat or to provide extensive reservoirs of moisture with which the air can interact.

4.1 THE COMPOSITION OF THE ATMOSPHERE

Early spectroscopic studies by Adams and Dunham, working at Mount Wilson Observatory during the 1930s, failed to detect oxygen and also revealed there to be little or no water vapour in the atmosphere. Carbon dioxide certainly had not been detected at that time, and the assumption being made most widely was that the main constituent was probably nitrogen, while theory predicted that argon might also be a minor but significant constituent (see Kuiper, 1952). Gerard Kuiper made the first positive identification of carbon dioxide as recently as 1947, and it was not until the mid-1960s that the composition and atmospheric pressure became known with any degree of certainty.

The earliest reasonably accurate estimates of atmospheric pressure were those of Owen (1966) and Belton *et al.* (1968), who arrived at very low values that were confirmed later by Kliore *et al.* (1969) during the Mariner 6 and 7 missions. The average surface pressure has been found to be only 8 mbar – less than 1/100th that of the Earth – which means that a person standing at a level near to datum on Mars, would experience a pressure close to that found at an altitude of 30 000 m above terrestrial sea level. There is, furthermore, a range in elevation of over 30 km on Mars; consequently, because the pressure varies with the altitude, there is a range in surface pressure approaching an order of magnitude. Thus, at the summit of Olympus Mons (27 km above datum) the air would be exceedingly thin and would render the safe landing of a Viking-type spacecraft an equivocal matter, as the atmospheric braking mechanism might not function sufficiently well to avoid high-velocity impact.

Because of the rarified atmosphere and the low temperatures characteristic of the planet, liquid water becomes unstable and freezes on the surface. Another manifestation of the low atmospheric pressure is the relative ease with which particulate material can be transported across the Martian surface and raised into the air. Clouds of dust so formed may coalesce into global dust storms which become a common feature during certain seasons.

Table 4.1 illustrates how the Martian atmosphere is composed mainly of carbon dioxide (CO_2), with nitrogen (N_2) and Argon (Ar) next in abundance; very little oxygen is present. While amounts of water vapour are very small indeed, for night-time temperatures the atmosphere is close to saturation for this component almost everywhere and theory predicts that water may be locked in the subsurface as ice which may lie at quite shallow depths at latitudes higher than 40°.

Table 4.1 Composition of the atmosphere at the surface
(Owen *et al.*, 1977)

Component	Proportion
Carbon dioxide	95.32%
Nitrogen	2.7%
Argon	1.6%
Oxygen	0.13%
Carbon monoxide	0.07%
Water vapour	0.03%
Neon	2.5 p.p.m.
Krypton	0.3 p.p.m.
Xenon	0.08 p.p.m.
Ozone	0.03 p.p.m.

4.2 ATMOSPHERIC PRESSURE VARIATIONS

The abundances of minor constituents, like H_2O, O_2 and O_3, vary with the season and also the geographical location; the total surface pressure also varies with the season. The latter was measured at both Viking lander sites, where a long enough run of data was obtained to show that during southern winter there was a gradual decrease in the surface atmospheric pressure which was then reversed as spring approached (Fig. 4.1). At the Viking 1 site the pressure ranged from 6.7 mbar during northern summer to 8.8 mbar at the commencement of northern winter (Hess *et al.*, 1977; Ryan *et al.*, 1978); at the Viking 2 site the equivalent data were 7.4 mbar and 10 mbar, the higher values here probably reflecting the somewhat lower elevation of this site.

It seems reasonable to assume that this behaviour reflects the seasonal condensation of CO_2 on the southern polar cap, thereby reducing the atmospheric mass during southern winter. With the onset of spring, CO_2 would be released again and produce an atmospheric mass gain. Because winters in the southern hemisphere currently are both longer and colder than those in the north, the south cap is larger than its northern counterpart and therefore incorporates a correspondingly greater volume of atmospheric CO_2 during southern winter. Thus it is that variations in the southern hemisphere tend to dominate the seasonal pressure cycle at the present time.

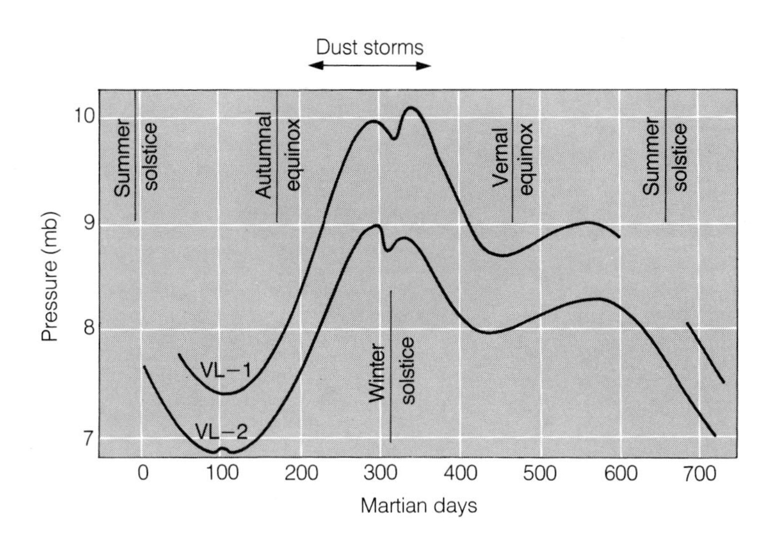

Fig. 4.1 *Mean diurnal pressure variation at the two Viking lander sites for one Martian year (after Hess et al., 1980).*

4.3 ISOTOPIC ABUNDANCES IN THE ATMOSPHERE

Valuable clues to the evolutionary history of the Martian atmosphere can be gleaned from study of certain isotopes, in particular those of nitrogen, argon, neon, xenon, oxygen and carbon. The noble gases, in particular, are useful indicators, since being inert, they are not removed from the atmosphere by chemical reactions. The present atmosphere of Mars may be quite different from the original one, both in terms of the relative proportions of the components and of the actual amounts. For instance, much of the original CO_2 and H_2O may now be locked into the subsurface rocks and regolith, while exospheric processes in the upper reaches of the atmosphere may have stripped away a significant fraction of the original air.

Now nitrogen, oxygen and carbon all have the potential to escape from the Martian gravity field, although the process is a very slow one. The potential for escape is biased towards the lighter isotopes of these elements (e.g. ^{12}C, ^{14}N and ^{16}O), which means that should these escape from the upper levels of the air mantle, the lower levels should become relatively enriched in the heavier isotopes (e.g. ^{13}C, ^{15}N and ^{18}O). In 1976, Nier *et al.* reported that the naturally occurring heavy isotope of nitrogen (^{15}N) is enhanced by a factor of 1.7 over the terrestrial value. Subsequently, this was confirmed for the Viking 1 samples by Owen *et al.* (1977). On the basis of such an enrichment McElroy *et al.* (1977) estimated that a significant proportion of the nitrogen inventory of Mars must have been lost, with a corresponding loss of original water to the extent of 130 m, averaged over the whole planet. There is, surprisingly, no similar enrichment in ^{18}O, and McElroy and his colleagues attributed this fact to the existence of a large reservoir of oxygen in close proximity to the Martian surface; this could be in the form of H_2O or CO_2. Such a source would readily be interchangeable with oxygen in the atmosphere, and if large enough, could lead to dilution of the fractionation effects, even after 4.5×10^9 a of Martian evolution. On the assumption that the oxygen isotopes are fractionated by about 5%, they estimated that the reservoir – possibly composed mainly of water – could be equivalent to a layer of water 13 m deep, over the entire surface of the planet. However, the actual degree of fractionation may be considerably less than this, with the consequence that the amount of degassed water could be far higher. Should this be so, then the estimates relating to nitrogen isotopes (a 130 m layer of water) could be nearer the truth.

Owen and his co-workers also made determinations of carbon and oxygen isotopes, finding them to exhibit terrestrial values, at least insofar as they could tell within the error limits of their measurement procedures ($\pm 10\%$). They also discovered ^{36}Ar, and established that the ratio of $^{36}Ar:^{40}Ar$ was only about one tenth the terrestrial value. When the Viking 2 lander was successfully deployed, they made further isotopic measurements, and established that ^{129}Xe was much more abundant with respect to the other xenon isotopes (^{131}Xe and ^{132}Xe) in the Martian atmosphere than in the terrestrial.

What is the significance of these data? In an attempt to synthesize all the chemical information, Anders and Owen (1977) argued that two particular elements are important in terms of the way certain elemental groups behaved in the solar nebula during the formative stages of planetary development. These two, potassium (K) and tellurium (Tl), provide keys to the groups of elements which condense at temperatures between 1200 K and 600 K, and <600 K, respectively; the latter (Tl) therefore provides clues to the group that contains carbon, nitrogen, chlorine, bromine and the noble gases, and naturally is of particular significance to studies of Mars. After examining abundances of these two elements in meteorites, on the Moon and on the Earth, Anders and Owen estimated that the Martian atmosphere contains 100 p.p.m. K and 0.14 p.p.b. Tl, their quoted figure for Tl being estimated from the measured ratio of $^{40}Ar:^{36}Ar$. Now the terrestrial value for Tl (4.9 p.p.b.) is substantially greater than the Martian, and they argue that the high ratio for $^{40}Ar:^{36}Ar$ cannot be explained away by more effective outgassing of ^{40}Ar, because this would also release ^{36}Ar; consequently they assert that amounts of ^{36}Ar must always have been low, and along with it, all other elements of the Tl group.

Since the present abundance of ^{36}Ar should approximate to the amount that has been outgassed (little should have escaped), it seems reasonable to suggest that Mars' outgassing efficiency could only have been about 0.27 that of the Earth. With this figure in mind, Anders and Owen estimated the amount of CO_2 which had degassed and, assuming all of this had been held in the atmosphere at the same time, suggest it would have exerted a surface pressure of 140 mbar. Had Martian outgassing been as efficient as Earth's, this figure should have been as high as 520 mbar.

Since Anders and Owen completed their work, fresh data derived from the Pioneer Venus mission appears to have rendered some of their arguments invalid. For instance, it appears that the non-radiogenic rare gases actually behave independently of other volatiles during planetary accretion, and Pollack and Black (1979) argue that the abundances of the rare gases within the atmospheres of the inner planets reflect the original nebular gas pressure during accretion, while abundances of other volatiles provide a more accurate measure of outgassing efficiency. Their calculations indicate that the nebular gas pressure in the vicinity of Mars was between one fifth to one twentieth that of the Earth, and they estimate that a volume of water equivalent to a uniform layer between 60 m and 160 m deep must have outgassed. This is comfortingly close to the figures arrived at by McElroy *et al.* (1977) after analysis of the oxygen and nitrogen isotopic data. The general consensus is, therefore, that because Mars' atmosphere is deficient in most volatiles with respect to the Earth, its degassing is significantly less complete, which means that the present atmosphere can represent only a small fraction of the total amount of volatile elements outgassed by the planet.

4.4 ATMOSPHERIC CIRCULATION PATTERN

The absence of large expanses of open water on Mars means that the entire surface responds quickly to solar heating. In this respect, therefore, the circulation of the Martian atmosphere is simpler than that of the Earth. Topographic effects, on the other hand, are more significant and complicate the pattern to a greater degree than they do on Earth. Modelling of the Martian circulation has been undertaken by several groups, utilizing techniques first developed for the Earth (Leovy and Mintz, 1969; Blumsack, 1971; Pollack *et al.*, 1976; Webster, 1977).

In broad terms, warm air on Mars rises over the summer hemisphere and descends over the winter one; the global circulation is, therefore, driven by seasonal temperature gradients. The peculiarly Martian phenomenon which sees approximately 30% of the air condensing on the winter pole, generates a high-latitude region of relatively low pressures. The marked pressure gradient set up by this seasonal activity induces a strong global circulation towards whichever pole is growing by the condensation of CO_2. Such condensation flow tends to dominate wind directions at all latitudes. At mid to high latitudes

Fig. 4.2 *Cyclonic weather system near the northern polar cap. Viking orbiter frame 738A27.*

in the winter hemisphere, westerly winds prevail and both cyclonic and anti-cyclonic weather systems move across the planet (Fig. 4.2). The Martian weather at these latitudes in winter is much like that on Earth. Summers in each hemisphere are different, both from winter and from each other. Thus northern summers see quiet conditions and only a small east-west zonal air flow; cyclonic systems generally do not develop. Southern summers, on the other hand, see frequent dust storms developing; these sometimes affect the entire planet and disrupt the general circulation pattern set up by the global north-south temperature gradients. Viking recorded 35 such storms during 1977, of which two were global in extent. Such storms coincide with the retreat of the southern polar cap, which produces large pressure gradients between the freshly exposed surface and the shrinking cap itself. During spring and autumn, quiet conditions prevail in both hemispheres.

The extreme thinness of the Martian air means that it has a low heat capacity, it therefore cools and heats up much more quickly than our own air. Furthermore, since it is largely composed of carbon dioxide, which is a good IR transmitter, large diurnal temperature ranges are experienced in the lower levels of the atmosphere. When there is little or no dust suspended in the atmosphere, Martian air absorbs little solar energy directly, its temperature profile being controlled principally by heat transfer from the ground, either by conduction or convection. The consequence of this is that whereas there may be diurnal fluctuations in temperature of 50 °C at the surface, these peter out quickly with increasing height (Fig. 4.3); tidal winds therefore are weak since only the near-surface layer is affected. However, in the presence of dust, solar radiation is absorbed and the air heats itself directly. As a result, the atmosphere cools more efficiently at night, with the result that the temperature profile becomes more nearly isothermal (Fig. 4.3). Under these conditions there will be narrower variations of temperature near the ground, but wider ones at greater altitude; tidal winds may become stronger under such 'dirty' conditions.

More local winds are the result of topography. This is much more pronounced on Mars because of the atmosphere's more rapid response to local ground temperatures. Because temperatures are mainly a function of reflectivity and the local cycle of insolation, they may be much the same at the top of a large volcano like Olympus Mons, as at its base. The large horizontal temperature gradients so produced, generate quite strong slope winds. Such winds are downslope at night and upslope during daytime.

4.5 CLOUDS

As has already been mentioned, despite the small amount of H_2O in the Martian atmosphere, it is always close to saturation in water vapour; consequently, the formation of clouds and fogs is a common feature of Martian weather. Larger cloud masses give rise to transient brightenings over whole regions of the planet and frequently have been recorded by Earth-based telescopic observers.

One of the most extensive regional cloud masses is that which forms over the north polar cap during autumn. The northern autumnal 'hood' is pervasive and may extend as far south as latitude 50°N; thus far only scattered clouds have been recorded at the southern cap at this season. An extensive hood

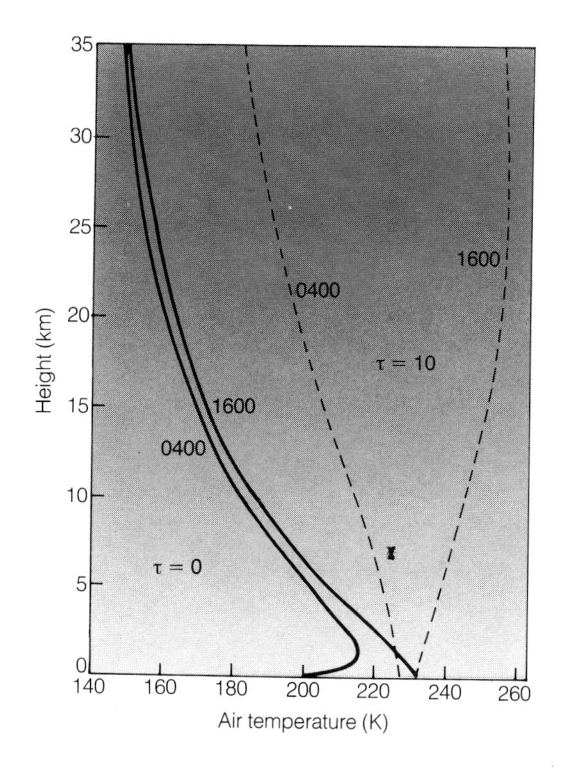

Fig. 4.3 *Temperature as a function of altitude. The large near-surface fluctuations die out with increasing height.*

Fig. 4.4 *Wave clouds formed over the Tharsis Montes volcanoes. Viking orbiter mosaic P-55A.*

does, however, form around the south cap during early spring. The polar hoods appear to represent thick hazes of water ice (and possibly carbon dioxide ice).

Hazes and fogs are also frequently seen to form in low-lying areas at both dawn and dusk. These probably are composed of water ice and are produced by the melting of ground frost due to solar vapourization. Early morning fog is common along the canyon floors of Valles Marineris, particularly amongst the interconnecting canyons of Noctis Labyrinthus.

Clouds also form in the lee of major topographic features such as impact craters. Such wave clouds also develop around the polar caps when the hood is developed. During spring and summer, moisture-laden air forced to rise over major features such as the Tharsis Montes and Olympus Mons, results in the growth of orographic clouds (Fig. 4.4). These tend to develop slowly during the morning and reach a maximum development during the afternoon. Convective clouds also develop in elevated regions, usually around midday. Areas such as Syria Planum are affected in this way. The formation of such clouds appears to be the result of atmospheric instabilities created by strong surface heating. Finally, regional or even global obscuration of Mars may be effected by dust storms. These are the yellow clouds recorded frequently by Earth-based observers.

THE INTERIOR OF MARS

5

5.1 INTRODUCTION

Despite knowing a considerable amount about the surface of Mars, relatively little is known about the interior with any degree of certainty. This is largely a function of the very limited geophysical and geochemical data that is available. What is known is the mean density which is $3930\,kg\,m^{-3}$, and the mass which is only about one-tenth that of the Earth. When the density value is corrected for the effect of self-compression in Mars' gravity field, the figure is seen to be smaller ($3730\,kg\,m^{-3}$) than that for either Venus or the Earth. This density deficit traditionally has been explained in one of two ways: either Mars always contained less iron than the Earth (Urey, 1952), or the total amount of iron was roughly the same for all the terrestrial planets (bar Mercury) and the carbonaceous chondrites, but on Mars it was in a more oxidized state (Ringwood, 1966).

The moment of inertia – a property which allows for modelling of the internal distribution of density within a planet – can be calculated. In the Earth's case, because both the Sun and Moon exert torques on the equatorial bulge, its moment of inertia is known with a high degree of accuracy. Clearly this does not apply to Mars, and moment of inertia calculations are based largely on the effects Mars has upon spacecraft trajectories and also upon its larger moon, Phobos.

Much of our information relating to the Earth's interior structure is derived from geophysical data, in particular that relating to seismic activity. Further information, particularly concerning the upper mantle, derives from geochemical analyses of exotic blocks brought to the surface in volcanic vents and diatremes, and from experimental work on igneous rocks. Thus far little such petrological data exists for Mars, while seismic data is also minimal; the best that can be said is that Mars appears to be much less seismically active than the Earth. An estimate of the thickness of the Martian crust, derived from two possible seismic events recorded at the Viking 2 lander site in Utopia, gives a figure of 16 km. Current knowledge of the internal structure of Mars largely stems from what has been learned of its figure and gravity field and from modelling exercises, most of which assume Mars to have been formed from primordial material of approximately Type 1 carbonaceous chondrite composition.

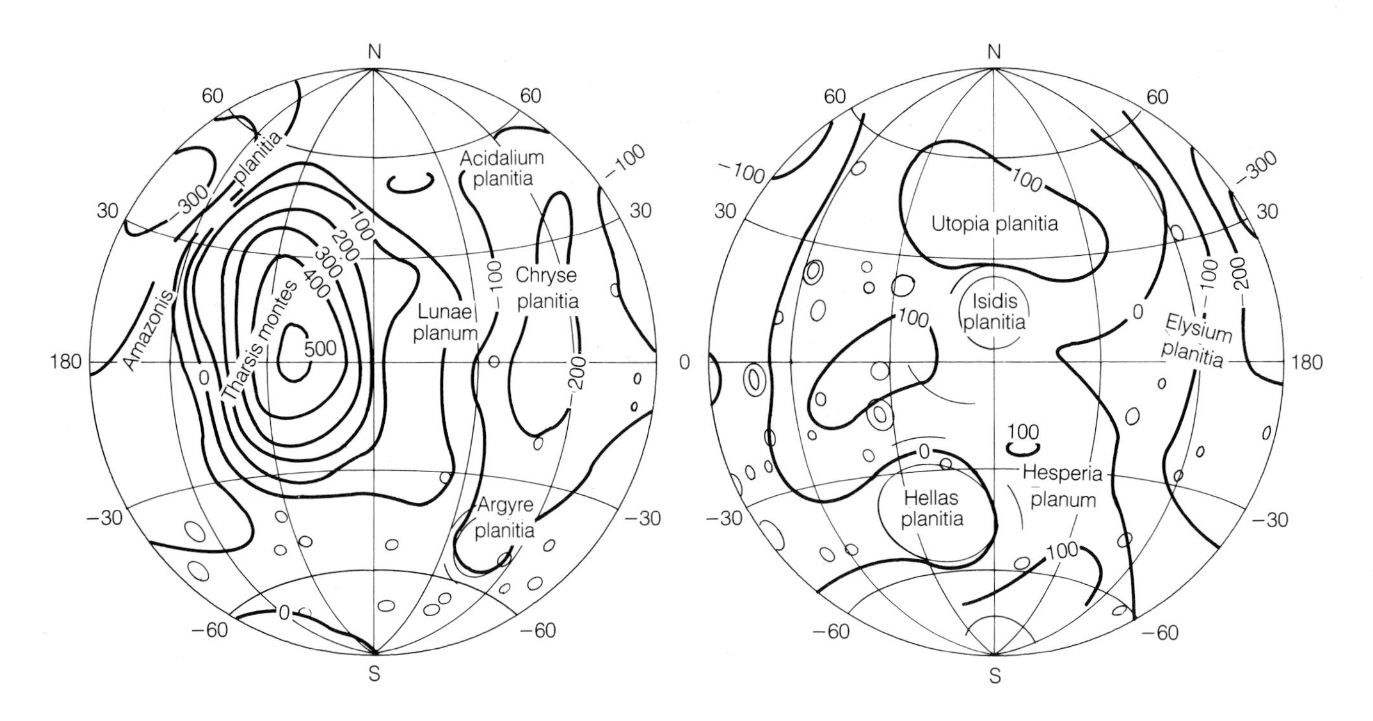

Fig. 5.1 *Map of long-wavelength free-air gravity values (after Sjogren).*

5.2 THE FIGURE AND GRAVITY OF MARS

Careful measurements of the shape of the planet reveal the mean equatorial radius to be 3396.6 km and the polar radius 3376.7 km; this gives an ellipticity of 0.005 (Earth is 0.003). The precise determinations that have been made of the orbital parameters and rotational period of Mars, and of its two small moons, allow for calculation of the mass (6.418×10^{23} kg) and the hydrostatic approximation of dynamical flattening (5.24×10^{-3}). The latter has consistently been reported as being smaller than the optical flattening (difference between equatorial and polar radii, divided by the polar radius – 5.86×10^{-3}), which implies that the planet cannot be in isostatic equilibrium (they should be equal). The difference between the two is about 25 km and from this figure Urey (1950) suggested that Mars might have an equatorial belt of isostatically compensated low mountains, producing the observed discrepancy without requiring large non-hydrostatic stresses.

Using data on the figure of the planet, calculations of the Martian moment of inertia coefficient (I/MR^2) have been made; they suggest a value of between 0.365 (Reasenberg, 1977) and 0.3759 (Cole, 1978) – values which imply that Mars is not a homogeneous body. The former value was based on the assumption that the huge load of Tharsis was largely uncompensated and was supported by the innate strength of the interior. A similar value was arrived at by Kaula (1979), whose model made the assumption that non-hydrostatic contributions to the moment of inertia are axially symmetrical with respect to the principal axis in the direction of Tharsis. The latter of the two quoted values derives from an equation which assumes the planet to be in hydrostatic equilibrium; interestingly, it is very close to the figure of 0.3752 resulting from an equation involving the dynamical behaviour of Mars (Cole, 1978). The closeness of these two values means that the interior of Mars must be approximately in hydrostatic equilibrium.

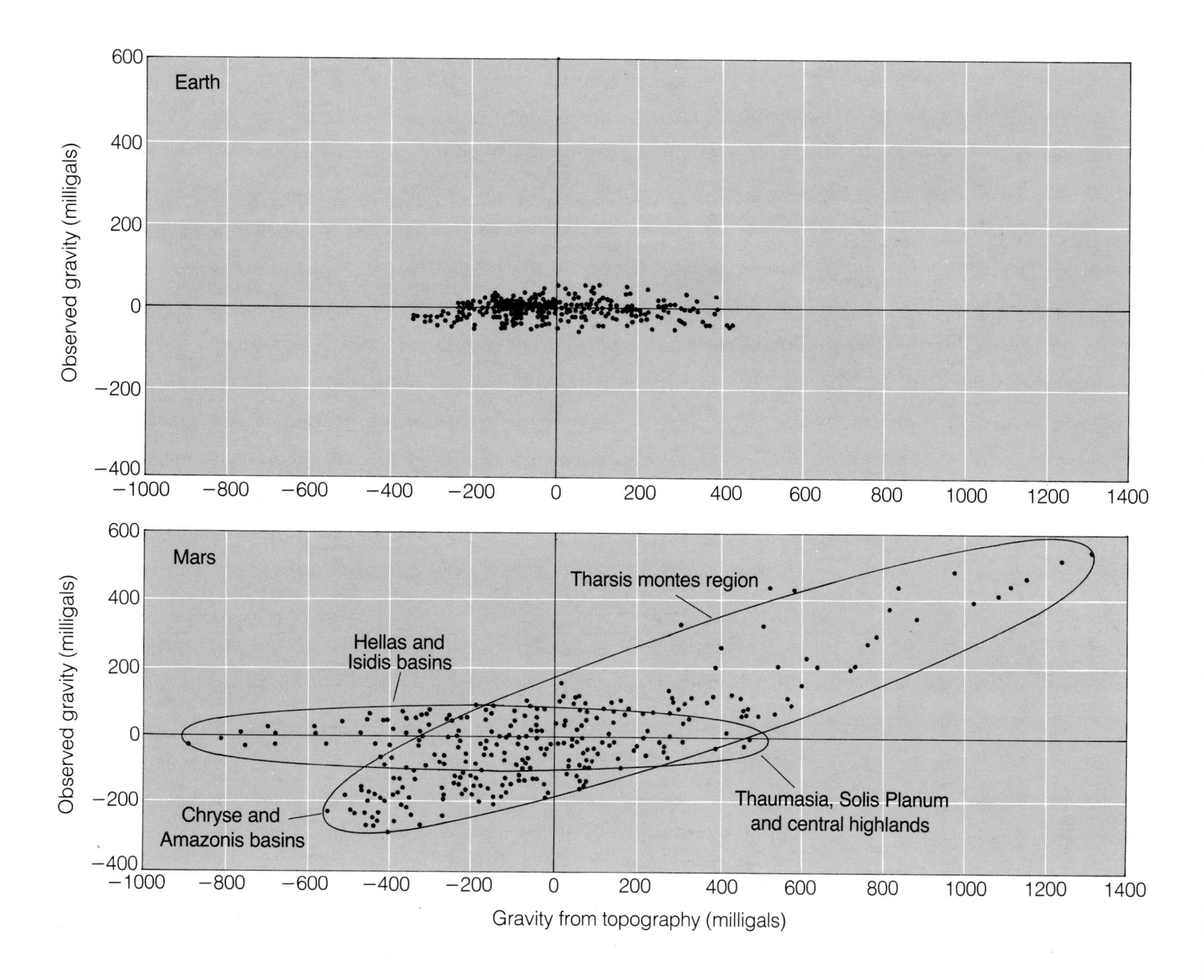

Fig. 5.2 *The relationship between gravity and topography.*

Most geophysical modelling exercises presuppose that Mars, like Earth, has a crust, mantle and core; furthermore, generally they assume the crust to be a shell of approximately uniform density and thickness, not in hydrostatic equilibrium, and making but a small contribution to the total planetary mass and moment of inertia. The non-equilibrium state of the Martian crust was confirmed by the Mariner 9 spacecraft which experienced marked variations in acceleration as it orbited the planet. The gravity measurements made by the Mariner and Viking spacecraft indicate that the elevated region of Tharsis and the adjacent basins of Chryse and Amazonis are assocated with large free-air gravity anomalies (Fig. 5.1). The long-wavelength free-air positive gravity anomaly over the Tharsis region is approximately 500 mGal, while large negative anomalies exist over Chryse and Amazonis (Lorrel *et al.*, 1972; Sjogren *et al.*, 1975). Because of the distinct regional correlation between gravity and topography (Fig. 5.2) – something which does not happen on Earth – it is

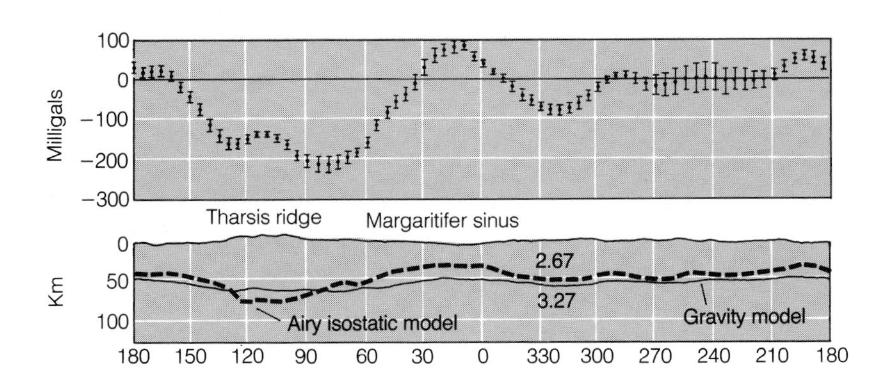

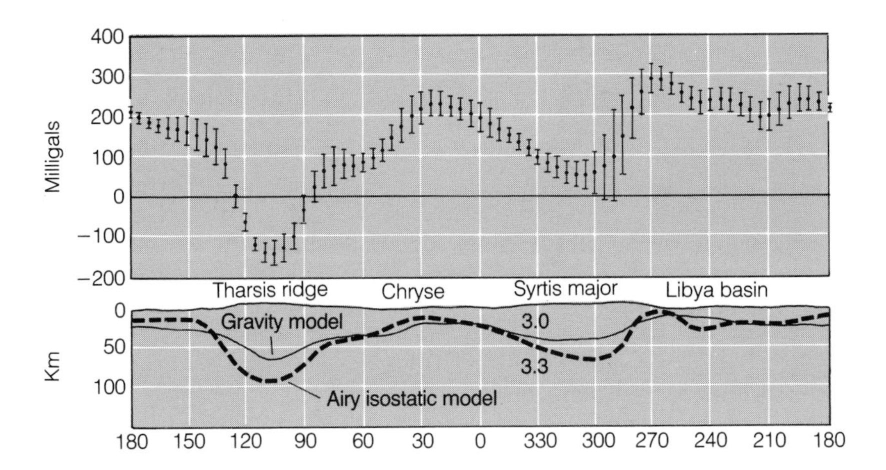

Fig. 5.3 *Bouguer gravity data for Mars.*

generally accepted either that there is incomplete isostatic compensation over these extensive regions, or that compensation is achieved at depths greater than 1000 km (Phillips and Saunders, 1975). Elsewhere, gravity and topography do not correlate and it has to be assumed that compensation is achieved at relatively shallow depths.

By calculating the gravitational effects of the observed topography and subtracting these from the gravity as measured by spacecraft, it is possible to arrive at Bouguer anomalies. Phillips *et al.* (1973) and Phillips and Saunders (1975), in undertaking such calculations, established there to be a large negative Bouguer anomaly beneath Tharsis, and corresponding positive anomalies beneath the adjacent lowland plains (Fig. 5.3). The obvious inference here is that at least partial compensation has been achieved, and Phillips and his co-workers suggest that the anomaly associated with Tharsis could be accounted for by variations in crustal thickness across the region. They suggest that crustal thickness may range from 20 km beneath basins to 130 km beneath Tharsis. Subsequent geophysical modelling by Comer *et al.* (1985), who determined the thickness of the elastic lithosphere of Mars at different sites by analysing the tectonic effects of major volcanic loads, gave a range of 20 to 150 km over the Tharsis region, and a figure in excess of 120 km beneath Isidis.

5.3 INTERNAL DENSITY PROFILE

As we have seen, the mean density of the planet is 3930 kg m^{-3} – about 75% that of the Earth – implying that, compared to the Earth, Mars is apparently deficient in heavy elements such as metallic iron. Nevertheless, the density at the centre of Mars is likely to attain a value close to 8000 kg m^{-3}. Early work by Sir Harold Jeffreys (1970) assumed that the interior of the planet was composed largely of the mineral olivine, with only a small iron core, and that density variations with depth largely are the result of olivine/spinel phase changes. The pressure at the centre of Mars must approach 1011 Pa which is well above the pressure at which terrestrial-type materials would show phase changes of this type.

Since the magnetometer on board Mariner 4 was unable to detect a magnetic field, it has to be assumed either that Mars lacks a metallic Ni/Fe core or that one is present but is convecting too slowly to act as a dynamo and thus generate a field. Various geochemical, geophysical and geological constraints suggest the latter to be the more likely. Calculations by Cole (1978) based on the moment of inertia, suggest that if an iron core does exist, its radius must approach 0.33 that of Mars; this would give it about 6% of the planetary mass, compared with 32% for the Earth's iron core. Johnston and Toksoz (1977) and Solomon (1979) respectively calculate the core to have a radius of 1400 km and 2000 km, accounting for between 7 and 21% of the total volume. Geophysical and geochemical constraints also mean that the mantle of Mars must be enriched in FeO compared with the Earth, perhaps by as much as three times (McGetchin *et al.*, 1981), with a density of 3500–3600 kg m^{-3} (Phillips, 1990). A density profile for Mars, based on the arguments of Cole (1978), is presented as Fig. 5.4.

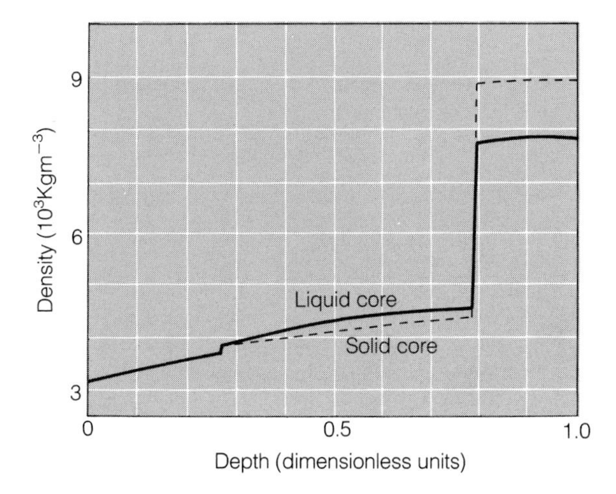

Fig. 5.4 *Density profile for Mars (after Cole, 1978).*

5.4 COMPOSITION OF THE INTERIOR

Numerous models have been proposed for the Martian interior, most having been developed since the 1960s at which time the relationships between high-pressure mineral phases became better understood (Kovach and Anderson, 1965; Binder, 1969; Johnston and Toksoz, 1977). Ringwood and Clark (1971) investigated the properties of various bulk compositions similar to dehydrated Type 1 carbonaceous chondrites, with iron in the oxidized state. The most plausible of their models contained 60% olivine +19% pyroxene +21% magnetite, a composition which, at $P-T$ conditions appropriate to the Martian interior, showed two major series of phase changes, at depths of 1200 and 2000 km. Since the calculated density of a homogeneous Mars of this composition was 3970 kg m^{-3} – in good agreement with observation – this at first seemed an appropriate model; however, the moment of inertia coefficient for such a composition turned out to be 0.391, much greater than that observed. Consequently Ringwood and Clark modified the model to assume that there had been extensive melting and differentiation, with the result that the high-pressure form of magnetite could segregate into a core of radius 1640 km (Fig. 5.5). By doing this, the calculated density of 3940 kg m^{-3} and inertia coefficient

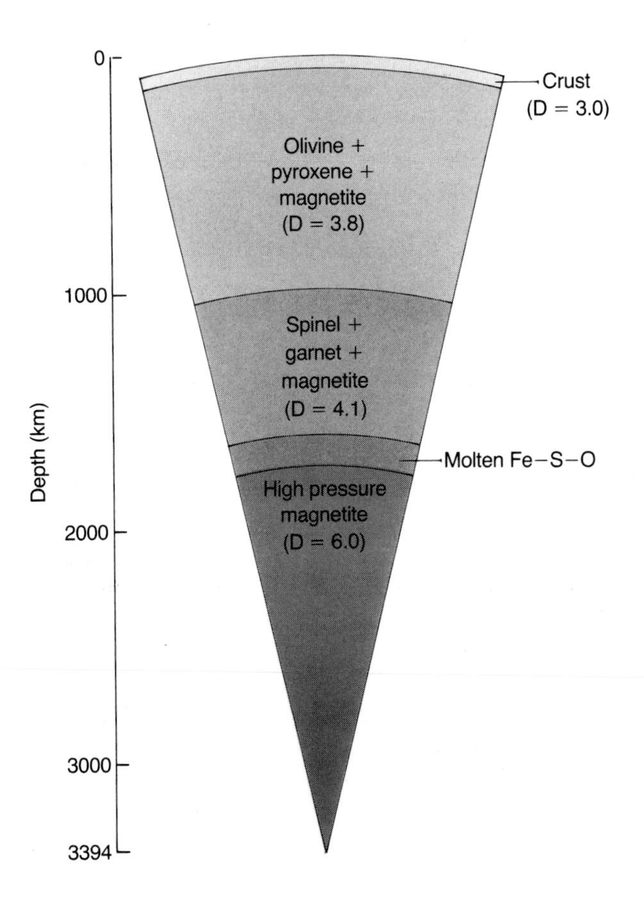

Fig. 5.5 *Model for Martian core segregation (after Ringwood and Clark, 1971).*

of 0.373, fell into excellent agreement with observation. While this may not be the only plausible or even the actual model for the interior of Mars, it is capable of accounting for the density and moment of inertia, and has the essential characteristic of a Martian mantle which contains substantially more FeO than Earth's. For this reason it has become widely accepted that Mars is significantly more highly oxidized than the Earth.

More recently, studies of SNC meteorites – now widely attributed to Mars – have led to newer experimental and modelling exercises, some of which have been reviewed by Kerridge and Matthews (1988), and also by Holloway (1990). Such work indicates that the parent magmas of SNC-meteorites must have been low in Al_2O_3, high in CaO or CO_2 and depleted in light REEs, while multiple melting events have to be invoked to account for the phase relationships seen. The bulk compositions of the mantle/crust and core of Mars, derived from SNC-meteorite studies are shown in Table 5.1.

Table 5.1 Bulk composition of Mars derived from SNC meteorites (after Chicarro *et al.*, 1989)

Mantle/crust (%)		Core (%)	
SiO_2	44.4	Fe	77.8
Al_2O_3	3.02	Ni	7.6
FeO	17.9	Co	0.36
MgO	30.2	S	14.24
CaO	2.45	Core mass = 21.7%	
TiO_2	0.14		
Na_2O	0.50		
P_2O_5	0.16		
Cr_2O_3	0.76		
K(p.p.m.)	305		
Ni(p.p.m.)	400		

5.5 DIFFERENTIATION OF THE MARTIAN CRUST

The Viking samples are not particularly helpful when it comes to modelling how Mars may have fractionated chemically, producing the magmas that reached the surface. Geophysical and geochemical constraints (including those from SNC meteorites) indicate that the mantle of Mars must be relatively enriched in iron when compared with the Earth, perhaps by as much as a factor of three (McGetchin *et al.*, 1981). This means that the first 1–2% of partial melt to form from a composition consistent with all available data, will be relatively enriched in iron and depleted in magnesium compared to a similar melt on Earth (Holloway and Bertka, 1989). The result is that Martian magmas are likely to be very low in silica and alumina and very rich in iron. This makes for lavas with very low viscosity and high density.

Several workers have suggested that primitive Martian magmas (i.e. those which were generated during terrestrial Archaean times) were like terrestrial komatiites, which have similar properties to the above and, in addition, are rich in sulphur – like the finds at the Viking lander sites. It may well be, therefore, that komatiite-type magmatism was prevalent on Mars during its early history and, indeed, since the SNC-meteorites have a 1.3 billion year crystallization age, may have persisted into more recent times on Mars than it did on the Earth (Burns and Fisher, 1989). This remains an area of very active research at the present time.

6 THE ANCIENT CRATERED TERRAIN

6.1 INTRODUCTION

While the heavily cratered upland plains which lie south of the line of dichotomy bear some similarities to the lunar highlands, they show widespread evidence for a much more extensive history of modification than has occurred on the Moon. Early impact cratering left a very obvious imprint on both worlds while subsequent impact erosion severely degraded the more ancient craters and basins. On Mars, however, fluvial and aeolian activity, volcanism, tectonism and non-impact erosion each has played an important part in bringing the highland plateau to its present geomorphological state. Thus, while a superficial look at a few Viking images of the upland hemisphere suggests a lunar-like surface, closer inspection reveals not only a difference in the nature of the cratering record but also in the morphology of the craters themselves (Fig. 6.1).

6.2 THE MARTIAN CRATERING RECORD

Our knowledge of the distribution of the different kinds and sizes of interplanetary object presently moving around within the solar system comes largely from studies undertaken in the vicinity of Earth (Dohnanyi, 1972). It appears that within the solar system today there are solid objects which range in mass from less than 10^{-9} kg (micrometeorites) to about 10^{20} kg (asteroids). Due to the shielding effect of the Martian atmosphere, at the present time only objects with masses more than about 1 kg produce craters on its surface (Gault and Baldwin, 1970). Compared with the Moon, therefore, there should be a relative paucity of the smaller crater sizes on its surface.

Our knowledge of the cratering history of the Moon is fairly well constrained since it has been possible not only to derive crater ages for its different surfaces, but also to date returned lunar samples radiometrically; for Mars we do not have this luxury. It has become clear that about 4×10^9 a BP the cratering rate for the Moon was very high indeed and that subsequently it declined sharply around 3.9×10^9 a BP. In consequence the lunar highlands – which formed before the decline in impact rate – are nearly saturated with craters, while surfaces which were produced after this reduction in the flux (e.g. the maria), are relatively sparsely cratered. Since very few surfaces exist on the Moon which exhibit intermediate crater densities (i.e. the crater density record is bimodal), it has been assumed that the impact flux fell away very

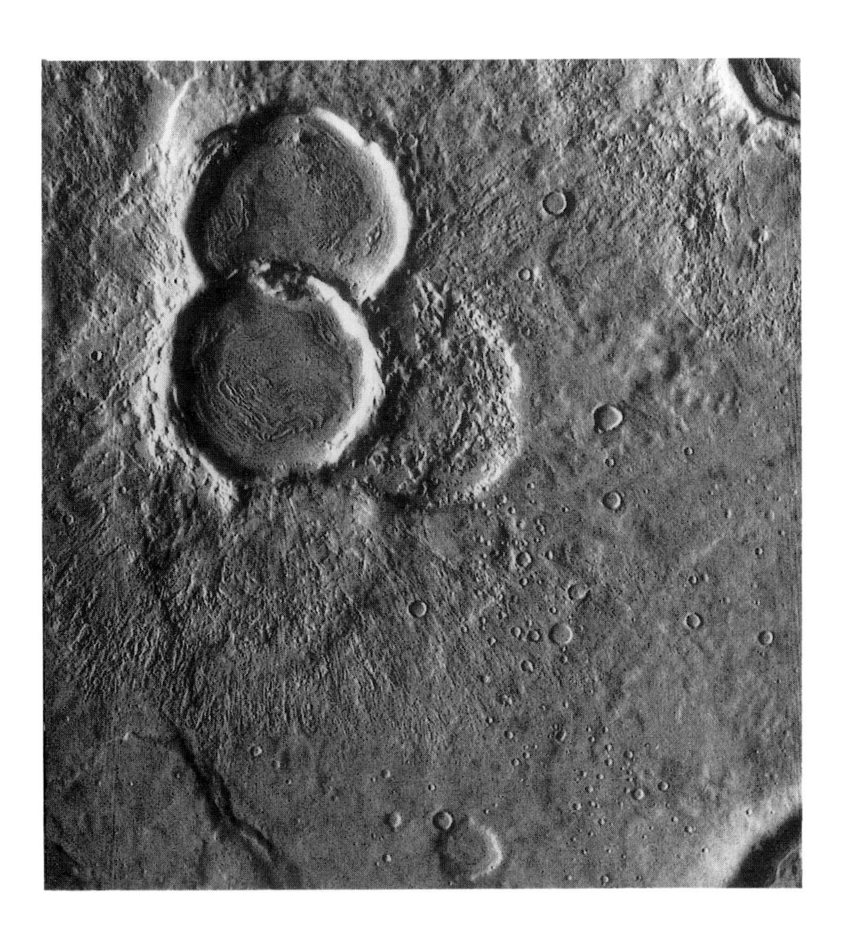

Fig. 6.1 *Craters and intercrater plains in the heavily cratered uplands northwest of Syrtis Major Planum. The pair of overlapping 15 km diameter impact craters show degraded walls and a radiating ejecta pattern; dunes may also be seen on their floors. The intercrater areas are relatively smooth and lack >5 km craters. Note the lobate feature and ridge towards the bottom left and the small moated crater at bottom centre. Viking orbiter frame 641A09. Frame width 60 km. Centred at 34.79°N, 309.14°W.*

rapidly. Soderblom *et al.* (1974) suggested that a similar situation pertained to Mars – a view which generally has been accepted; however, there is continued debate concerning, first, the time at which such a decline occurred, and second, the present cratering rate.

There are abundant statistical studies of lunar cratering but this is not the place to review them. The interested reader is referred to a listing given on page 55 of Mike Carr's excellent book on Mars (1981). It is sufficient here to realise that as a solid surface ages, the tally of superimposed impact craters increases until a point is reached at which the number of new craters being formed is exactly balanced by those being destroyed as a result of the ongoing impact process. The surface is then said to have reached an equilibrium termed **saturation**. The general principles are illustrated in Fig. 6.2. This appears to have occurred on the lunar highlands and over much of Mercury; however the situation on Mercury is complicated by the subsequent (and possibly penecontemporaneous) development of intercrater plains, many of which have a volcanic origin (Cattermole, 1990). The situation for Mars is somewhat similar to Mercury in this respect, though even more complex and possibly less well understood.

There is little doubt that the most densely cratered parts of the Martian upland plateau are representative of the most ancient surfaces on Mars, and may be comparable with the lunar highlands. When the crater statistics are studied, however, the size–frequency distributions for lunar and Martian

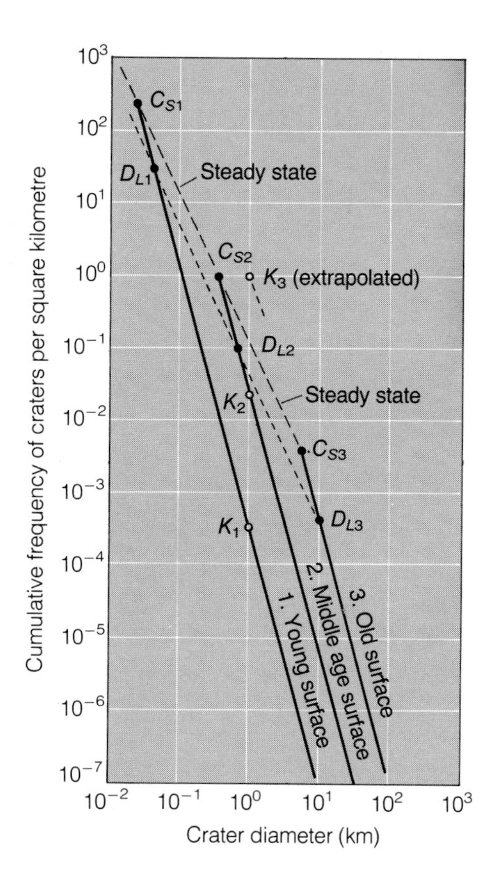

Fig. 6.2 Hypothetical cumulative frequency curves for craters on surfaces of three different ages. The distribution for each is composed of two parts: a steady-state distribution and a crater-production distribution. Crater-production curves for each of three surfaces of increasing antiquity intersect the steady-state curve at C_{s1-3} respectively.

craters are found to be markedly different, since there is a distinct paucity of craters below about 20 km diameter. The resultant sharp change in the slope of the size–frequency curve (Fig. 6.3) was noticed, among others, by the astronomer Ernst Öpik (1965, 1966), who attributed it to an obliterative process which he thought might have removed most of the smaller craters and certainly modified many of the larger ones.

Subsequent work by the US Mariner 6 and 7 teams also noted a bimodal distribution of craters, not only in terms of their size–frequency relationships, but also their morphology. Thus craters larger than about 5 km seemed to have relatively low ramparts and to be shallow; smaller craters were mostly bowl-shaped, like small lunar craters. The more definitive studies associated with the later Mariner 9 mission, led to the realization that the size–frequency curves consisted not of two, but three segments (Fig. 6.4): craters smaller than 5 km and those larger than 30 km gave curves with a slope often exceeding −2, while those between 5 and 30 km diameter fell on a curve of shallower slope verging on −1. The more recent Viking data has confirmed this. In seeking to explain this behaviour, Bill Hartmann (1973) surmised that craters larger than 30 km diameter had survived from an early period when the cratering and obliteration rates were most intense, giving a saturation curve. The middle segment on the size–frequency curve – that with the lower slope – he suggested was representative of an equilibrium curve, produced during the same early era, during which time the formation and destruction of craters exactly matched, the smaller craters being more readily destroyed than the

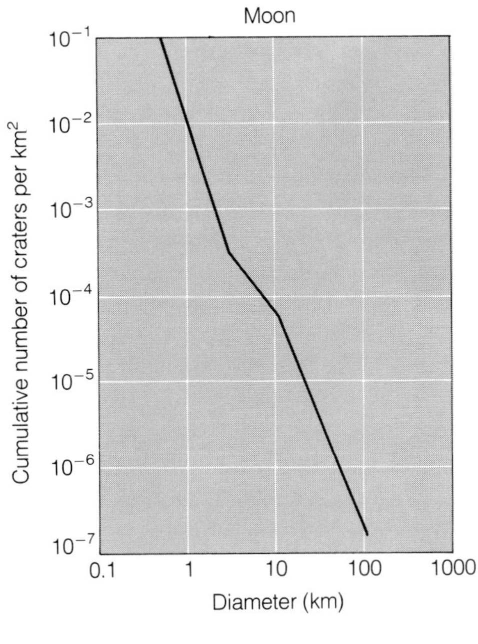

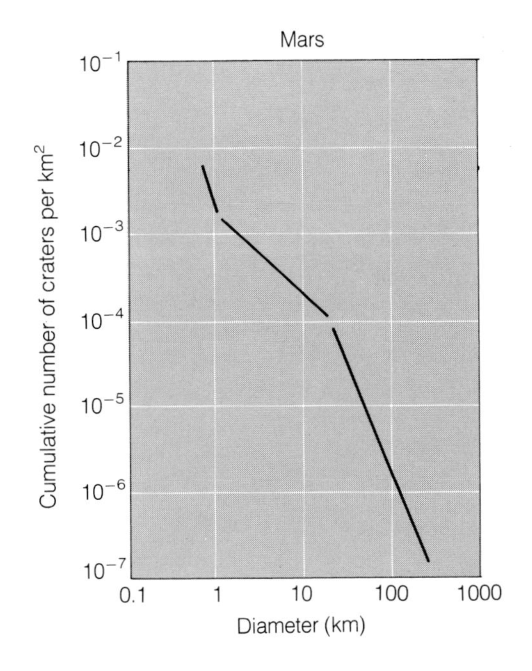

Fig. 6.3 Size–frequency crater curves for Mars and the Moon compared.

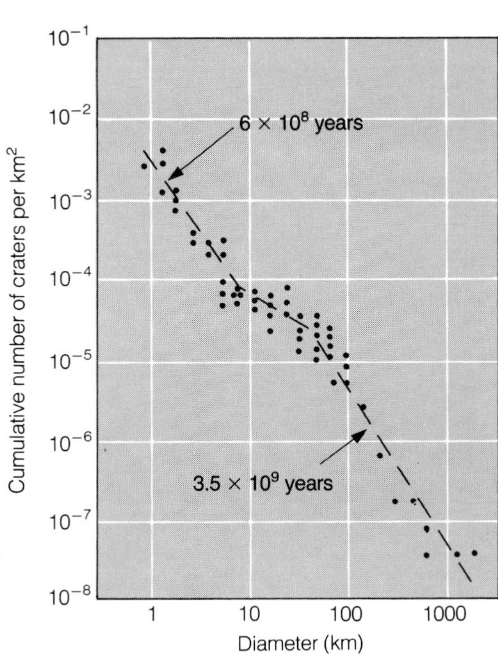

Fig. 6.4 Modified crater-frequency curve for Mars (Hartmann, 1973).

large ones. To explain the third segment on the curves, Hartmann postulated that the impact and obliteration rate then sharply declined, the population of smaller craters (<5 km) postdating this event.

In the mid-1970s, an alternative explanation for the cratering statistics was provided by Jones (1974) and by Chapman and Jones (1977). They suggested that during the early period of intense bombardment, rates of crater formation were much the same throughout the complete size spectrum; subsequently they declined to leave a crater distribution close to saturation for all sizes (thus far, their model is similar to that of Hartmann). They then go on to hypothesize that there was a phase of intense obliteration which destroyed all craters below about 5 km diameter and most in the 5–30 km class; it also greatly modified the larger ones. The new population of <5 km craters was formed after this event.

Regardless of which of these and similar hypotheses comes closest to the truth, it does appear that two populations of impactors affected the inner regions of the solar system. Thus a recent statistical study by Barlow (1988), identifies an older population which gave rise to the multi-sloped distribution curve, and which represents the phase of intense bombardment which affected Mercury, the Moon and Mars (producing the heavily cratered plains which cover about 60% of the latter's surface). She also identifies a younger population whose distribution follows a power-law function, and which primarily is recorded in the more lightly cratered plains regions of these same planets.

Naturally, if cratering were the sole process of crater destruction, then the 5–30 km crater population ought to have approached the lunar slope value of −2. The fact that it does not, strongly implies that other crater-destructive processes must have been at work; the more obvious of these would have been volcanism, fluvial and aeolian activity. There is ample evidence that intense volcanism was widespread during the early geological history of Mars, while the presence of fluvial channels on the upland plateau implies that the Martian atmosphere may have been considerably denser early on. Furthermore, the varied state of preservation of the channel networks themselves is most readily interpreted to have been the result of obliteration by the same process as that which modified the cratering record. Most estimates point to a date of 3.8×10^9 a BP for the decline in impact flux (e.g. Carr *et al.*, 1984).

Of the various attempts at providing an absolute Martian time scale, those of Neukum and Wise (1976), Soderblom *et al.* (1974), Hartmann *et al.* (1981) and Carr (1981) are most widely credited. Table 6.1 shows estimated absolute ages for selected regions of cratered plains using the scheme of Carr (1981), who takes 1 km diameter crater statistics and calibrates them against the ages from

Table 6.1 Ages of Martian plains as derived by Carr (1981)

Plains region	No. of craters <1 km/10^6 km^2	Age in 10^9 years	
		Best estimate	*Range in likely age*
Mare Acidalium	830	1.2	0.2–1.7
Sinai Planum	970	1.4	0.4–3.0
Utopia Planitia	1270	1.8	0.6–2.3
Noachis Planitia	1740	2.5	0.9–3.6
Amazonis Planitia	1940	2.8	1.0–3.7
Syrtis Major Planum	2053	2.9	1.2–3.7
Chryse Planitia	2100	3.0	1.2–3.8
Lunae Planum	2400	3.5	1.7–3.8
Hellas	2640	3.8	2.9–3.9
Hesperia Planum	2710	3.9	3.0–3.9

Hartmann *et al.* (1981). The large errors inherent in such methods are shown by the wide range of possible ages quoted.

6.3 MARTIAN IMPACT BASINS

As with the Moon, the oldest recognizable geological features on Mars are the circular impact basins. Of these Hellas – which arguably is the largest impact basin so far discovered within the solar system (1600 × 2000 km) – Argyre and Isidis are the most obvious on spacecraft images, but many other >250 km diameter ring structures have been identified by Wood and Head (1976) and by Schultz *et al.* (1982). The global distribution of these basins is shown in Fig. 6.5 and in Table 6.2.

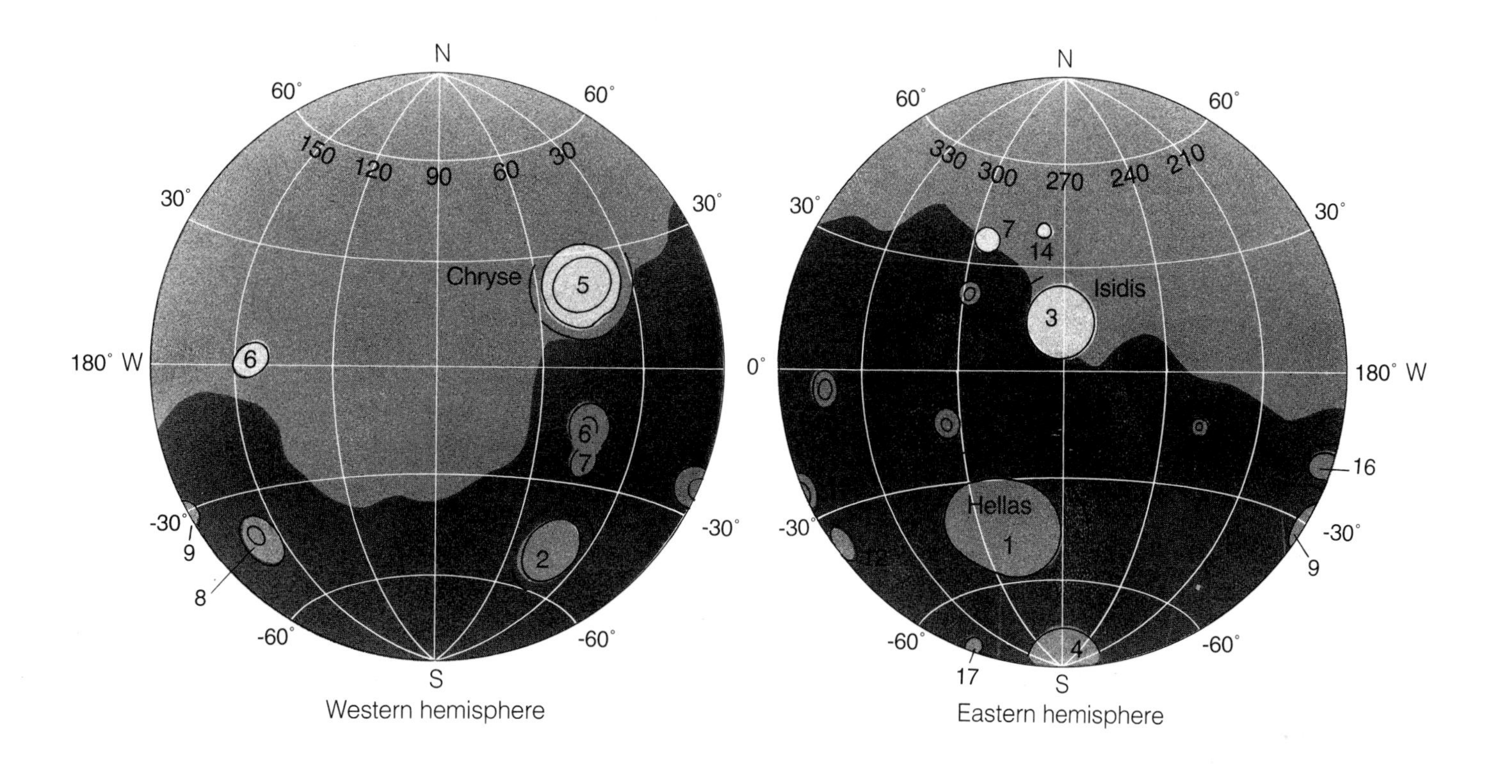

Fig. 6.5 Distribution of Martian basins.

Whereas lunar impact basins are characterized by several rings of scarplike peaks, those of Mars have a rather different morphology. The rim morphology is best seen around the Argyre basin, where a belt of closely spaced blocky massifs extends from about 300 to 800 km from the basin's centre (Fig. 6.6). The incomplete ring of the Isidis basin is closely similar, while that surrounding parts of Hellas is broader and lower. Some Martian basins have several rings. Schultz *et al.* (1982) mapped six for the Chryse basin, four for both the Ladon and Aram basins, and two for each of Argyre and Isidis. Hellas, Isidis and Argyre also have concentric fractures around their peripheries, in which respect they resemble the experimentally produced Snowball impact structure and some circular lunar basins. The marked difference between lunar and

Table 6.2 Impact Basins of Mars (diameter >250 km).
Data from Schultz *et al.* (1982) and Wood and Head (1976)

Name	Latitude	Longitude	Diameter (km)
Hellas	−43.0	291.0	2000
Isidis	16.0	272.0	1900
Argyre	−49.5	42.0	1200
S. Polar	−82.5	267.0	850
Chryse	24.0	45.0	800
Renaudot, S. of	38.0	297.0	600
Ladon	−18.0	29.0	?550
Sirenum	−43.5	166.5	500
Hephaestus Fossae	10.0	233.0	500
Schiaparelli	−3.2	343.5	470
Huygens	−14.0	304.2	460
Le Verrier, W. of	−37.0	356.0	430
Antoniadi	21.7	299.1	400
Nilosyrtis Mensae	33.0	282.5	380
Nr Newcomb	−22.5	3.0	380
Al Qahira	−20.0	190.0	300
Nr South	−73.0	344.0	300
Herschel	−14.6	230.2	290
Holden	−25.0	32.0	260

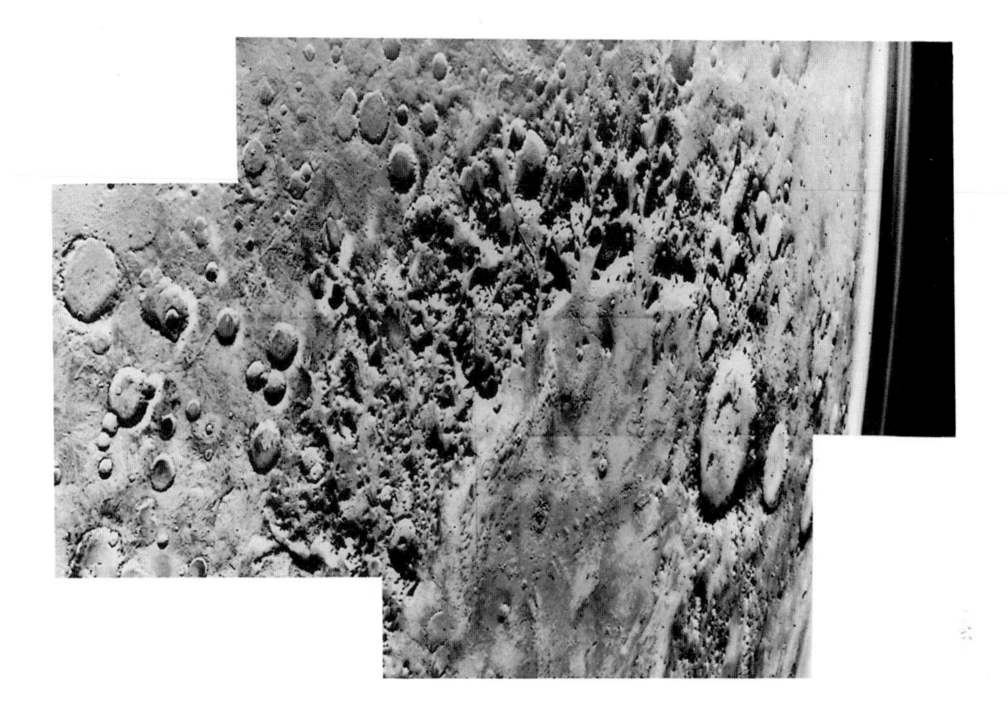

Fig. 6.6 *Viking mosaic of the Argyre impact basin. Note the blocky rim massifs. The large crater on the rim is Galle.*

Martian basins is undoubtedly due to differences in the target materials, both in terms of their volatile content and subsurface layering.

Unlike their lunar counterparts, particularly Orientale and Imbrium, Martian impact basins seldom show preserved ejecta facies; however, peripheral areas of enhanced erosion may focus on original circum-basin ejecta that became saturated with water at some stage in the distant past, when the atmosphere of the planet was far denser than at present. The volatile-saturated deposits are presumed subsequently to have undergone sapping.

Martian multi-ringed basins all have suffered a complex history of burial and exhumation which has led to many of them having remained unrecognized until relatively recently. However, their ultimate recognition has led not only to a fuller understanding of the early impact record, but also to an appreciation of the way in which basin rim massifs apparently have controlled subsequent geological developments. On the Moon it is patently obvious that impact basins have exerted strong controls upon surface geology. Although such a deduction was not made for Mars during the reconnaissance work undertaken on Mariner 9 and Viking data, subsequent detailed studies have shown that basin structure has often controlled the development of both fracturing and channel networks within the cratered plains of Mars. Thus, the older ring structures commonly exhibit extensive narrow valley systems on their rim massifs, while large outflow channels may originate along one of their component rings (Plate 6). Such an association is unlikely to be fortuitous and implies that while their original form is now obscure, the deep-seated concentric fracture belts generated during basin excavation and subsequent tectonic readjustment, have exerted an important control on geological features.

The degree to which such control has affected global-scale Martian geology has been discussed by Schultz *et al.* (1982), who suggest that the Chryse Basin, although experiencing an early stage of lava infilling, subsequently was reactivated – perhaps by regional Tharsis-related volcanism – such that there was a complex interplay between it, the adjacent Aram and Ladon basins, major outflow channels and Valles Marineris. In particular, there is a broadly arcuate belt of disruption which extends from the Aram basin, through the northern rampart of the Ladon Basin and the offset canyons associated with the central regions of Valles Marineris, to Kasei Vallis. This appears to be a manifestation of an arcuate zone of instability along the outer rim of Chryse (Fig. 6.7) and, if so, underlines how ancient basin structure, resurfacing, flood volcanism and subsequent basin reactivation may all be interrelated subtly. In particular, it seems likely that impact-induced fractures, giving rise to unstable zones surrounding basins, may have provided pathways for magmas rising from the Martian mantle, much in the same way as basin-related volcanism occurred on the Moon.

6.4 MORPHOLOGY OF IMPACT CRATERS

It was clear even from the early Mariner 6 and 7 images, that Martian impact craters were shallower than those on the Moon. This characteristic is a function of the complex erosional and depositional history of the planet, with partial infilling of the interiors by windblown material and of rim destruction by erosion (Fig. 6.8). Wall slumping, similar to that seen in larger lunar craters, leads to cavity enlargement in much the same way as it does on the Moon, but

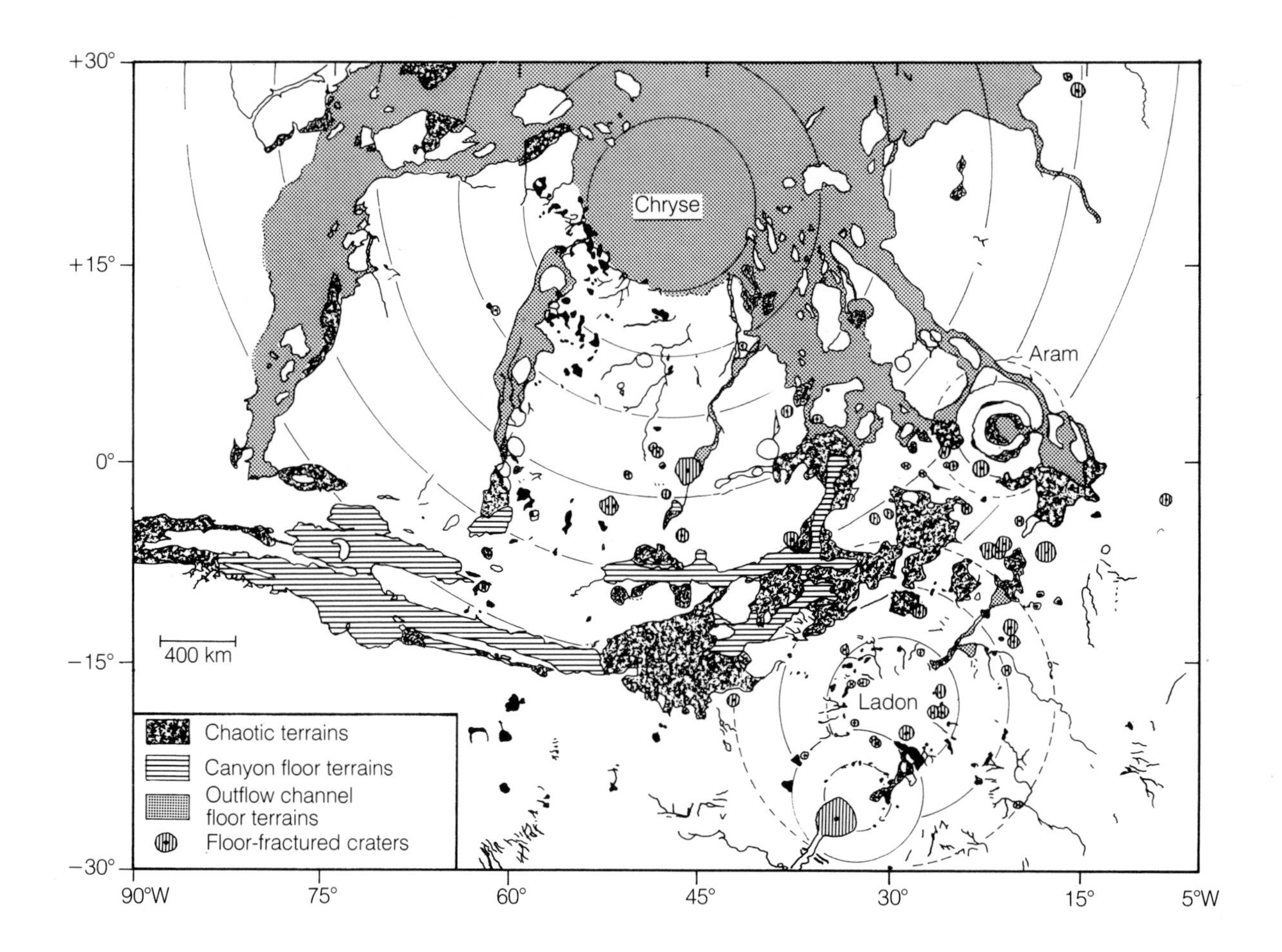

Fig. 6.7 *Geological sketch map showing the zone of instability surrounding Chryse (after Schultz et al., 1982).*

the transition from simple to complex crater morphology occurs within the diameter range 3–8 km, as opposed to about 20 km on the Moon (Pike, 1979; 1980a). Certainly the larger Martian craters are relatively shallower than their lunar counterparts, but deeper than those on Earth. Since gravity is the predominant factor in controlling this parameter (Pike 1980a), this is hardly surprising, but one further factor which may have led to an enhancement of the enlargement process is the presence of volatiles within the Martian regolith (suggested by the fluidized ejecta patterns shortly to be described). By analogy with the terrestrial Prairie Flat crater, Boyce and Roddy (1978) indicated that if large volumes of volatiles were locked up in the regolith, Martian craters should be proportionately shallower than those on both Mercury and the Moon. Furthermore, many craters within the size range 30–45 km contain central pits (Wood *et al.*, 1978) or peaks with summit pits on them. Chuck Wood and his co-workers assert that the presence of ground ice, which would volatilize from the core of a crater's central uplift, could reasonably explain this particular characteristic, too.

Dick Pike (1980a) also notes that with increasing crater size a series of changing morphological features appears sequentially: thus within the 3–4 km diameter range, flat floors are typical; as depths become proportionately shallower with increasing diameter (4–5 km) central peaks appear; at around 6 km scallopped rims are typical, while around 8 km, wall terracing develops.

Fig. 6.8 Degraded impact craters at latitude 35°N. The large crater, 65 km across, has gullied walls, a floor with at least two levels partially filled with aeolian debris and/or volcanic flows and little or no evidence of a surrounding ejecta blanket. The smaller craters to the south and west exhibit dunes and laminated deposits on their interiors. Viking orbiter frames 192S19–21 (parts). Centred at 24.5°N, 301°W.

He suggests that shallow depth of excavation and some unspecified rebound mechanism, rather than centripetal collapse and deep-seated slippage, gave rise to the central peaks and, in turn, engendered rim collapse. Pike (1980b) also notes that the transition from simple to complex craters is not the same on the ancient heavily cratered terrain as on the less heavily cratered plains. The onset diameter for complex craters is about 10 km on the latter, but is only about 3 km on the former. The most simple explanation for this particular behaviour is terrain dependance, whereby stronger materials typical of the younger plains support larger bowl-shaped cavities than the weaker, heavily impacted ancient crust.

In general terms, therefore, because of extensive modifying processes, the original forms of most large Martian impact craters are poorly preserved. However, this cannot be said to be true of their associated ejectamenta, whose distribution and form have provided important clues as to the constitution of the planet's regolith.

6.5 CRATER EJECTA MORPHOLOGY

It is the ejecta patterns surrounding Martian impact craters which are their most distinctive characteristic. Lunar craters are surrounded by an inner zone of continuous hummocky ejecta, often with coarse blocky debris along the rim, which extends outwards for about one half the crater radius. This is encircled by a zone of discontinuous dune-like ejecta, or radial ridges and grooves, that extend outwards for about one crater radius. Beyond this the ejecta is dis-

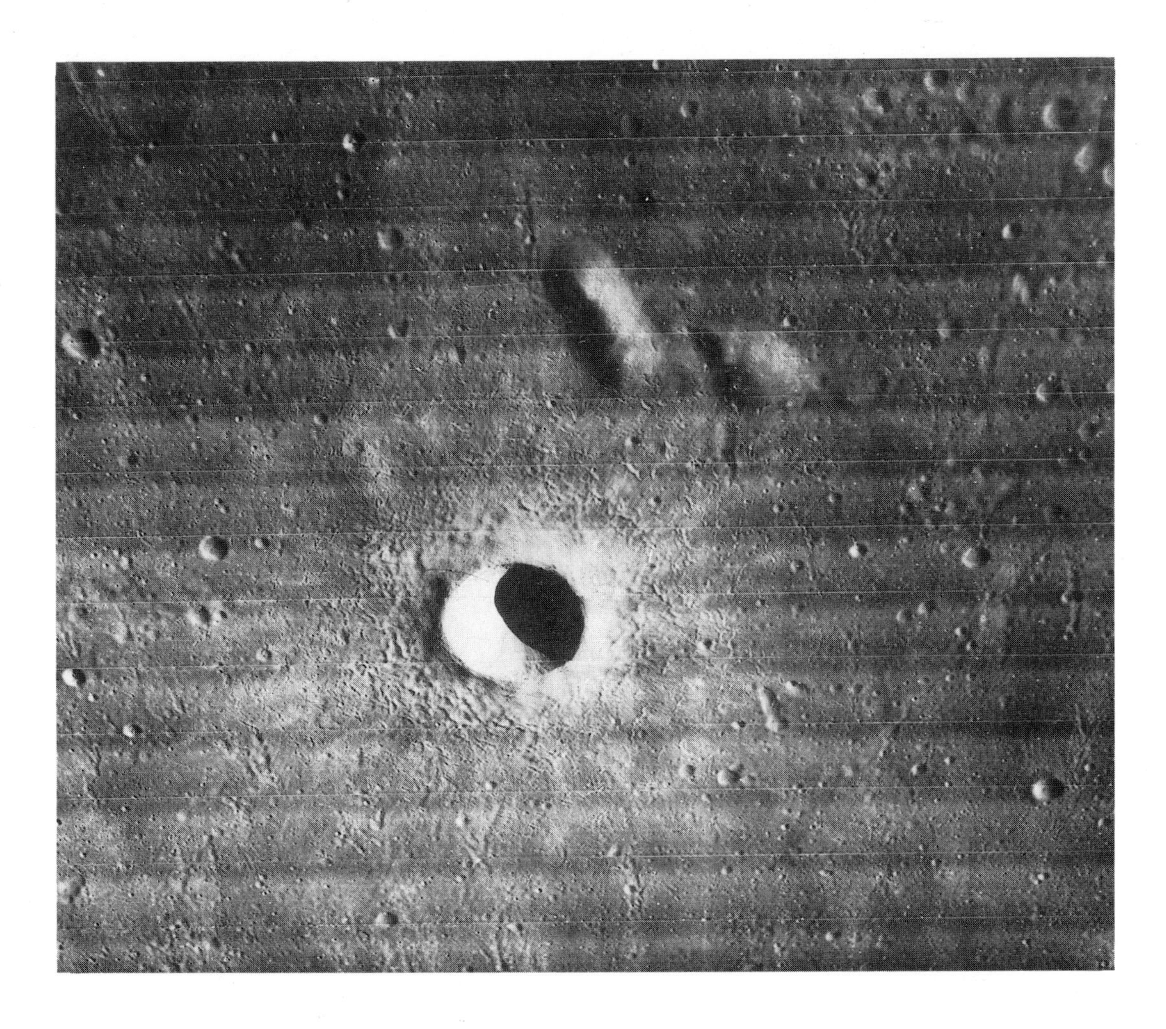

Fig. 6.9 Lunar impact crater Mösting-C and associated ejecta zones. Lunar Orbiter frame III-113 M.

continuous and finally merges with a zone of secondary cratering and bright rays (Fig. 6.9). All of this is due to deposition from ballistic trajectories.

On Mars the situation clearly was different. Below a diameter of about 5 km, crater ejecta blankets appear to have been laid down with ballistic trajectories. However, above this size, most ejecta patterns show indisputable signs of having been laid down by flow across the surface (Carr *et al.*, 1977). This kind of fluidized ejecta emplacement appears to be uniquely Martian. What have widely been described as 'Yuty-type' or rampart craters – named after the 18 km diameter crater Yuty (22°N, 34°W) show a series of overlapping sheets of ejecta with lobate margins, which extend outwards for about two crater diameters from the rim, and each of which has a raised rim or rampart along its outer margin (Fig. 6.10). A second type of occurrence sees a continuous sheet of ejecta forming an annulus, often with concentric ridges on its surface, which overlies ejecta with a strongly radiating texture (Fig. 6.11). Boyce (1979) suggested that because only craters with a diameter >5 km exhibited such ejecta patterns, this critical diameter may have been a function of the minimum depth at which the impacting bolide encountered volatiles trapped below the Martian surface.

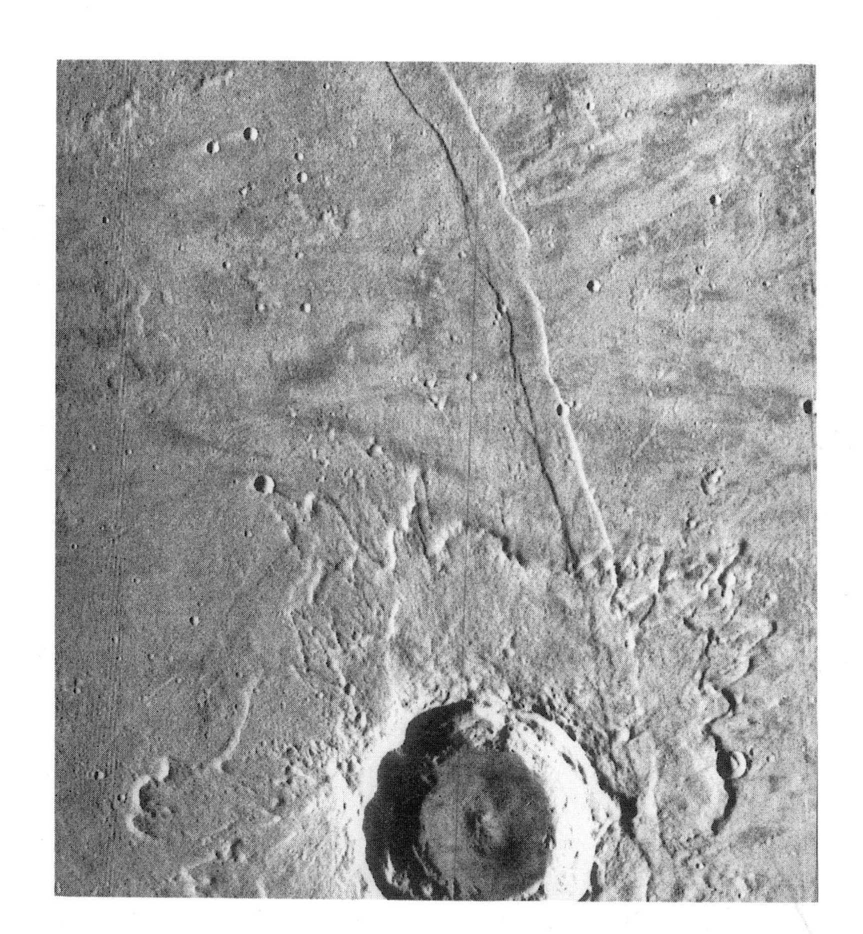

Fig. 6.10 Yuty-type crater, showing lobate ejectamenta with raised edges. Viking orbiter frame. Centred at 30.28°N, 124.12°W.

Fig. 6.11 12 km diameter impact crater with striated ejecta pattern. Viking orbiter frame 653A22. Centred at 28.35°N, 184.72°W.

Experiments conducted by Don Gault and Ron Greeley into viscous target materials support the hypothesis that ejecta fluidization can occur and that the mobilized ejecta probably behaved as a Bingham fluid (Gault and Greeley, 1978). Furthermore they showed that decreasing viscosity promoted post-depositional ejecta flow and increased the radial extent of the continuous ejectamenta. Other workers, including Alex Woronow (1981), after measuring a large number of rampart craters, concluded that all such craters >6 km in diameter were surrounded by ejecta of approximately the same thickness (40–50 m). Mutch and Woronow (1980), assuming that the ejecta initially was ejected ballistically, proposed that if the deposit exceeded some critical thickness, it would fail under its own weight; the laharic-type deposit so formed would then give rise to a flow-lobe ring. Further rings could be emplaced as ejecta continued to be excavated, the weight of overlying deposits periodically exceeding the weight that could be supported by the lower sheet. Depending on the mode of failure assumed for the mobile ejecta as it traversed the Martian surface, Woronow (1981) felt his data implied a total volatile content of between 16 and 72%. Whatever the precise figures may have been, it is clear that compared with the Moon, Martian ejecta emplacement was complicated by both entrainment of volatiles and atmospheric drag, such that locally varying conditions could affect the details of individual ejecta deposits.

6.6 CHANNELLING ON THE CRATERED PLATEAU

Dissection of the ancient terrain of Mars varies in its level of development. Over very large regions the cratered surface is incised by branching valley networks (Fig. 6.12). These may be very well defined or barely discernible. Branching valleys are typical of less heavily impacted areas between large craters, while stubby valleys frequently are seen cutting through the crater ramparts and extending partway across their floors. There seems little doubt that these ancient valleys are of fluvial origin and were formed at a time when the Martian atmosphere was denser and richer in volatiles than it is today. Their variable state of degradation strongly suggests that these, like craters, were subject to early attack by meteorites during the stage of heavy bombardment. The better-preserved networks on this basis, probably are the younger. They are more fully described in Chapter 10.

6.7 INTERCRATER PLAINS

Intercrater plains are defined as those less heavily cratered deposits which occur between the large craters and on their floors. In places the level floors of these craters clearly are made from eroded layered deposits; elsewhere they are covered by plains units crossed by curvilinear ridges; then again, some floor-mantling units are pitted by hundreds of very small pits. The latter are particularly difficult to accommodate in the light of how active has been the surface of the planet. The intercrater plains units partially submerge the plateau surface over large areas. They are described in detail in Chapter 8.

Fig. 6.12 Valley networks incised into ancient cratered terrain, south of Isidis. Viking orbiter frame 067B71. Frame width 250 km. Centred at 6.10°S, 271.30°W.

6.8 VOLCANOES OF THE CRATERED PLATEAU

While central volcanism generally is associated with the northern hemisphere, there are several ancient volcanic structures in the south, particularly around the Hellas basin. Plescia and Saunders (1979) referred to these as highland paterae. The most equivocal is Amphitrites Patera, which lies close to the southern border of Hellas. It comprises several 100 km diameter rings, with little vertical relief, and associated radiating ridges, some of which extend on to the Hellas plains. Hadriaca Patera is very similar, though somewhat better defined, and has at its summit a caldera 60 km across (Fig. 6.13). The flanks have a smoothed appearance and are incised by radiating channels. The low shield has a diameter of about 300 km. It is very reminiscent of the mantled parts of Alba Patera, and may have been built mainly from pyroclastic deposits.

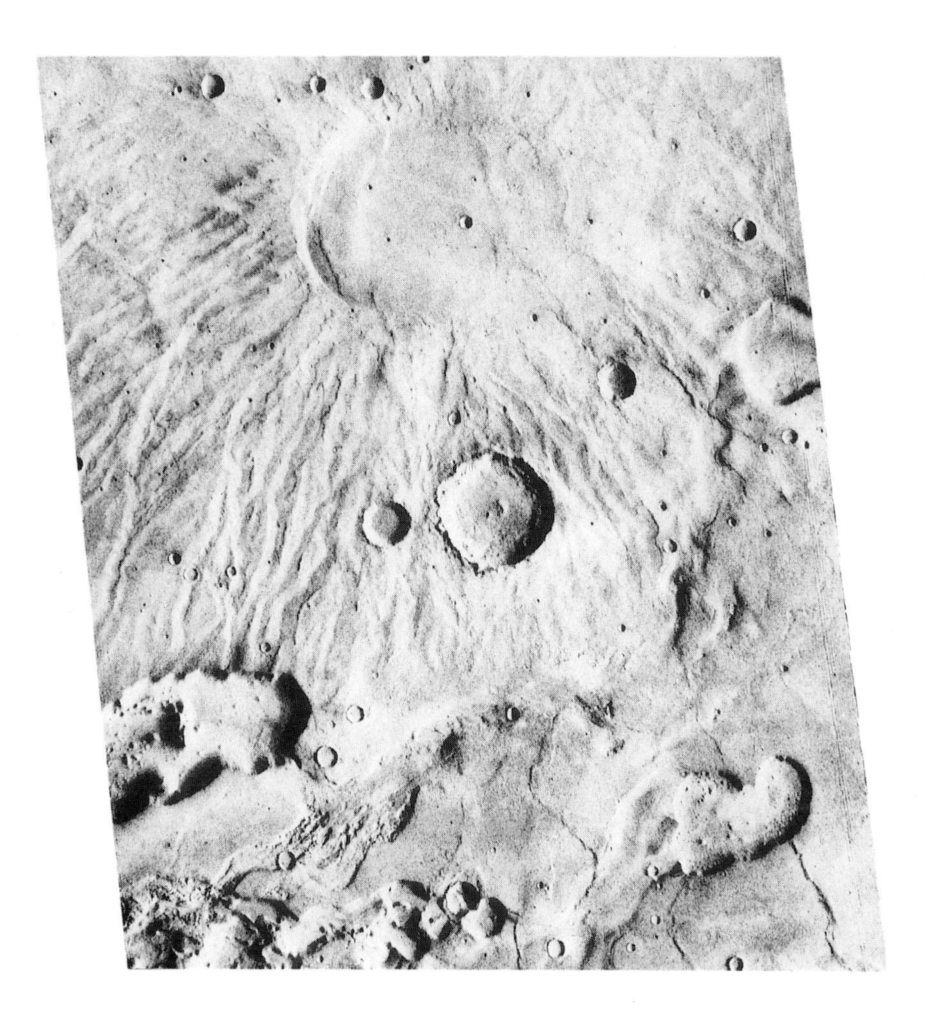

Fig. 6.13 *The highland patera volcano, Hadriaca Patera, showing the 60 km diameter caldera, gullied flanks and adjacent channels. Viking orbiter frame 625A18. Centred at 30°S, 266°W.*

Tyrrhena Patera has received closer study than either of the former, particularly by Greeley and Crown (1990). Situated northeast of Hellas, there are two sets of ring fractures surrounding the summit, the inner one defining a 50 km diameter region in which is an off-centre depression (Fig. 6.14). A broad channel leads off from this. Although there is a lava unit on the southwest flank, the lower parts of the shield are highly dissected and embayed by younger units; they have a higher albedo than these. This led Pike (1978) and Greeley and Spudis (1981) to surmise that it was composed of ash. Smooth units surrounding the shield are also interpreted by Greeley and Crown to be pyroclastics; they envisage Tyrrhena Patera to have been built from phreatomagmatic eruptions, more or less contemporaneously eroded by water, wind and mass wasting. A similar origin probably can be supported for several other structures around Hellas.

Because the ash deposits are very widespread (they extend from between 300 and 600 km from the summit), it is unlikely that they were air-fall in origin. More likely is the notion that they were laid down by ash flows, capable of being emplaced at velocities of up to 300 m s^{-1} on the Earth (Sparks *et al.*, 1978). Greeley and Crown calculate that initial flow velocities of up to 650 m s^{-1} would be necessary to emplace the smooth units surrounding the shield,

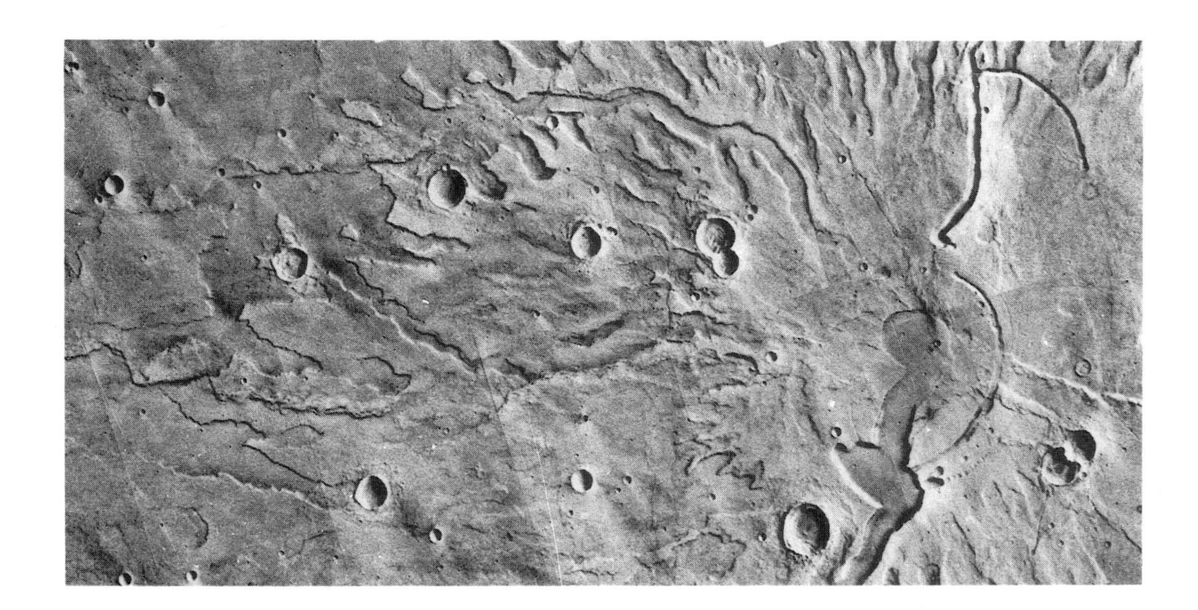

Fig. 6.14 Synoptic view of the ancient volcano, Tyrrhena Patera, showing summit faulting, caldera and channelled flanks. Viking orbiter mosaic.

should they also be ash. This seems entirely plausible and is in accord with the climatic changes believed to have affected Mars during this early stage (Clifford *et al.*, 1988). That ash eruptions appear not to have been a feature of more recent volcanism, i.e. in the northern hemisphere, further supports the idea that the Martian climate changed as time progressed. Thus the conditions under which the ancient cratered terrain, intercrater plains and ancient paterae developed, were undoubtedly very different from those pertaining during the emplacement of the northern plains and the Tharsis and Elysium shield volcanoes.

6.9 THE GEOLOGICAL STORY AS REVEALED BY THE UPLAND ROCKS

The oldest rocks that can be discerned in the upland hemisphere are the upstanding remnants of impact basin rims. These are assigned to the Lower Noachian epoch of Martian history. Perhaps the most characteristic of these are the massifs of Nereidum and Charitum Montes which represent the elevated rim materials produced during the Argyre impact. These blocky mountains form an annular belt up to 700 km wide and in places 1–2 km above the interior of the basin. Similar massifs would have been produced for each of these early impact events and the associated and widely dispersed basin ejecta would have covered the surrounding cratered crust (Fig. 6.15a). The latter, now completely buried by younger deposits, doubtless represented the primitive material – formed by solidification of molten materials from within Mars – which subsequently was bombarded by myriad meteorites to generate the ancient cratered terrain.

During the Middle Noachian, cratering was still intense and the Martian crust became battered by impacts that generated not only cavities but ejectamenta that constructed a complex interdigitating succession of rocks (Fig. 6.15b). Subsequently, mainly in Upper Noachian times, intercrater plains were

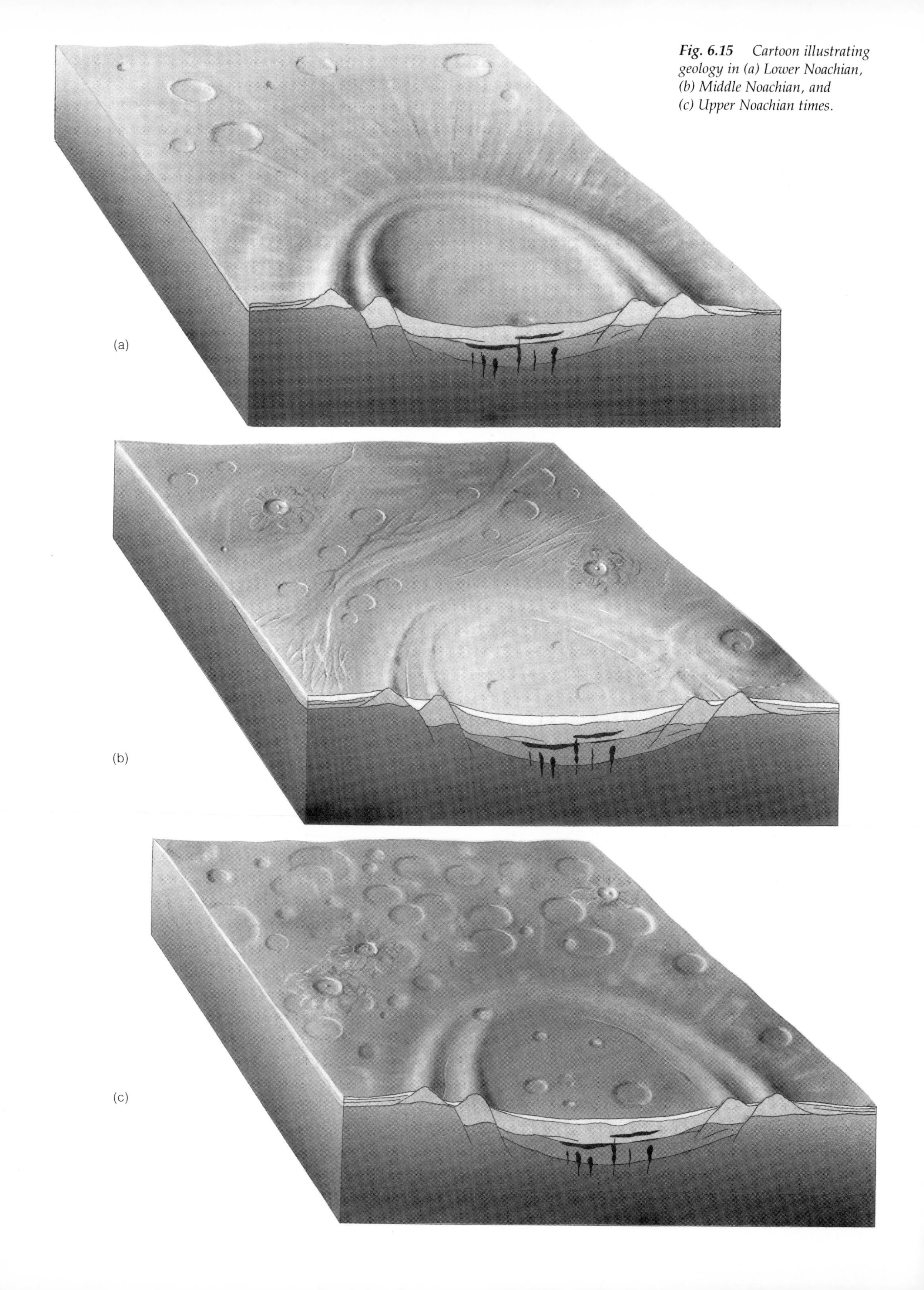

Fig. 6.15 *Cartoon illustrating geology in (a) Lower Noachian, (b) Middle Noachian, and (c) Upper Noachian times.*

emplaced widely; these now overlie or embay older cratered crust. The smooth surfaces of these plains, the presence of wrinkle-type ridges, and the occurrence of volcanic flow lobes, suggests that these were generated by early volcanism. That volcanism did develop at this early stage is confirmed by the presence of several very large, low-profile paterae, particularly surrounding the Hellas basin. These may have been of the explosive type, throwing out enormous quantities of ash, some as ash-flows, as well as extruding fluid lava. At this time the atmosphere of Mars apparently was denser than it is today and the climate may have been significantly milder, since large numbers of valley networks with well-developed tributary systems were incised into the inter-crater areas and rims of large impact craters. These are assumed to have been formed by fluvial processes (Fig. 6.15c).

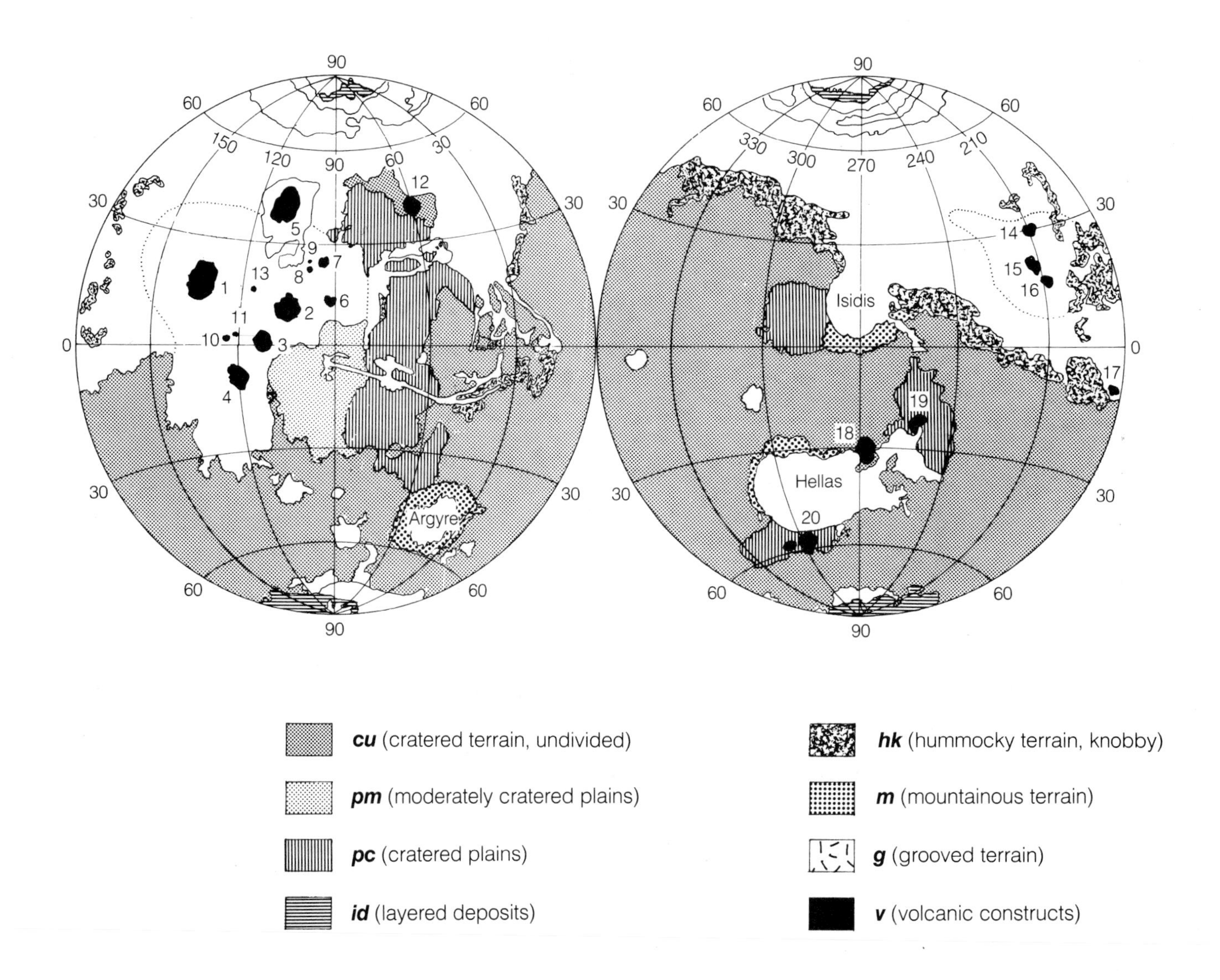

Fig. 7.1 *Distribution of the larger central volcanoes on Mars. Individual centres are shown in solid black. Key: 1. Olympus Mons, 2. Ascraeus Mons, 3. Pavonis Mons, 4. Arsia Mons, 5. Alba Patera, 6. Tharsis Tholus, 7. Ceraunius Patera, 8. Ceraunius Tholus, 9. Ulysses Patera, 10. Biblis Patera, 11. Pavonis Patera, 12. Tempe Patera, 13. Jovis Tholus, 14. Hecates Tholus, 15. Elysium Mons, 16. Albor Tholus, 17. Apollinaris Patera, 18. Hadriaca Patera, 19. Tyrrhena Patera, 20. Amphitrites Patera.*

THE CENTRAL VOLCANOES OF MARS

7

7.1 INTRODUCTION

The earliest major outbreak of volcanism on Mars which left a record visible on spacecraft images, emplaced the plateau plains of Upper Noachian age. This was succeeded by the even more extensive ridged plains of lower Hesperian age. Both were largely emplaced by flood lava eruption (see Chapter 6 and also Tanaka, 1986). Centralized volcanic activity appears to have begun with the formation of the enigmatic structure, Amphitrites Patera, during the Lower Hesperian epoch, and this was followed in Upper Hesperian times by generation of a small number of large, low-profile ash or mixed lava and ash volcanoes in the southern hemisphere, near the borders of the Hellas impact basin, at Hecates Tholus in the region of Elysium, and in Syrtis Major Planum. The characteristic phreatomagmatic activity of this period was a manifestation of the rather different climatic conditions proposed to have been a feature of Mars at that time, as a result of which significant volumes of volatiles became entrained in magmas rising through the Martian subcrust.

Towards the close of Hesperian times, another episode of flood volcanism began constructing the volcanic plains of the northern hemisphere. This was accompanied by the growth of major volcanoes in the region of Elysium at Albor Tholus, and in northern Tharsis, where activity became focussed on Alba Patera. While some explosive activity continued, there was a gradual change towards effusive volcanism, with the eruption of very large volumes of low-viscosity mafic magmas. The later stages in Mars' volcanic history saw the growth of vast shield volcanoes along the crest and margins of the Tharsis Bulge and the construction of Elysium Mons. The culmination of this uniquely Martian shield-building activity was the growth of Olympus Mons, located some distance west of the crest of Tharsis.

While the earliest central volcanism may have been controlled by deep-seated fracturing produced during the excavation of Hellas, this is less likely to have been the case for the activity in Tharsis and Elysium. More likely, volcanism here was in some way related to the growth of the major crustal upwarps which exist there or to the internal processes which generated these. Considerable research has centred around trying to explain the Tharsis 'Bulge'. For instance, in an attempt to explain the gravity data for the Tharsis region, Sleep and Phillips (1979) proposed a model which assumed the Martian crust to be thinner beneath Tharsis than elsewhere and the mantle less dense. (A lesser density might be expected in a region of active volcanism.) On the basis of such a model, they were able to achieve isostaic compensation under Tharsis at depths of only 300 km.

7.2 VOLCANO DISTRIBUTION

The global distribution of Martian central volcanoes is shown in Fig. 7.1. The less linear global distribution pattern of the Martian volcanoes compared with that for Earth, implies the absence of a segmented lithosphere on Mars. Martian large shields are presumed to be related to a number of long-lived mantle hot spots.

The greatest number of volcanoes and also the youngest individual structure are located in the Tharsis province, the region of a major tumescence in the planet's crust (the Tharsis Bulge), which lies between 10 and 11 km above datum. It is roughly the size of Africa south of the River Congo, extending 4000 km from north to south and 3000 km from west to east. Its slopes are, however, low and it is really a broad, gentle rise. On the north flank, slopes range between 0.2° and 0.4°, but on the south side they are only about half this. This asymmetry affects also its extent, for not only is it steeper but it is also more extensive to the north of its crest due to its straddling the line of dichotomy. Strangely, most of the large shield volcanoes are sited either near the crest of the bulge or on its northwestern flank; there are none on the southern flank.

The most prominent shield group – the Tharsis Montes – comprise Arsia, Pavonis and Ascraeus Montes. These are spaced approximately 700 km apart and aligned in a southwest-northeast direction along the crest of the bulge. Since major fractures pursue the same trend northeastwards and southwestwards, even beyond the Tharsis volcanic province, it can be assumed that they have developed along a major fracture zone, now buried by the products of Tharsis volcanism. Several smaller shields and steeper-sided volcanoes or tholi lie close to the continuation of this line to the northeast of Ascraeus Mons; others lie to the east and west (Fig. 7.2). Twelve hundred kilometres northwest of Tharsis Montes lies Olympus Mons, the youngest of the Martian shields, while at a slightly greater distance north of Ascraeus Mons, on the extreme edge of Tharsis, lies the vast Alba Patera, an older low-profile volcano surrounded by prominent circumferential fractures.

Elysium Mons, Hecates Tholus and Albor Tholus are located on another broad rise in Elysium which has a diameter of approximately 2000 km and a mean height of 5 km. The nature of the Elysium volcanoes is somewhat different from those of Tharsis and there is evidence that their development may have involved not only lavas but also pyroclastic rocks. Several major volcanic centres are located around the Hellas impact basin. Of these, Hadriaca and Tyrrhena Paterae are almost certainly mixed lava and ash volcanoes. The only other major volcano is Apollinaris Patera, which is located southeast of Elysium on the lowland hemisphere at 10°S, 185°W, that is, just north of the line of dichotomy. Its morphology also suggests an origin in both effusive and explosive activity. Among more equivocal structures are low shields in Syrtis Major and volcano-tectonic features in Tempe.

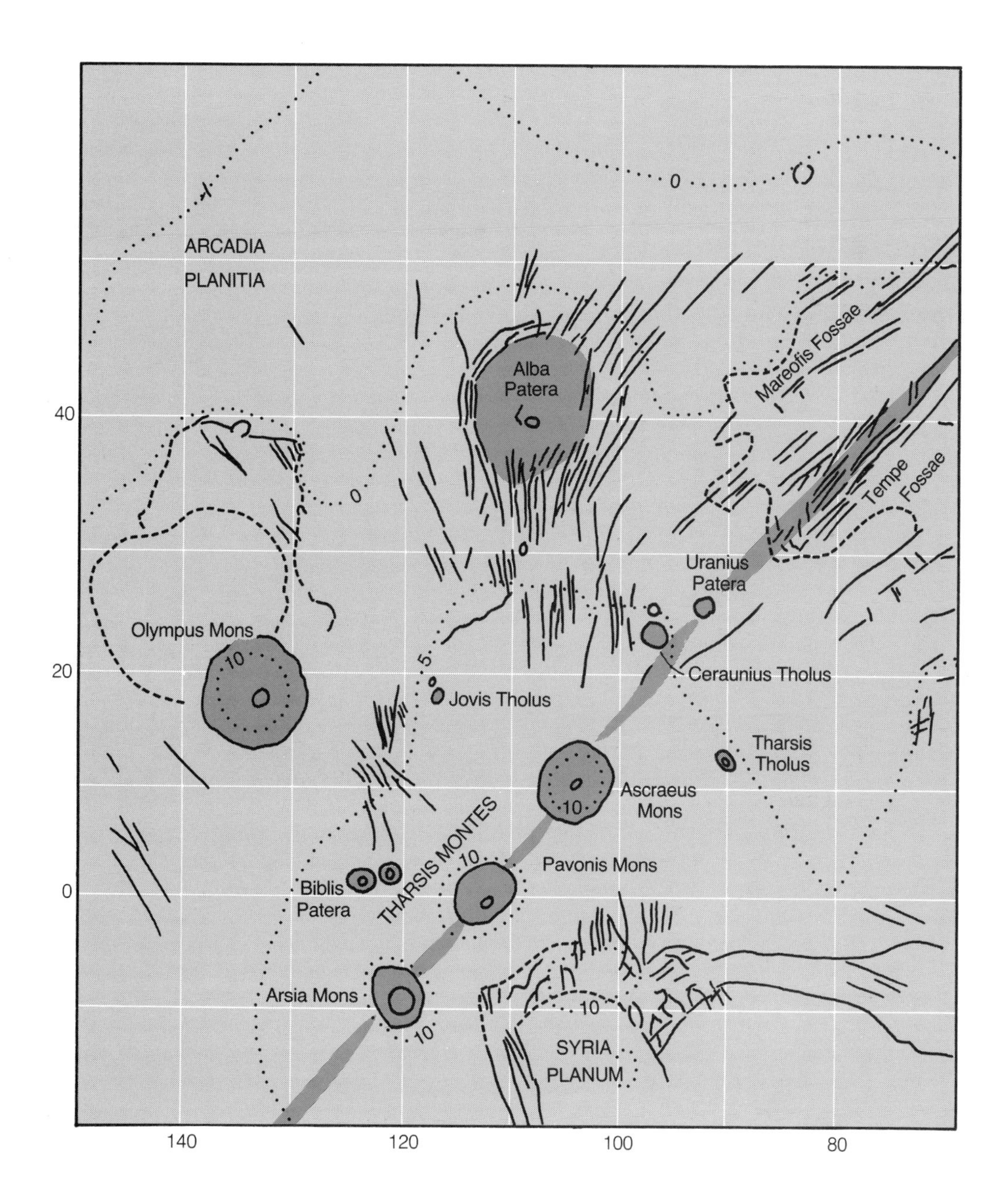

Fig. 7.2 Map of the western hemisphere of Mars between 60°N and 60°S showing the position of the main Tharsis volcanoes (black areas) and principal fractures and ridges. Volcanoes: 1. Alba Patera, 2. Uranius Patera, 3. Ceraunius Tholus, 4. Olympus Mons, 5. Ascraeus Mons, 6. Pavonus Mons, 7. Arsia Mons, 8. Biblis Patera, 9. Tharsis Tholus, 10. Jovis Tholus. (Modified from J.B. Plescia, 1979.)

7.3 CLASSIFICATION OF VOLCANO TYPES

The large volcanoes have been classified into three main types: (1) shields, which are built from thousands of individual flows, have summit calderas and overall low profiles characterized by steeper upper regions and gentler lower flanks; (2) tholi or dome volcanoes, which are similar to the former but have somewhat steeper slopes that may be a function of more viscous lava, lower eruption rates or increased pyroclast content; and (3) paterae which are

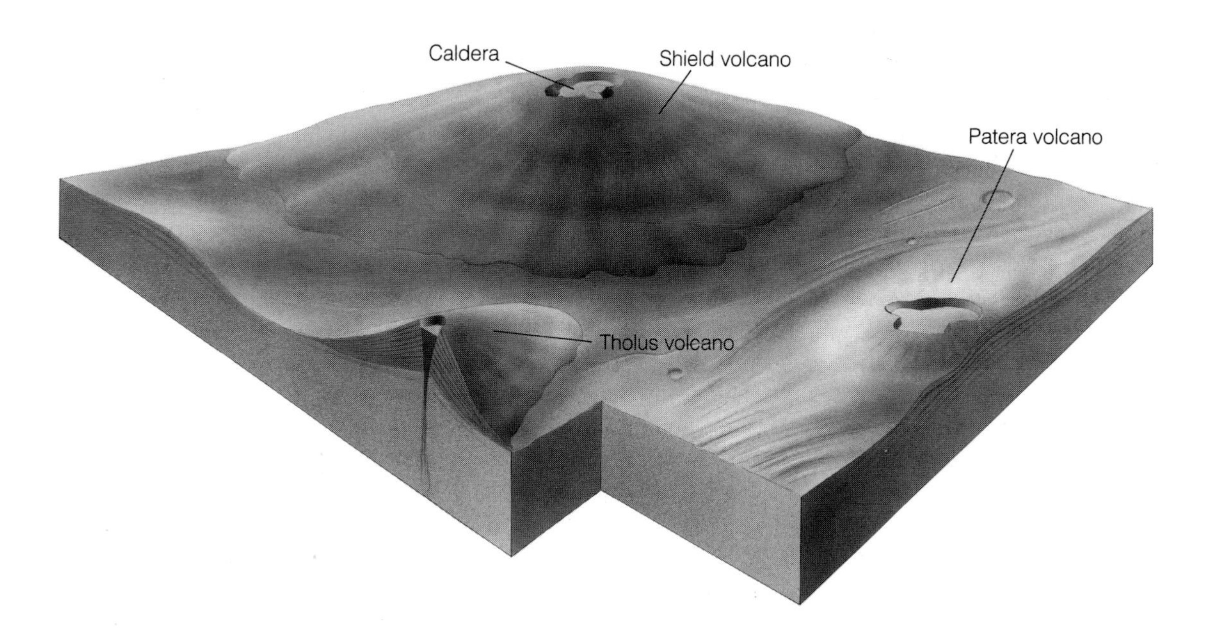

Fig. 7.3 The three principal volcano types on Mars: (a) low-profile patera, (b) shield, and (c) tholus (dome-type).

two kinds: (i) lowland paterae (e.g. Alba Patera, Uranius Patera), which are northern hemisphere lava shields characterized by extremely low profiles and complex summit calderas, and (ii) highland paterae (e.g. Tyrrhena Patera), which are located mainly in the southern hemisphere, have very low profiles and summit caldera complexes and also may be incised by channels (Plescia and Saunders, 1979). Most of subgroup (ii) are believed to be mixed lava and pyroclast edifices. A fourth group – volcano-tectonic depressions – may also be recognized, including certain suspected volcanic structures in Syrtis Major, in the region south of Hellas and amongst the volcanic plains around Tharsis. Figure 7.3 illustrates the broad characteristics of the three main volcano types.

7.4 VOLCANO AGES

In the absence of Martian returned samples there can be no radiometric dates, as there are for the Moon. Therefore, only crater-counting can be used to establish a relative timescale. Plescia and Saunders (1979) provided crater counts for each of the major volcanoes, divided into their types, with Lunae Planum as a datum (Table 7.1). 'Absolute' ages, derived by putting these data into the chronologies of Neukum and Wise (1976) and Soderblom (1977) give an idea of the sequence of events in the context of Martian geological evolution. However, it should be noted that individual structures did not form at any one instant but evolved often over very lengthy periods, episodes of activity being interspersed with periods of inactivity measurable in tens or hundreds of millions of years.

7.5 HIGHLAND PATERAE

Several of the more ancient volcanic structures on Mars are located near the borders of the Hellas Basin. Plescia and Saunders (1979) termed these highland

Table 7.1 Number of craters greater than or equal to 1 km in diameter per 10^6 km^2 for individual Martian volcanoes, compared with absolute ages derived from the chronologies of Neukum and Wise (1976) and Soderblom (1977) (after Plescia and Saunders, 1979)

Volcanic centre	Number >1 km of craters per 10^6 km^2	Implied absolute age in 10^9 a (Soderblom, 1977)
Olympus Mons	27	0.03
Ascraeus Mons	110	0.1
Pavonis Mons	350	0.3
Arsia Mons	780	0.7
Apollinaris Patera	990	0.9
Biblis Patera	1400	1.3
Tharsis Tholus	1480	1.38
Albor Tholus	1500	1.4
Hecates Tholus	1800	1.7
Alba Patera	1850	1.7
Jovis Tholus	2100	1.95
Hadriaca Patera	2100	1.95
Elysium Mons	2350	2.2
Tyrrhena Patera	2400	2.25
Uranius Patera	2480	2.3
Uranius Tholus	2480	2.3
Ceraunius Tholus	2600	2.4
Ulysses Patera	3200	3.0
Tempe Patera	4300	3.4
Lunae Planum	2500	2.3

paterae and, on the basis of crater counts, all appear to have formed during the same general phase in Martian history, 3.7–3.1 × 10^9 a ago. Potter (1976), Peterson (1977) and King (1978) made the first studies of these during geological mapping. The Viking imagery they studied suggests that some provide evidence for the earliest explosive volcanism on the planet and appear very similar to some terrestrial ash shields. Also, Peterson (1977) suggested that Hadriaca and Tyrrhena Paterae are both located on inferred rings of the Hellas multi-ringed impact basin, one of which extends 4260 km from the basin centre.

The most equivocal of these ancient structures is Amphitrites Patera, situated close to the south rim of Hellas and comprising several 100 km diameter ring structures with associated radiating ridges, some of which extend into the basin itself. The rings appear to have negligible vertical relief. The overall structure does, however, suggest some kind of volcanic origin (Fig. 7.4). Hadriaca Patera, located on the northeast rim of Hellas, is better defined and has at its summit an obvious flat-floored caldera 60 km in diameter. It also appears to have little vertical relief – though more than Amphitrites Patera – but there is possibly as much as a 2 km height difference between the caldera and the foot of the channellized shield which extends 300 km from the caldera rim. The smooth, ridged morphology of the flanks – which appear to be devoid

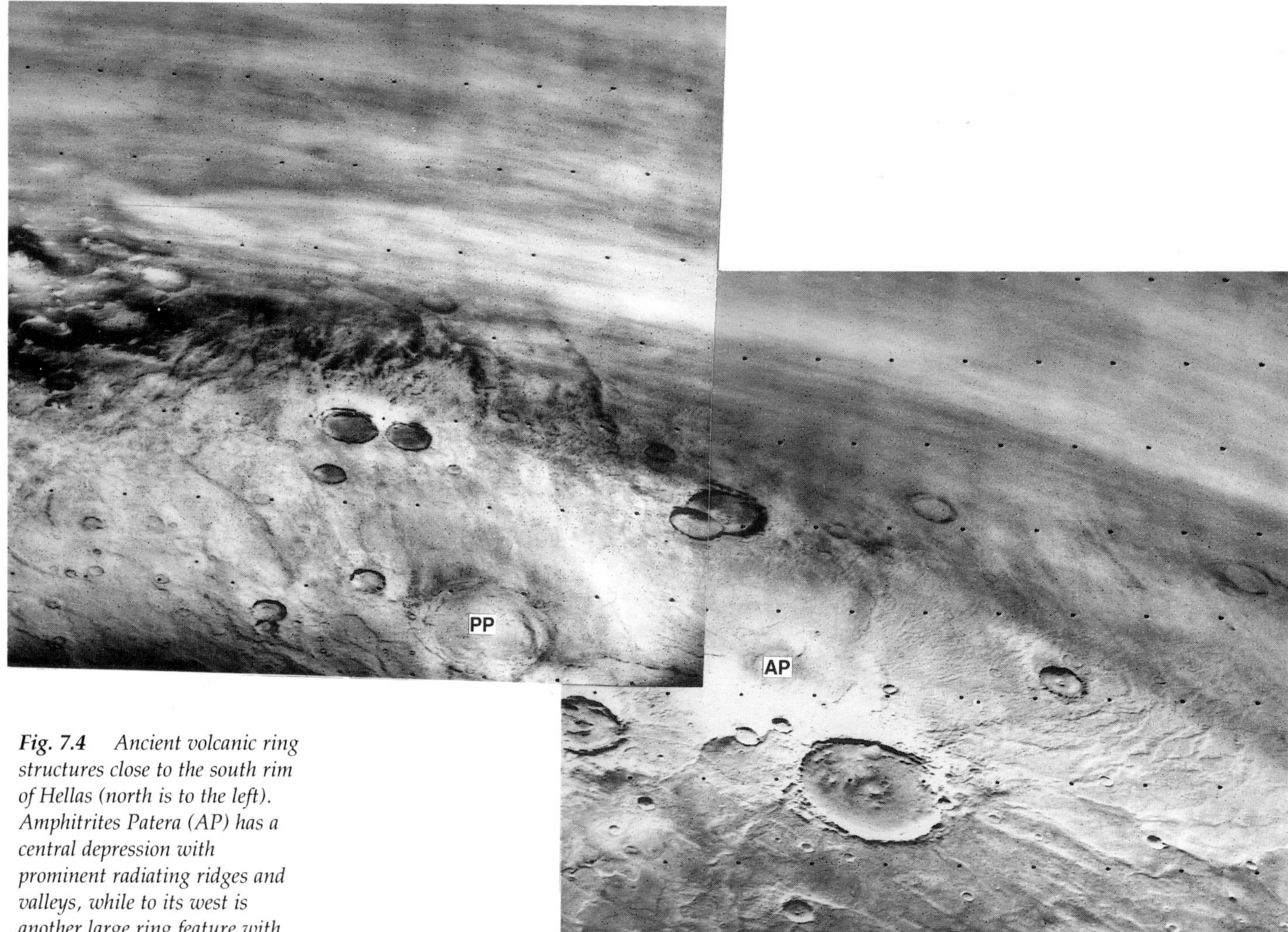

Fig. 7.4 *Ancient volcanic ring structures close to the south rim of Hellas (north is to the left). Amphitrites Patera (AP) has a central depression with prominent radiating ridges and valleys, while to its west is another large ring feature with concentric graben, Peneus Patera (PP). Frame width approximately 2000 km. Viking orbiter frames 079B17–18.*

of lava flow lobes and scarps – is rather reminiscent of the mantled parts of Alba Patera which elsewhere are suggested to have a pyroclastic origin. Interestingly, a large channel, which begins its southwestward course in a large depression on the lower southeast flank, eventually merges with the floor of Hellas, some 800 km distant.

Tyrrhena Patera, the third of the major highland structures, has received closer study than the others, largely because it has benefitted from better image coverage. Located northeast of Hellas, it presents a markedly eroded appearance (Fig. 7.5). At the summit are two sets of ring fractures, the innermost defining a region 50 km in diameter inside which is an off-centre caldera depression. Leading off southwestwards from the caldera is a prominent broad channel which appears to be volcano-tectonic in origin; two others commence lower down the flanks. Greeley and Crown (1990), after a recent study of the region, recognized the presence of five geological units, of which the two older are a basal shield unit that extends 340 km to the south and over 300 km to the

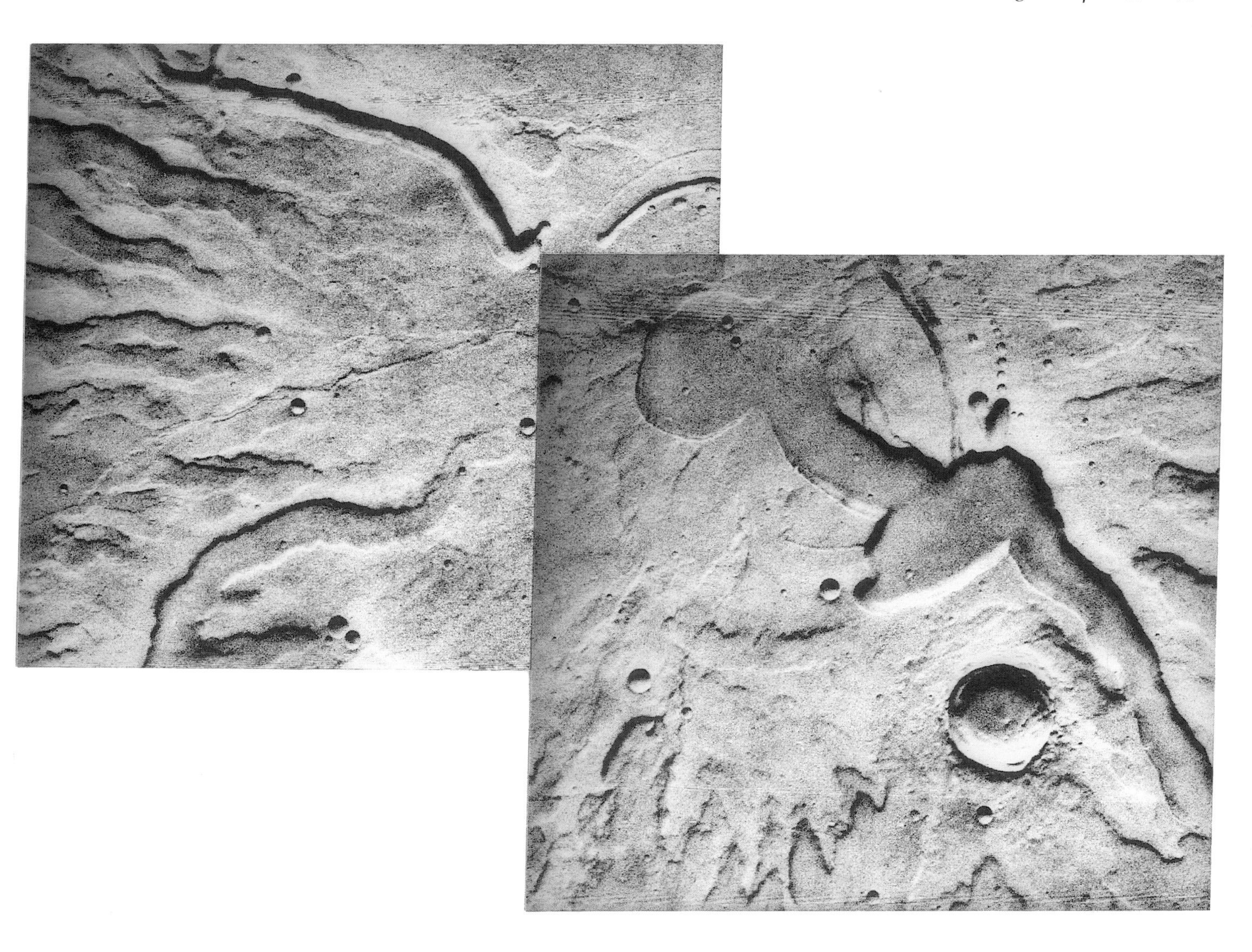

north and west of the summit caldera, and a slightly less old summit shield unit, which, on average, has a diameter of 200 km. The margins of both are embayed and dissected by younger units. It is these dissected, eroded deposits of relatively high albedo which were attributed by Pike (1978) and Greeley and Spudis (1981) to volcanic ash. Extensive smooth plains which differ morphologically from the adjacent ridged plains also are believed to be composed of volcanic ash. On the southwest side of Tyrrhena Patera is a younger, fan-shaped southwest flank flow unit whose apex focusses on the caldera and which is composed of a plethora of narrow lava flows, often with levees and flow channels. At the summit there is a caldera-filling unit, believed to represent late-stage ponded lavas.

Fig. 7.5 *The summit region of the eroded volcanic structure, Tyrrhena Patera (north is to the left). The central ring-faulted depression has been filled by later deposits and subsequently incised by volcano-tectonic depressions/channels. Mosaic width 140 km. Viking orbiter frames 445A53–54.*

The radial texture and etched appearance of much of the basal and summit shield materials (Fig. 7.6) is interpreted by Greeley and Crown to be the result of erosion by water, wind and mass wasting of early ash deposits generated more or less contemporaneously by phreatomagmatic eruptions here and at a

Fig. 7.6 *High-resolution image of a part of the summit shield margin of Tyrrhena Patera, northwest of the volcano summit. Note the erosional scarps at the margin of the prominent channel towards the upper margin of the image and the faint lineations on the channel floor. Frame width 11.5 km. Viking orbiter frame 794A01.*

number of other centres surrounding Hellas. Such activity is seen as the result of eruptions through water-charged regolith that is believed to have been widespread in the southern hemisphere and whose former presence is indicated by the extensive fluvial channels incised into the Noachian-age plateau plains (Greeley and Guest, 1987). Subsequent to ash deposition, the degree of internal heating appears to have increased, whereupon flood lava eruptions emplaced the surrounding plains, which partially buried the flanks of the volcano. The sinuous channels that originate close to the summit may have continued to supply lavas to the lower flanks, after infilling most of the summit caldera.

After careful study, Greeley and Crown conclude that the ash deposits cannot be air-fall in origin due to their very widespread dispersal (300–600 km from the caldera region). Gravity-driven pyroclastic flow could, however, account for the observed distribution, the requisite energy to drive these being supplied by explosive eruptions rich in magmatic volatiles or groundwater (Crown and Greeley, 1990). Theoretical analysis of terrestrial ash-flow eruptions has indicated that large flows may have been emplaced at velocities of up to 300 m s^{-1} (Sparks *et al.*, 1978), while other theoretical considerations indicate that initial velocities as high as 400–600 m s^{-1} may be possible (Sparks and Wilson, 1976). Their calculations suggest that initial flow velocities of between 325 and 450 m s^{-1} would be required to emplace the basal shield unit and an initial velocity of around 250 m s^{-1} could have emplaced the summit shield unit. If the smooth plains units are also accepted as being ash-flow deposits, initial velocities of around 650 m s^{-1} would be necessary. An ash-flow origin therefore seems highly plausible.

On the basis that eruption volumes were comparable with large terrestrial eruptions (10^2–10^3 km^3) – probably an underestimate bearing in mind what we know about Mars – Greeley and Crown estimate that between 110 and 1100 basaltic eruptions could account for the entire edifice of Tyrrhena Patera. On the additional basis of calculations relating to the rate at which subsurface water could accumulate versus the magma volume per eruption, they conclude that the volume of water required to drive the proposed explosive volcanicity could, in fact, accumulate rapidly (in tens of years) and that phreatomagmatic activity is quite consistent with the climatic changes suggested to have affected Mars during this early epoch (Clifford *et al.*, 1988). The fact that climatic conditions changed as time progressed may explain why ash eruptions do not appear to have characterized the younger shields of Tharsis.

7.6 SHIELD VOLCANOES AND PATERAE

7.6.1 *General form*

Shield volcanoes are broad, gently sloping cones. Usually, they have a shallow caldera at their summit, while smaller pits and spatter cones are often concentrated along lateral rift zones. Patera volcanoes have very low profiles and may be of even larger areal dimensions; they were first identified on Mars. In general form they closely resemble upturned saucers, being characterized by flank slopes as low as 0.25°. Most have complex summit calderas and, like shields, often have well-defined rift zones.

7.6.2 *Terrestrial shield volcanoes*

A detailed description of terrestrial shields may be found in another of the author's books (Cattermole, 1990). For this reason only a brief account of such structures is given here.

Terrestrial shields show a range of sizes from small, low-profile shields such as those of Iceland to the very large volcanoes of the Galapagos Islands and Hawaiian–Emperor Chain. Those of Mars are of even greater size, some being more than an order of magnitude larger than any terrestrial counterpart. They appear to have been constructed largely from fluid basalt-like lava flows. While some paterae are similar in this respect to shields, others may have been largely due to ash eruptions, or a mixture of both explosive and effusive activity. Terrestrial lava shields are sited within some of the most volcanically active regions on Earth and are a manifestation of volcanicity associated with long-lived mantle hot spots. The same appears to be true of their Martian counterparts.

The volcanic shields of Hawaii attain immense proportions; thus Mauna Loa rises approximately 10 km from the Pacific floor and has a basal diameter of about 400 km. However, despite its volumetric immensity (425 000 km^3), the flank slopes are less than 6°, and in consequence it is not a striking landform. For a variety of reasons, not the least of which is the longer time over which they grow, the detailed geomorphology of large shields is generally more

Fig. 7.7 *Oblique view across South Pit (foreground) and Mokuaweoweo on the plateau-like summit of the Hawaiian shield, Mauna Loa. The larger caldera measures approximately 4 × 5.6 km across and on its floor, just beyond South Pit, may be seen the 1940 cinder cone (US Air Force photograph, 1966).*

complex than that of small ones. Thus the former have plateau-like tops which often are surmounted by cinder cones, small subsidiary shields and complex collapse depressions (Fig. 7.7).

In general, shield volcanoes are built from basalts or olivine basalts, but there are also a few andesitic, alkaline and ignimbritic shields. Hawaiian eruptions are typically basaltic and while they may emanate from the summit region, much activity originates from fissures and rift zones. Vigorous fire-fountaining usually accompanies the onset of eruptive activity, lava often reaching heights of 500 m and temperatures of 1165 °C (Swanson, 1973). Most lavas emerge as thin flows, averaging about 4 m thick, a very high proportion being emplaced via tubes and channels (Greeley *et al.*, 1976). The distribution pattern of flows associated with central vents active over lengthy periods tends to be radial, while fissure-related activity produces a more polarized distribution. The range in average eruption rate for Hawaiian shields is between 4 and 300 $m^3 s^{-1}$ (average 60 $m^3 s^{-1}$) but during the initial stages of an eruption lava may emerge from vents at much higher rates; for instance 560 $m^3 s^{-1}$ was recorded at the vent during the 1984 Mauna Loa eruption.

Geophysical data suggest that the oceanic crust immediately below the Hawaiian chain is between 15 and 20 km thick – over twice the thickness typical of oceanic crust in adjacent regions. The greater thickness is believed to be a result of isostatic subsidence related to the transfer of crustal and mantle materials during volcano development.

7.6.3 Caldera formation and summit activity at shields

On the basis of research conducted in Hawaii, caldera formation used to be considered symptomatic of a particular stage in a shield's evolution. In the light of more recent work it now seems more likely that calderas form over most of a volcano's active life. Unfortunately few historic caldera collapse events have been documented but vital evidence comes from one of the Galapagos shields where, on Isla Fernandina, June 11th 1968, a small seismic disturbance and large cloud of vapour was followed four hours later by generation of a much larger ash cloud. This was followed by a major explosion recorded at infrasonic stations throughout the hemisphere. The violence of the explosion was such that ash expelled from the vent area fell at locations 350 km distant. Further seismic events were recorded for the next ten days or so (Simkin and Howard, 1970). Prior to the 1968 eruption, the floor of the 4 km × 6.5 km caldera lay 800 m below the rim, but subsequently it subsided in a series of short drops focussed along steeply inclined elliptical boundary faults (Fig. 7.8). The principal question posed by the Fernandina events concerns the whereabouts of the displaced magma, for the only recorded contemporaneous effusion was a flow emplaced on the eastern flank during May 21st, which fell far short of the displaced volume. Since no related submarine activity was recorded, it must be assumed that the magma either must have been intruded as dikes into the volcano's substructure or withdrawn at greater depth by tectonic movements.

The calderas of Hawaiian shields are usually elliptical and may contain subsidiary pits which, during the volcano's history, may be filled partially, filled completely or devoid of ponded lavas. Thus the present caldera of Kilauea measures 4 km × 3 km and ranges in depth from 120 m in the north-west to just a few metres at the southern rim. In the southwestern part of the

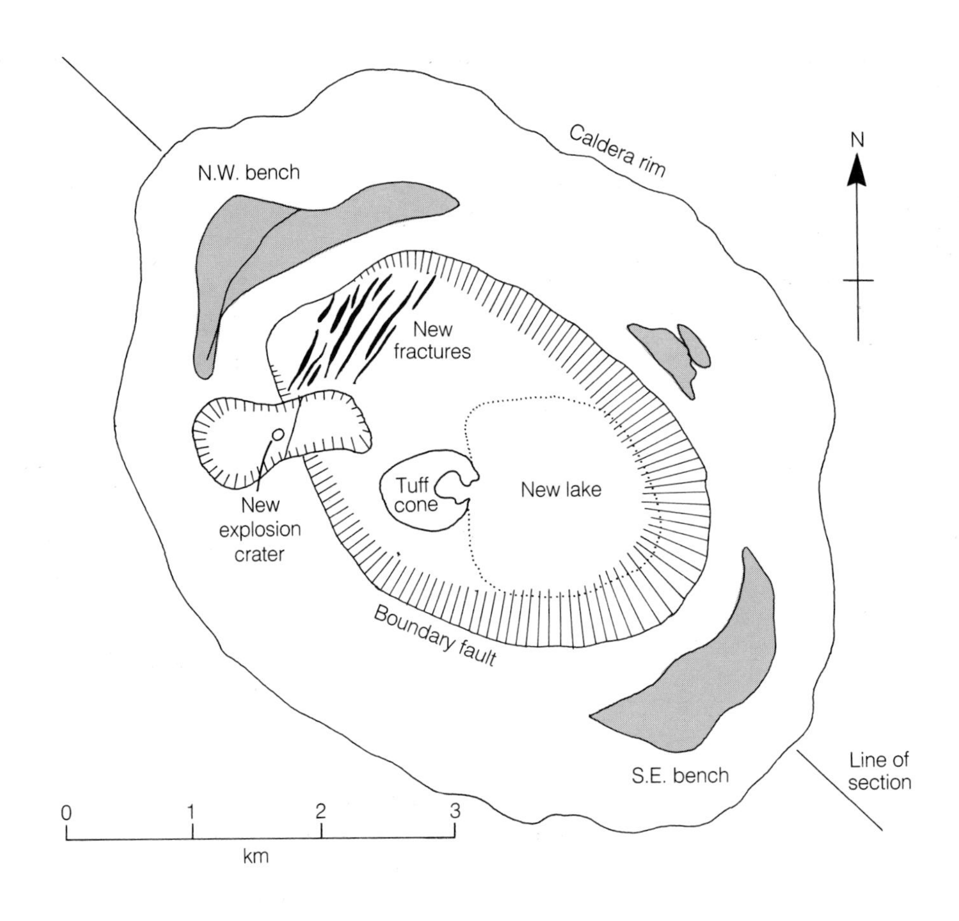

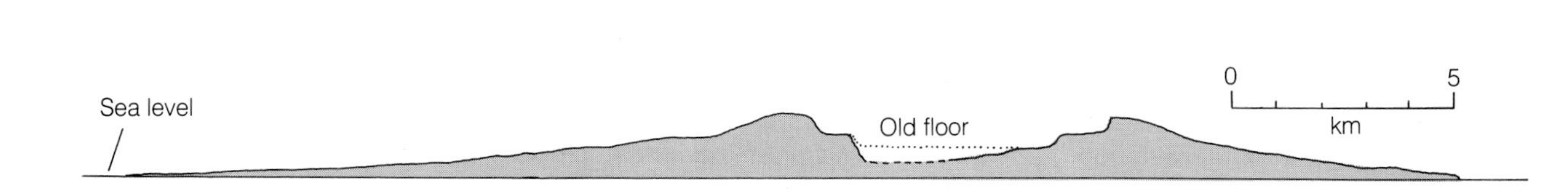

Fig. 7.8 Sketch map of the caldera of Fernandina, Galapagos Islands (above) and cross-section across the summit region (below) (after Simkin and Howard, 1970).

caldera is an 800 m diameter pit, Halemaumau, which resides at the crest of a low shield which has grown on the caldera floor (Fig. 7.9).

A major explosive eruption is known to have occurred at Kilauea in 1790 but it is still not entirely clear whether the modern caldera subsided prior to this event or was partly associated with it. What is known, however, is that the summit of the volcano has experienced alternating episodes of inflation and collapse over long periods and that the numerous pit craters have associated coarse tephra deposits that suggest past phreatic eruptions. In 1894 there was subsidence in Halemaumau and magma was withdrawn from that area. After a long period of inactivity the pit was filled to the brim with lava in both 1919 and 1921, and several flows spread out on to the main caldera floor. The pattern of almost continuous summit activity which was characteristic of the volcano during the nineteenth century and the early part of the twentieth, eventually ceased in 1924, when there were several violent phreatomagmatic

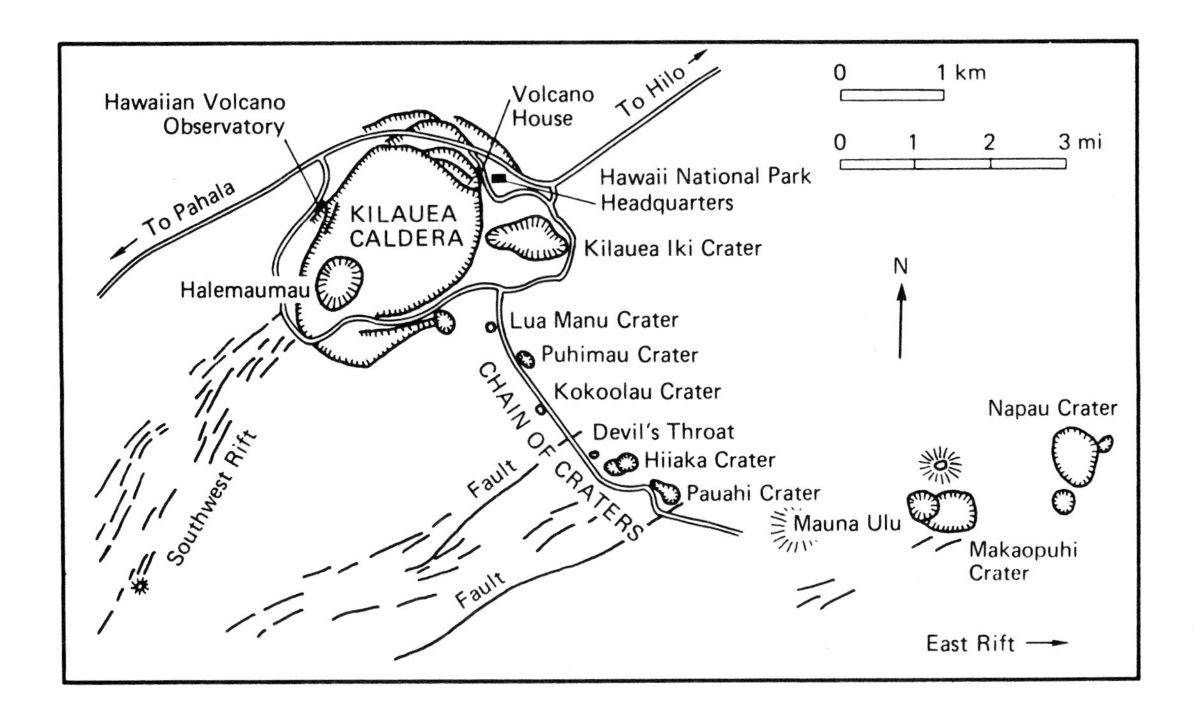

Fig. 7.9 *Location map for features at and near the summit region of Kilauea volcano (after Stearns and MacDonald, 1946).*

eruptions after which the walls of Halemaumau collapsed, leaving a pit 400 m deep. This change in the pattern of eruption appears to have followed a major submarine effusion from the East Rift Zone that coincided with summit collapse. It seems clear, therefore, that withdrawal of magma from a shallow reservoir beneath the summit of Kilauea was responsible for summit subsidence and that the magma withdrawn was both intruded into the volcano's substructure as dikes along the East Rift Zone and extruded as basalt flows on this flank. Episodic activity has continued at the summit since 1952.

Activity along rift zones is another characteristic of shields. For instance, along the upper parts of Kilauea's East Rift Zone there are several pit craters which owe their origin partly to magma stoping and partly to its withdrawal. The latter is a response to eruption lower down the rift, causing the evacuation of a magma chamber beneath the surface and collapse of the crust immediately above it. Sections eroded through rift belts associated with older volcanoes such as Oahu and Lanai, reveal dense swarms of near-vertical dikes. Orientation of active rift zones is evidently largely a function of local gravitational stresses since they form in those relatively unsupported sectors of shields that are not butted against adjacent volcanoes.

Summit eruptions usually are accompanied by shallow earthquakes; rift eruptions are characterized by less regular seismicity. During periods of prolonged rift activity the summit region deflates, presumably because magma is withdrawn sideways into reservoirs on the flanks. The forceful injection of magma into such zones may produce cracks and fissures at the surface and result in the intrusion of swarms of sub-parallel dikes. Such intrusions contribute significantly to volcanic construction and it is clear that while a shield may grow substantially by summit extrusions and dike emplacement, flank growth also occurs and is responsible for the elongated morphology and low profiles of most Hawaiian shields.

7.7 MARTIAN SHIELDS

Although Martian shield volcanoes share many similarities with those of Hawaii, they are very much larger, were active over extremely long periods and show quite a different global distribution pattern. The first and second of these imply a structural and thermal stability of the Martian lithosphere that probably has never been matched on Earth, while the third evidently is a reflection of the absence of plate tectonics from Mars.

7.7.1 The Tharsis Montes and environs

The Tharsis Montes epitomize the large volcanoes of Mars and bear a striking morphological resemblance to terrestrial large shields. By implication, therefore, they are likely to be the products of long-term eruption of large volumes of fluid, basalt-like lavas. The most northerly shield, Ascraeus Mons, rises to the greatest height (26 km) and has the largest relative height difference with respect to the surrounding plains (17 km); both Pavonis and Arsia Montes reach about 20 km above Mars datum. Each is between 350 and 400 km in diameter, with a summit caldera complex significantly larger than any known terrestrial analogue. However, flank slopes are relatively low, averaging less than 5°, while the summit region and lower flanks tend to be less steep than the intervening mid-flank zone. The prominent radial texture seen on Viking imagery is due to hundreds of narrow (<3 km) lava flows, many of which have apical channels; a large proportion of these can be traced upslope to the rims of the summit calderae, while a further significant proportion appear to have emanated from prominent embayments made up of numerous coalescing pits, and situated adjacent to the calderae in the southwest and northeast sectors of each shield (Fig. 7.10). Those near the summit of Arsia Mons are very large and striking, the embayment on the southwest side having erupted a vast shoulder of fan-like flows (Fig. 7.11). Successively older flows are exposed as the distance from the main shields increases, a characteristic common to many Martian shields.

Arsia Mons has a simple 120 km diameter caldera bounded by arcuate faults. A row of low domes connects the embayments in each wall, suggesting eruptive activity associated with the major southwest-northeast fracture line continued after the latest caldera subsidence had taken place. However, as is the case with all three Tharsis Montes, little or no evidence of intra-caldera constructional volcanism is to be found. On the caldera rim are numerous graben which extend outwards from the rim for about 60 km and whose spacing ranges between 1 and 12 km; (Fig. 7.11). The narrow (0.5–1.3 km) lava flows on the northwest slopes predate the faults, having been emplaced prior to fracturing of the caldera rim. The lavas themselves can be traced to a source in a major graben depression situated on the rim, approximately 12.5 km from the edge of the modern floor.

Arsia Mons' caldera is relatively shallow compared with that of Ascraeus Mons, whose deepest pit lies 3.15 km below the rim (Mouginis-Mark, 1981). It is but one of eight major depressions that form its nested caldera complex (Fig. 7.12). Study of the summit region by Peter Mouginis-Mark (1981) shows that some of the collapse events were preceded by major slumping of the caldera backwalls, these being approximately contemporaneous with the formation of circumferential graben. Originally the summit must have boasted

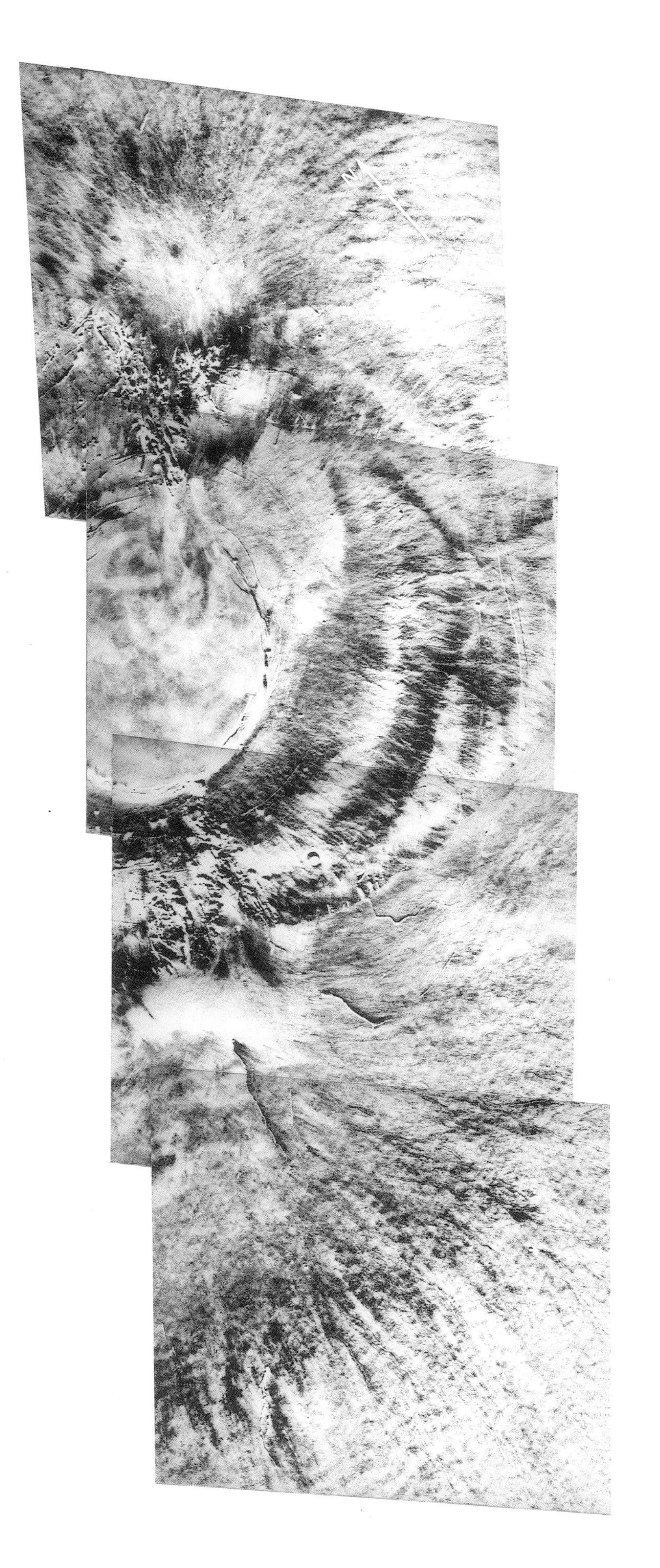

Fig. 7.10 *Mosaic showing the summit region of Arsia Mons. The 120 km diameter caldera has associated with it annular graben and from its backwalls hundreds of lava flows fan out down the flanks. Note that these form broad terraces close to the summit. The sources for many flows are the prominent embayments on both the southwest and northeast flanks. An enlarged view of the southwest embayment in shown in Fig. 7.11. Mosaic length 650 km. Viking orbiter frames 052A02–8.*

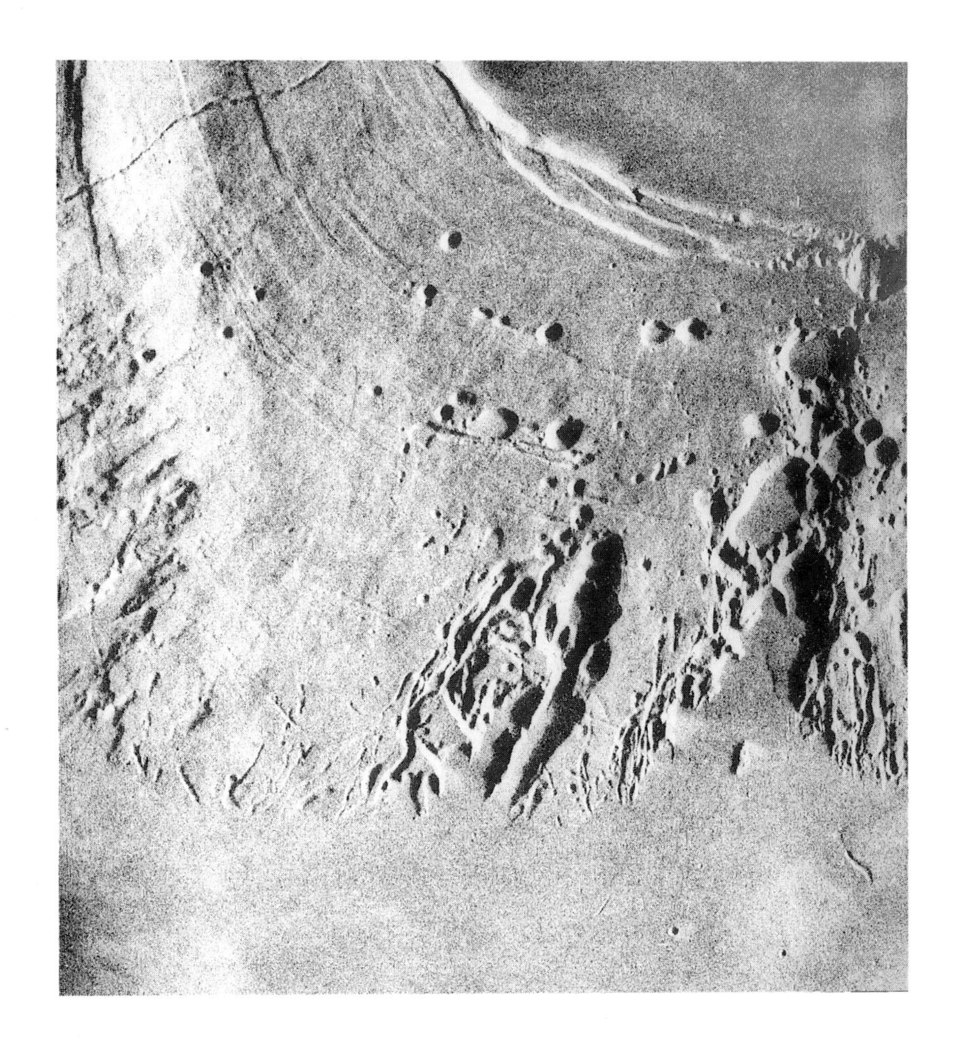

Fig. 7.11 *Enlarged view of southwest flank of Arsia Mons, showing prominent embayment composed of coalescent pits and circumferential grabens. Frame width 115 km. Viking orbiter frame 204A08.*

several smaller pits, but the latest collapse saw the production of the large 40 km diameter depression that now unifies the caldera.

Along the entire caldera rim it is possible to discern a large number of narrow lava flows and some sinuous channels; these radiate down the volcano's upper flanks. The flows are generally less than 1 km wide and between 10 and 20 km in length, while channels are between 100 and 200 m wide and up to 18 km long. Gerry Schaber and his colleagues estimate that most flows are no more than 10 m thick (Schaber *et al.*, 1978). Because these flows and channels can be traced right up to the caldera backwall, it is clear that effusive activity continued at the summit until the final collapse occurred and proves that late-stage explosive activity was not important in the shield-building process. However, source vents for flows are not visible, even on high-resolution Viking images, neither can traces of the flows be discerned on slump terraces within the caldera. This implies, first, that each collapse depression experienced resurfacing after its formation, and, second, that the source vents for the lavas originally were located further up the shield than they can now be traced (Fig. 7.13).

Pavonis Mons has a single, 45 km diameter caldera, about 4.5 km deep, surrounded by arcuate fault terraces that define a shallow summit depression

Fig. 7.12 *Summit region of Ascraeus Mons, showing the nested summit caldera complex, radial flows and flank embayments. Frame width 210 km. Viking orbiter frames 224A88–91.*

Fig. 7.13 *High-resolution mosaic of the south rim of Ascraeus Mons' summit caldera. The narrow summit lava flows are truncated by the caldera backwall, a characteristic they share with terrestrial lava shields such as Mauna Loa. Several flows are seen to be leveed, while others show apical channels. North is to the left. Frame width 25 km. Viking orbiter frames 401B16–18.*

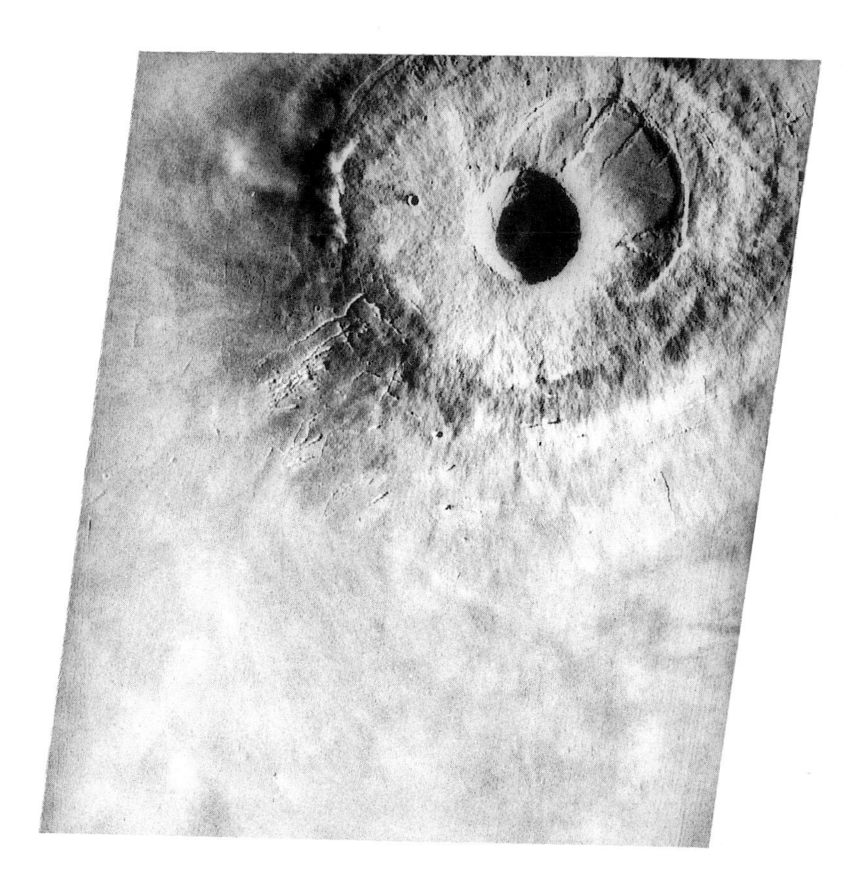

Fig. 7.14 The summit of Pavonis Mons, showing the 45 km diameter caldera within its larger, shallow depression. Arcuate graben surround this. Note also the prominent wrinkle ridges that traverse the otherwise smooth, dark floor of this depression and the embayments on the southwest flank. Frame width 300 km. Viking orbiter frame 643A27.

approximately 100 km in diameter. Annular grabens also occur lower down the northeast, east and southeast slopes, commencing about 120 km from the summit (Fig. 7.14). These are similar to those developed on both Ascraeus and Arsia Mons. Several sinuous rilles can be seen to cross these faults and are considered to be major lava channels incised by turbulent flow. Some of these apparently emerged from circumferential fractures, implying that the latter were the sources for the flows that subsequently poured down the volcano's flanks.

Rheological studies of flows associated with the main shield of Ascraeus Mons indicate that they were of low yield strength and low viscosity, consistent in the main with basalt-like composition (Moore *et al.*, 1978; Zimbelman, 1985). The average effusion rates calculated for these flows range between 18 and 60 $m^3 s^{-1}$ – towards the lower end of Hawaiian and Icelandic rates. While a basalt-like composition is implied by the morphology of the Tharsis Montes flows, they are not all exactly the same. On the upper parts of shield surfaces, out to radial distances of about 400 km, flows tend to be relatively narrow (less than 3 km), less than 150 km long (many being no more than 15 km in length) and often have a central channel, sometimes bounded by levees. Beyond this point they tend to widen (4–7 km) and become longer, some exceeding 400 km; central channels are common. At distances greater than 800 km, where regional gradients become lower, the flows broaden substantially and may be up to 50 km across at their distal ends. Such flows tend not to be channel-fed; some are at least 650 km long. Interestingly, the greater the source height of Ascraeus flows, the shorter they are; this appears to be the case for all three shields, and

Fig. 7.15 Mosaic showing landslide to the northwest of Arsia Mons. This huge lobe has a striated texture towards its margin, but is blocky inside this zone. Note that the striations cross both large sheet flows and small impact craters. Frame width 510 km. Viking orbiter frames 042B9–13, 33–36.

has been noted by the author at Alba Patera (Cattermole, 1988; 1989). This cannot be mere coincidence and almost certainly is a direct reflection of the greatest altitude to which lava can rise in response to the density contrast between the magma and rocks in its source region. Confirmation for this explanation is forthcoming from the observed concentration of shorter flows near shield summits, where eruption rates must have been less than on the lower flanks – a response to the much greater vertical distance through which they needed to be lifted before eruption.

It is hardly surprising, in view of their huge dimensions, that instabilities were set up from time to time. The most dramatic evidence for this comes from the west-northwest flank of Arsia Mons, where a lobate slump measuring 400 km × 350 km extends outward from the base of the shield. Within the slide unit are hundreds of small hills, while at its outer edge are closely spaced ridges which produce a strongly striated effect (Fig. 7.15). Many striations crosscut impact craters and other small landscape features without in any way modifying their outline, as though having been superimposed upon them. This is a curious phenomenon and the suggestion has been made that the slide formed while the landscape was mantled by ice (Williams, 1978). Not only would this have facilitated the flow of the debris, but its subsequent melting and removal would have superimposed the flow pattern on to the underlying topography. The strongly dissected terrain of the shield flanks above the slide's upper boundary suggests there was a zone of detachment along which the shield partly collapsed, slipping laterally as a gravity-assisted slide or indeed a series of slides that spread out over the surrounding plains.

It appears that the three Tharsis Montes followed a similar pattern of development. An initial stage of shield-building was achieved by the gradual accumulation of fluid lavas both from the summit area and from peripheral vents; then, after each shield had grown to its maximum height, effusive activity became concentrated along a major rift zone aligned in a southwest-northeast direction. Considerable lateral transport and supply of magma to vents and fissures along this zone over long periods of time resulted in repeated collapse of the shield summit region forming major embayments and small satellite calderae in which some lavas became ponded. Eruptions from these rift-aligned sources built out substantial shoulders on southwest and northeast flanks, from which large numbers of major eruptions disgorged high-volume flows. The great lateral extent of the latter implies that rates of eruption were high. Crater counts suggest that the construction of the main Arsia Mons shield terminated earlier than its two neighbours, but eruption both from southwest-northeast embayments and from within the caldera continued into the more recent past (Crumpler and Aubele, 1978).

7.7.2 *Olympus Mons*

Rising to a height of 27 km above Mars datum and at least 23 km above the surrounding plains, Olympus Mons is unquestionably the most spectacular of all Martian volcanoes. It has a diameter of at least 520 km, and a volume over fifty times that of any terrestrial shield. Furthermore, it is surrounded by a huge scarp that in places is 6 km high, and an aureole of peculiar blocky terrain extending for between 300 and 700 km beyond the scarp base (Fig. 7.16).

Like the Tharsis Montes, it has a nested caldera complex. This measures 80 km across and has suffered multiple collapse, the most recent events having formed a 3 km deep pit. Even on very-high-resolution images of the summit region, evidence for caldera-floor activity is not forthcoming, the only features discernible on the floor being several wrinkle ridges and arcuate graben, both landforms due primarily to tectonic forces (Fig. 7.17). In contrast to Ascraeus Mons, where the largest pit was formed last, here the reverse is true.

The flank slopes are generally rather low (mean slope is 4°), but the shield has a somewhat sinusoidal profile since the middle part of the structure is steeper than either the summit region or the lower flanks. Also, the surface of the volcano is built from a series of 15 to 50 km wide flow terraces separated by distinct breaks in slope and crossed by innumerable thin flows which have a roughly radial disposition. The flows themselves are difficult to see close to the caldera region but further away long, narrow flows (1–3 km wide), often with central channels and levees, are widely distributed, and except in the western sector, may be seen draping the face of the scarp before extending over the lower, flatter ground (Fig. 7.18). The fact that these radial flows extend

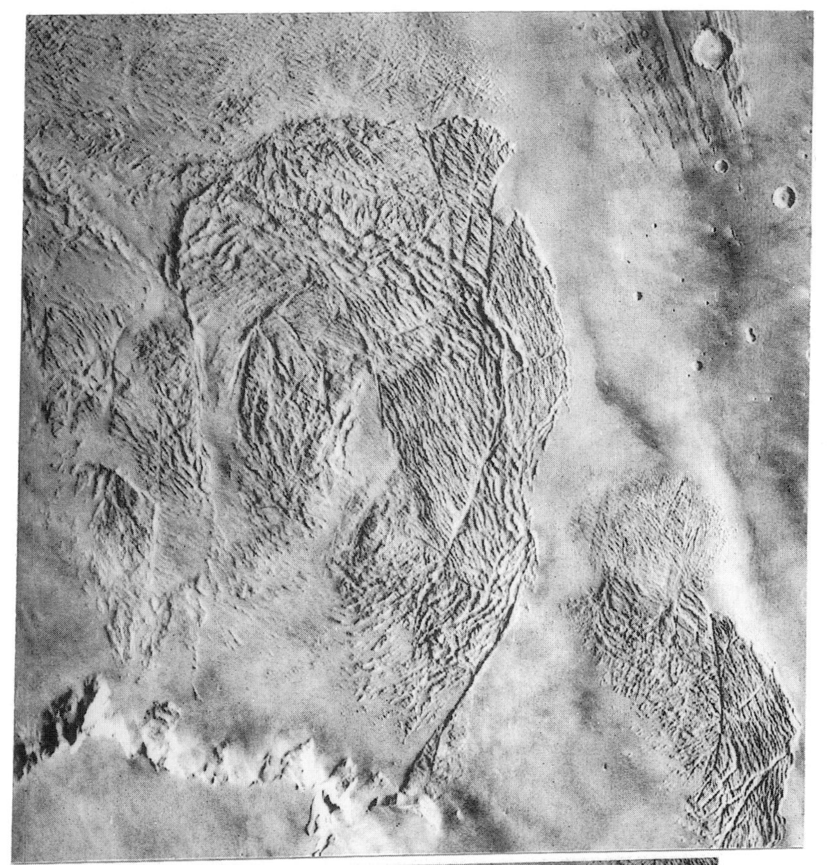

Fig. 7.16 *Synoptic view of Olympus Mons, showing the summit caldera, terraced flanks, basal scarp and aureole. Length of photo pair 1200 km. Viking orbiter frames 741A05 (above), 741A07 (below).*

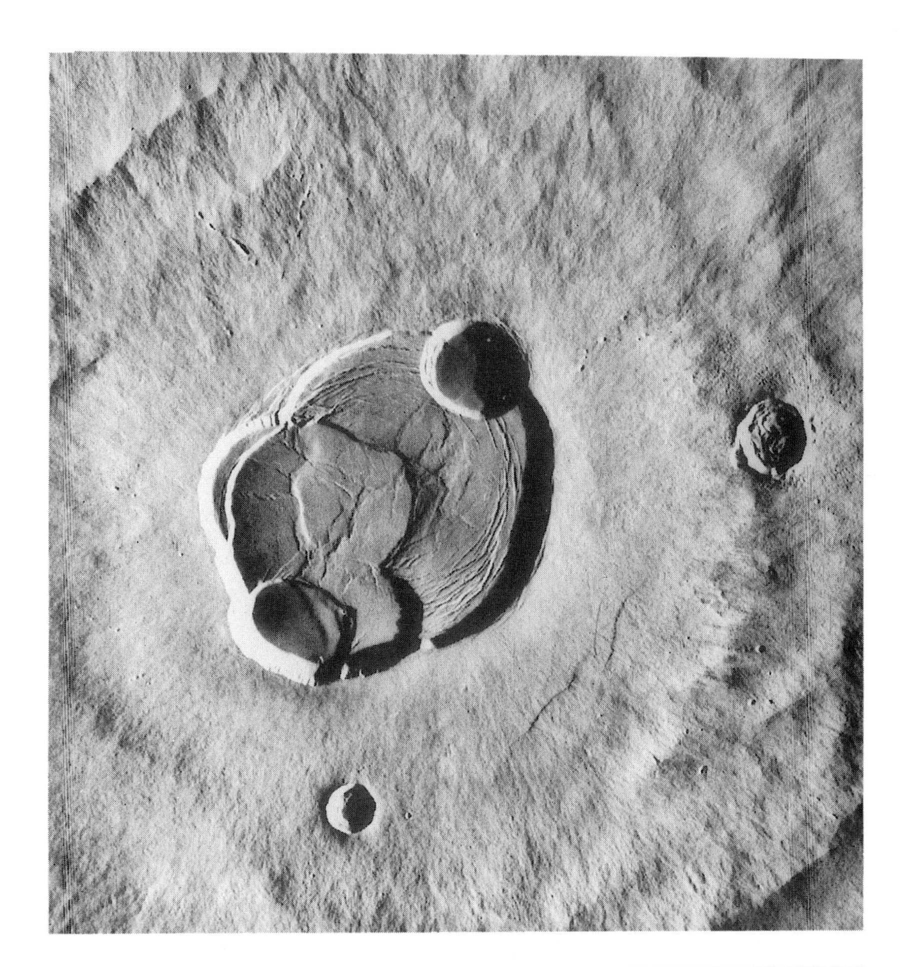

Fig. 7.17 The summit region of Olympus Mons, showing the nested caldera pits, radiating flow texture and broad flow terraces. Annular grabens occur on the floor of the oldest caldera component, while wrinkle ridges characterize the smaller, younger pits. Frame width 175 km. Viking orbiter frame 890A68.

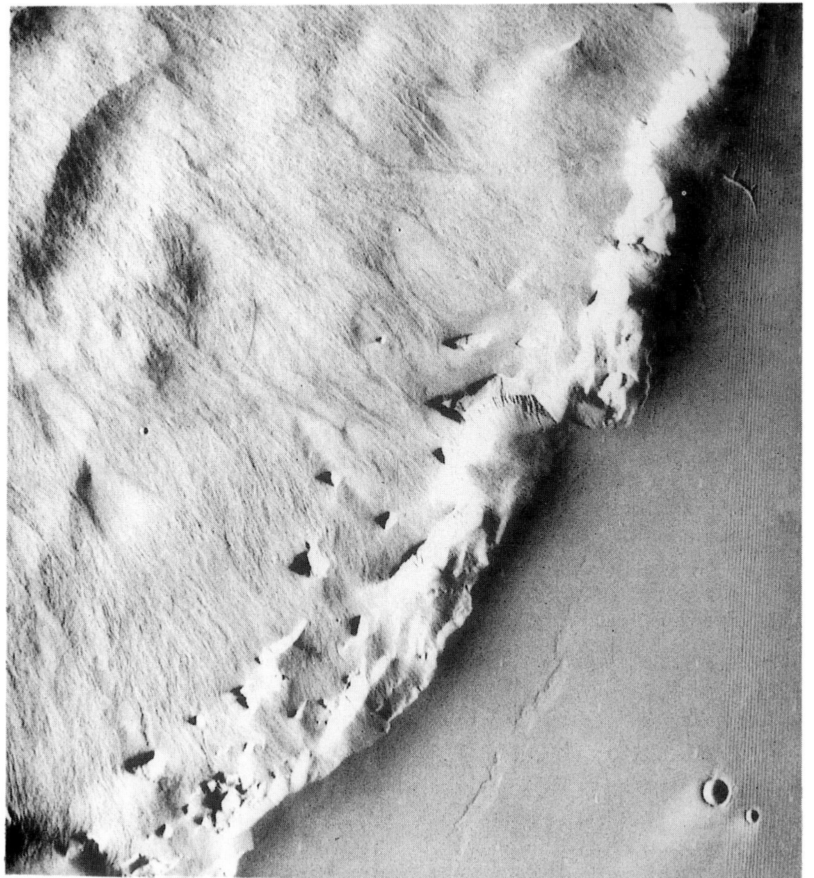

Fig. 7.18 Plethora of narrow volcanic flows draping the basal scarp on the northern flank of Olympus Mons. Most flows are less than 3 km across. Note the pre-shield blocky terrain emerging from beneath the scarp in the upper right. Frame width 130 km. Viking orbiter frame 222A64.

outwards for distances of about 350 km from the summit may mean that it is more realistic to consider 700 km as the real diameter of the volcano.

The scarp transects a broad pedestal of pre-shield material about which little is known. This is cut by several faults, seen in section in the scarp backwall. These may be related to the generation of the scarp. Where the scarp has not been inundated by lava flows there is evidence for several major landslides which have a hummocky texture except near their margins where a ridge-and-trough fabric prevails.

7.7.3 Olympus Mons aureole

In some respects the landslide texture is similar to that of the remarkable aureole, whose formation was either contemporaneous with or just subsequent to scarp development. It extends in places for at least 1000 km from the shield summit and comprises a series of terrain blocks made up from distinctively textured, closely spaced ridges (Fig. 7.19). The inner margin of each block is embayed by younger flows or aeolian deposits, while the outer boundary is more scarplike; the blocks form a complex of inwardly tilted prisms of intensely fractured terrain. The northwestern sector of the aureole is marked by a positive free-air gravity anomaly of several tens of milligals (Sjogren, 1979). Significantly, the number of superposed impact craters is very small, and although smaller ones (<1 km) might be removed by surface creep or land-slipping, the relative sparsity of 5 km craters – which are less likely to be removed in this way – implies the feature must be young (Schaber *et al.*, 1978).

Explanations for this unique but enigmatic landform are as numerous as they are diverse. Carr (1973) suggested that the aureole might be the remains of an older shield volcano, while Blasius (1976) suggested it was an unroofed intrusion. Neither of these early suggestions appears valid, however, since both would necessitate much higher rates of erosion than are seen on Mars. King and Riehle (1974), on the other hand, proposed Olympus Mons to have been a composite volcano, a significant amount of construction having been accomplished by ash generated during extensive explosive activity. The aureole blocks are considered by them to represent the eroded remnants of lava and tuff sheets, the latter being deposits of nuées ardèntes. Morris (1979) also suggested construction from tuff sheets, but in his hypothesis these were erupted from several vents distributed around the main shield. Neither of these ideas seems entirely appropriate, however, since there is a total absence of evidence for any pyroclastic deposits anywhere on the Olympus Mons shield or, indeed, on the Tharsis Montes.

In contrast, Hodges and Moore (1979) formulated the ingenious proposal that Olympus Mons bore similarities to certain Icelandic volcanoes which erupted under a cover of ice. According to this theory, the height of the basal scarp represents the thickness of an ice sheet which once covered the region and beneath which eruptions took place. However, no evidence for subaerial activity within the confines of the aureole exists and, furthermore, neither is there any evidence for the ice sheet.

Various workers have considered the possibility of major thrusting or sliding. For instance, Harris (1977) suggested that the aureole lobes represent huge gravity-assisted thrust sheets, an argument pursued also by Morris (1981). Carr *et al.* (1977) and Lopes *et al.* (1980) have concluded the aureole to be the product of large-scale mass movement. The Lopes group invoke gravity-

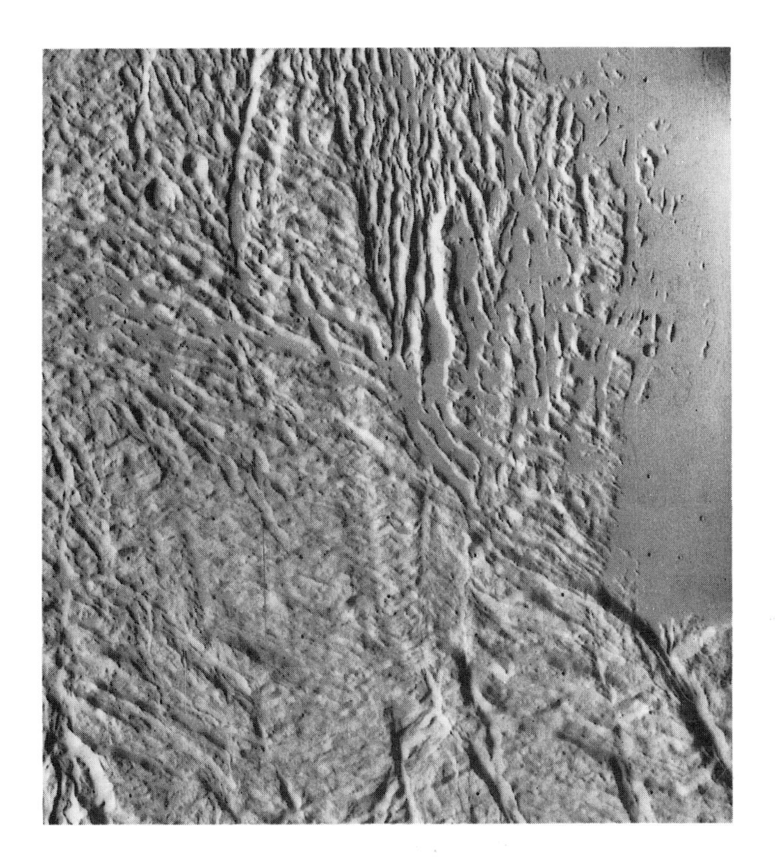

Fig. 7.19 Olympus Mons aureole texture. This consists of a series of blocks each composed of ridges interspersed with smoother plains units. Frame width 93 km. Viking orbiter frame 043B20.

assisted rockslides on an immense scale, an interpretation supported by a comparison between the volume of material available before scarp formation and the present volume of the aureole materials (Fig. 7.20). Furthermore, the greater extent of the aureole on the northwest side, where the shield slopes gently down to the surrounding plains (compared with the southeast, where it rises towards Tharsis Montes), clearly implies gravity control. Mass sliding of this order could have been aided had there been a permafrost reservoir in the

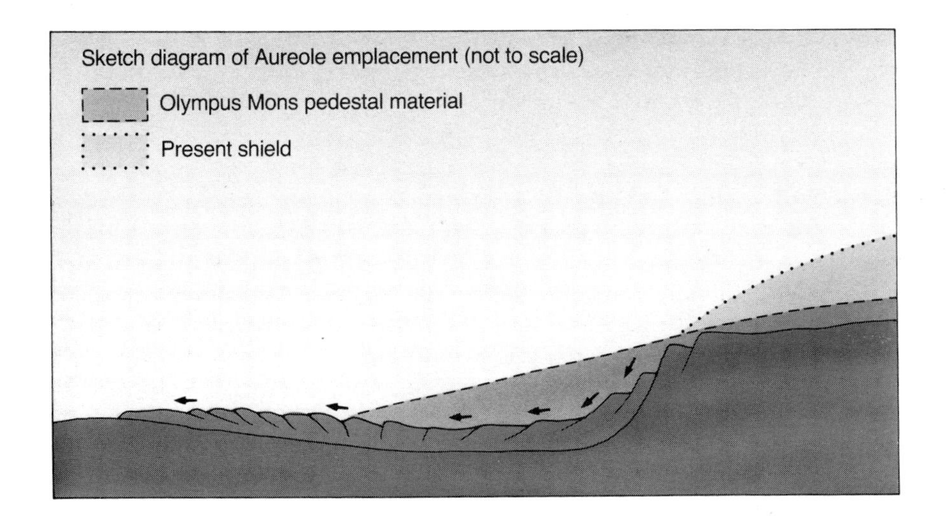

Fig. 7.20 Gravity-sliding mechanism for generation of Olympus Mons aureole blocks (after Lopes et al., 1982).

sub-shield pedestal. For this to have occurred the pedestal would, of necessity, have to have included a significant proportion of either brecciated or relatively porous material, such as volcanic ash or tuff. Although there is no evidence for this, it is not impossible that explosive activity preceded the building of the main Olympus Mons shield. If this had been so, the increased heat flow associated with the rise of magma to form the main shield might have set off melting of sub-shield permafrost. This, in turn, might provide a catalyst for the sliding out of the aureole blocks. While the author favours some kind of large-scale sliding mechanism, neither this nor any of the other suggestions explains the gravity data. At the present time, therefore, it has to be admitted that no single hypothesis is completely satisfactory and a solution to this enigma remains to be found.

7.8 OLDER THARSIS VOLCANOES

Elsewhere in Tharsis is a number of smaller central volcanoes all of which are older than the Tharsis Montes; some may be as ancient as paterae in the heavily cratered highland hemisphere (Plescia and Saunders, 1979).

Uranius and Ceraunius Tholi and Uranius Patera lie northweast of Ascraeus Mons (Fig. 7.21). The two tholi have steeper slopes (5° and 7°, respectively) than the patera (0.5°). Uranius Tholus has a diameter of 83 km and rises 3500 m above the surrounding lava plains. Its flat summit has a 14 km diameter pit set towards the eastern edge of a larger but very shallow 32 km caldera, largely infilled with ponded lavas. It is a relatively old structure and like the other two volcanoes, is older than the Hesperian flow plains which surround them. Ceraunius Tholus is somewhat steeper and larger (130 km × 92 km) and has a 23 km summit caldera with vestiges of an older, shallower pit on its north side. The flanks of both volcanoes are finely striated, the most obvious radial striations being narrow channels. In general appearance they are more reminiscent of the Elysian volcano Hecates Tholus, discussed in a later section, than other Tharsis shields. It could be that Plinian-type eruptivity played a role in their development.

A prominent 2 km wide channel runs from near the summit of Ceraunius Tholus down the north flank into an impact crater at the base of the shield (Fig. 7.22). Carr (1974) suggested that this larger channel and several others like it may have been due to lava erosion, but Reimers and Komar (1979) argued that the general appearance of both coarse and fine channelling on both tholi is reminiscent of the terrestrial stratovolcano Barceno, in Mexico, which, during 1952–1953, experienced large-scale explosive activity that generated fast-moving density currents – base surges and nuées ardèntes – which eroded arrays of channels in the volcano's flanks.

Biblis Patera and Ulysses Tholus, situated west of Pavonis Mons, both have simple calderas slightly over 50 km in diameter and numerous circumferential graben. The former measures 175 km × 105 km and rises 4 km above the plains, while the latter rises 3 km above the plains and has a diameter of 91 km. Both have flank slopes approaching 4°. Their general appearance is very reminiscent of the summit region of Arsia Mons. Jovis Tholus is a 55 km shield with a 27 km diameter summit caldera and is set somewhat apart from the rest, midway between Olympus Mons and Ascraeus Mons. Tharsis Tholus, east of Ascraeus Mons, rises 6 km above the plains and measures 155 km by 120 km; it too has a

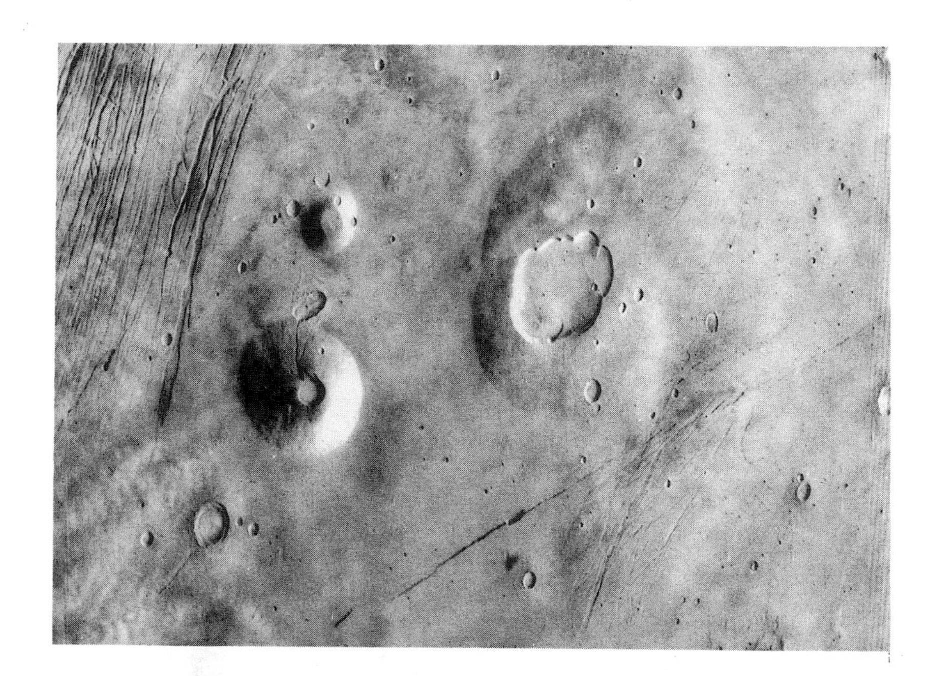

Fig. 7.21 Group of three volcanic shields situated east of Ceraunius Fossae (left). Uranius Tholus, the smallest, has a flattened summit region, while to its south is the larger, steep-sided Ceraunius Tholus. East of both is the volcano Uranius Patera, with a large summit caldera and much shallower flank slopes. Frame width 750 km. Viking orbiter frame 759A73.

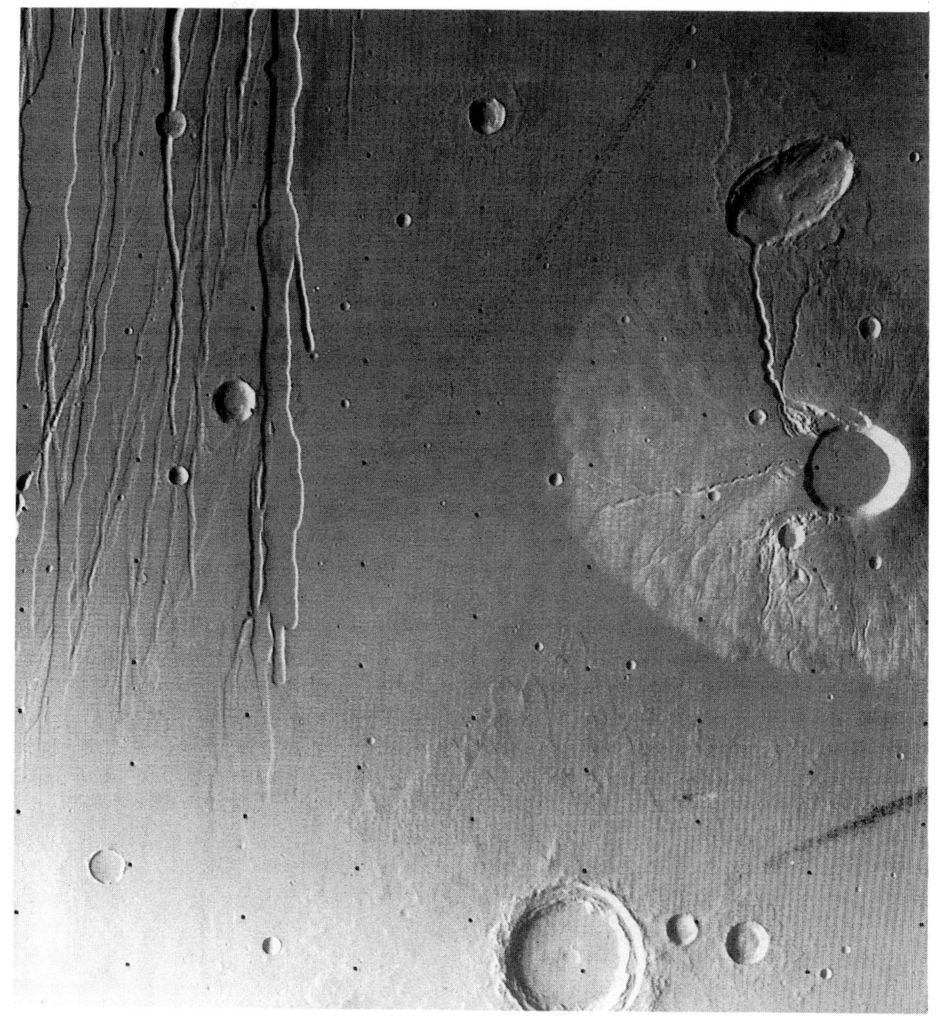

Fig. 7.22 JPL enhanced image of Ceraunius Tholus, showing the finely striated shield flanks and the prominent channels on the north side. The largest of these debouches into an elongate impact crater on whose floor it has deposited an apron of debris. Frame width 100 km. Viking orbiter frame 516A24.

caldera 62 km × 46 km across. In view of their resemblance to the upper parts of the main Tharsis shields, all are presumed to be partially buried centres that grew prior to the Upper Hesperian/Lower Amazonian effusive activity which saw the gradual building of major centres along the crest of the Tharsis Bulge. Whether they ever grew to the vast dimensions of their successors is not known.

The upper exposed part of Uranius Patera is traversed by well-defined, narrow and often levéed flows and measures 202 km × 184 km across. It has flank slopes of just 0.5°. The floor of the nested caldera complex is crossed by several prominent wrinkle ridges while the northwest part of the floor surface is inclined towards the centre of the depression, suggesting it subsided prior to the emplacement of the more or less horizontal ponded flows which occupy the centre. The very large size of the caldera implies that a large part of the shield has been buried by younger deposits.

7.9 ALBA PATERA

The volcano Alba Patera has associated with it some of the most extensive volcanic flow fields found anywhere in the solar system. Situated on the northern edge of the Tharsis bulge, its summit is located at 110°W, 40°N and lies at the centre of an oval ring-fracture zone measuring 550 km × 400 km across (Fig. 7.23). Crater ages suggest that its oldest flows predate those of Tharsis Montes, that its peak of activity may have occurred around 1725×10^6 years ago (it has a Hesperian/Amazonian age range), while its activity extended over a period of at least 1.5×10^9 a and may have spanned as much as 2.8×10^9 a (Cattermole, 1989).

The summit of Alba lies 7 km above Mars datum, and comprises a broad southwest-northeast trending ridge with the summit calderas situated towards the northeast end. The mean slope value for the patera is only about 0.5°, yet the volcanic flows extend at least 1350 km from the summit, giving it a diameter of 2700 km, an area approaching 2×10^6 km^2 and a volume of at least 1.4×10^9 km^3. The extreme length of these flows implies both low viscosity and high volume for the individual flows. Quantitative analysis of these flows indicates that the large sheet and tube-fed lavas had low yield strength and viscosity and were erupted at very high rates (Cattermole, 1987).

The summit region is the site of a double caldera complex which became the focus of extended effusive activity after an early period of flood lava activity. Volcanic flows which originated from the younger caldera partially bury the older, incomplete depression (Fig. 7.24). The former has at least five components, which together cover an area of 2065 km^2. The size of the individual collapse depressions decreased with time. Each is very shallow, the deepest being only 150 m deep, which contrasts markedly with the deeper pits on Ascraeus and Olympus Montes and indicates a significant difference in the dimensions of their subjacent magma reservoirs. As is the case with the Tharsis Montes, narrow and often levéed flows can be traced right up to the backwall of the caldera, indicating that Plinian-style activity was not important during the later stages of patera growth.

The general distribution of lava flows on the main part of the patera is radial about the summit, but flows are scarce over the uplifted block of Ceraunius Fossae and in the northwest sector they are either absent or mantled by

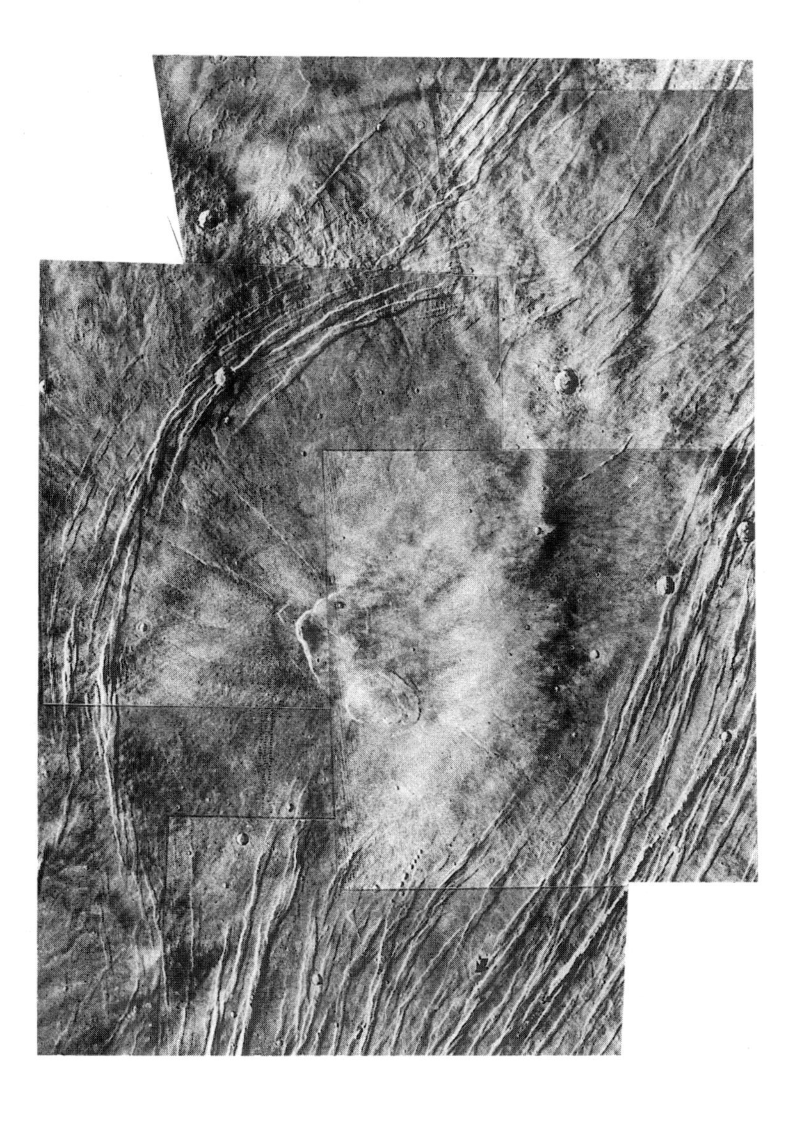

Fig. 7.23 *Synoptic view of Alba Patera. Viking orbiter frames 783A11–17.*

younger deposits (Fig. 7.25). On the lower ground to the northwest, large fields of tube- and channel-fed lavas have a generally northnorthwest-southsoutheast strike and appear not to have been related to summit activity. Detailed geological mapping by the author shows there to have been three principal episodes of patera growth (confirming earlier mapping by Scott and Tanaka (1980, 1986)), together with the generation of several large discrete flow fields whose development may have been related to major fractures on the volcano's flanks. The earliest phase in the development of Alba involved the widespread emplacement of fissure-fed flood lavas of Lower Hesperian age which now occupy distal locations. Subsequently, volcanism became more centralized, sheet flows and tube-fed lavas of large volumes being extruded, mainly from linear vents situated at or near the present summit, or from lower down the volcano's flanks (Fig. 7.26). These flows were responsible for the majority of patera construction. Quantitative measurements of these flows indicate volumes at least an order of magnitude greater than most Hawaiian flows, sheet flows having volumes in the range 1–110 km^3 and tube-fed lavas achieving volumes as great as 3500 km^3 (Baloga and Pieri, 1985; Cattermole, 1987).

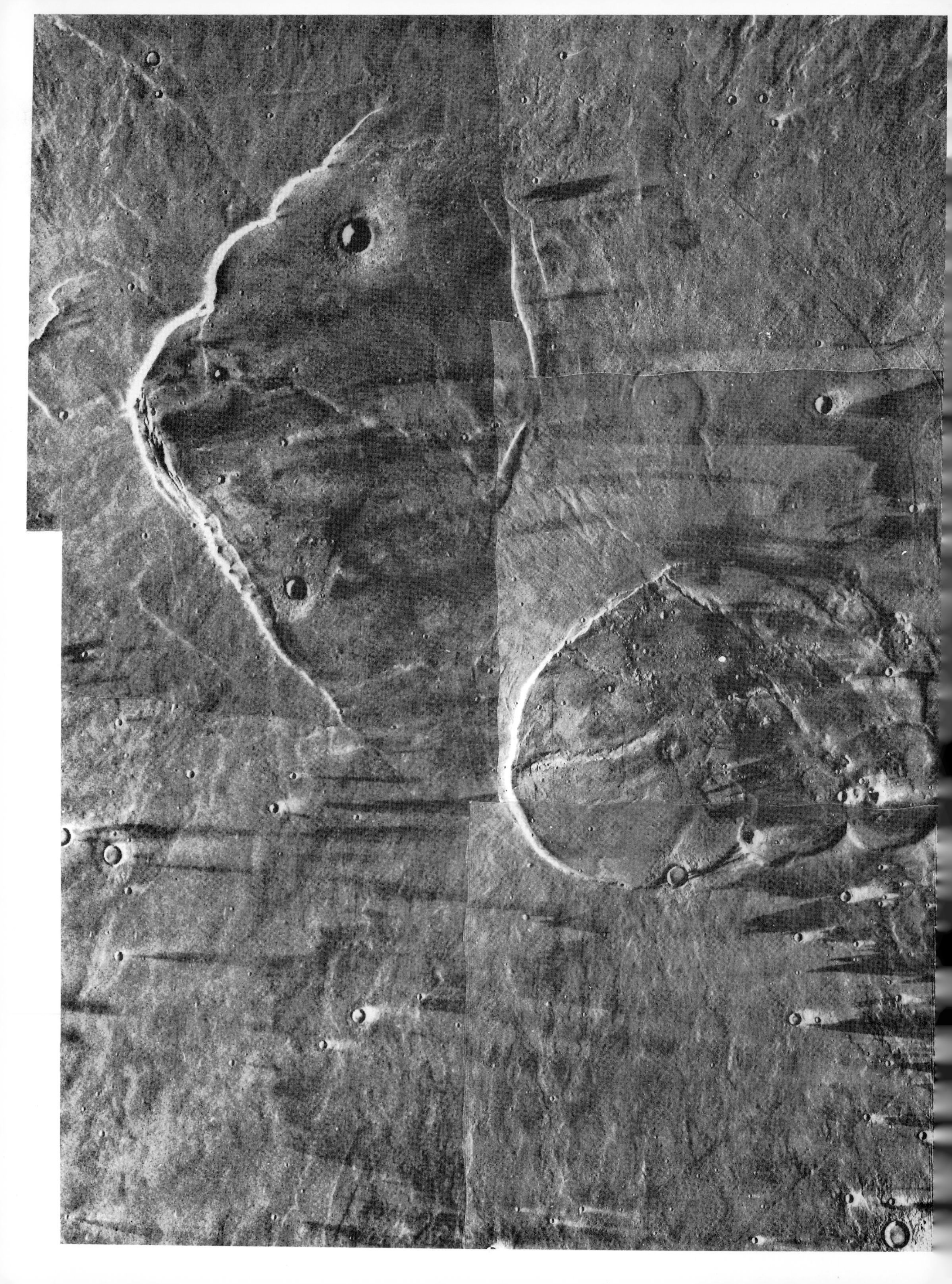

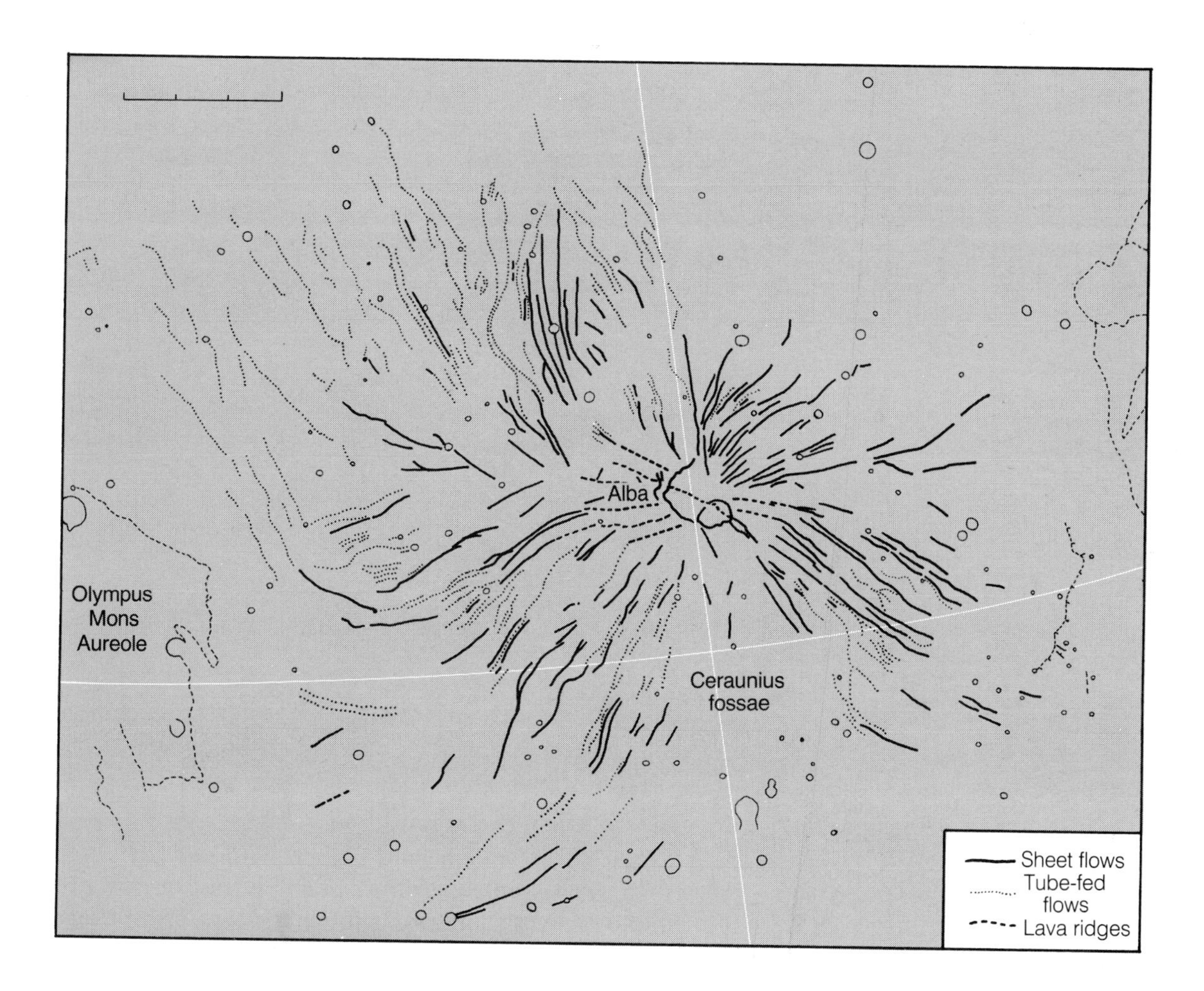

Fig. 7.25 *Distribution of different flow types at Alba Patera volcano. While there is a generally radial pattern, massive tube- and channel-fed lavas to the west and northwest of the structure do not appear to have a summit-related origin.*

Fig. 7.24 *(opposite)* *Alba Patera caldera complex. The younger complex measures 65 km × 48 km and is composed of at least five coalescent depressions. Narrow summit flows can be traced to the caldera backwall and many are seen to be leveed. These late flows partly bury the floor of the older caldera to the northwest. Viking orbiter frames 255S13–17.*

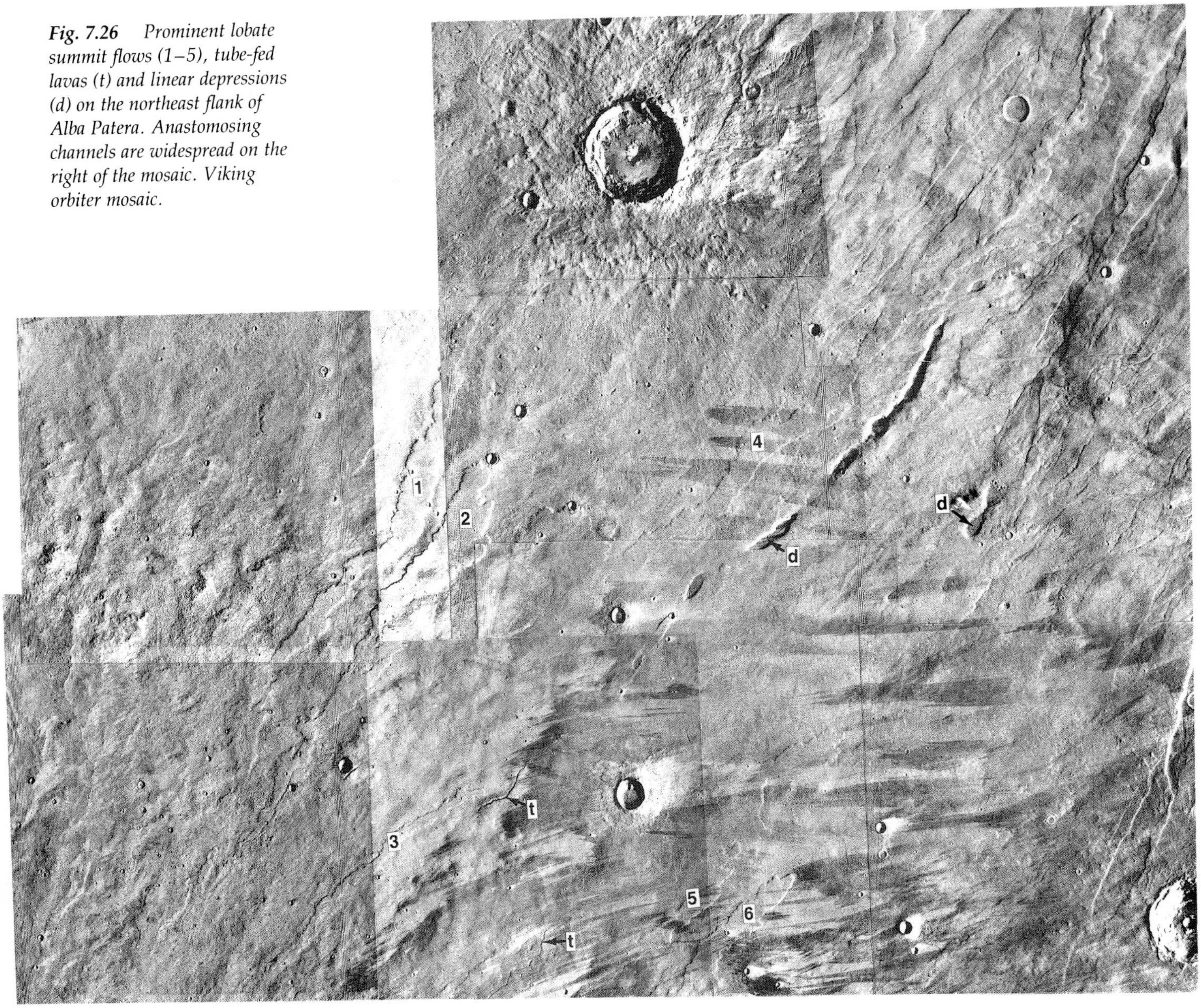

Fig. 7.26 Prominent lobate summit flows (1–5), tube-fed lavas (t) and linear depressions (d) on the northeast flank of Alba Patera. Anastomosing channels are widespread on the right of the mosaic. Viking orbiter mosaic.

A further characteristic of Alba Patera is the existence of anastomosing channel networks, mainly on the northern flank where lava flow lobes are absent. Channel incision appears to have separated the two main stages of lava shield growth, and a recent analysis of these by Mouginis-Mark *et al.* (1988) suggests, first, that the channels were of fluvial origin and, second, that they had been incised by a process of sapping induced by the release of non-juvenile water within relatively unconsolidated deposits on the volcano's flanks. They concluded that the fine-grained deposits themselves were of pyroclastic origin. Mouginis-Mark *et al.* (1988) discounted an air-fall origin on the basis that the channelled deposits extended too far (500–600 km) from the potential caldera source region and could only have been dispersed that far by extremely high eruption clouds. Since these are unlikely to have formed on Mars, they suggest that long run-out pyroclastic flows dispersed the fine-grained material.

If this conclusion is accepted – and there is no *a priori* reason why this should not be so – then eruption of the volatile-rich ash-flow materials can be shown to have post-dated the emplacement of high-volume sheet and tube-fed lavas on the lower flanks but to have preceded the final effusive phase of the volcano's evolution. As Mouginis-Mark and his co-workers note, the existence of pyroclastics within the lava pile certainly could have played a part in explaining its very low relief and also the form and distribution of the circumferential fractures.

Because the Alba flows are so well preserved, it is possible to consider certain volcanological implications. First, it is clear from the flow morphology that Hawaiian-style effusive activity characterized the growth of the main shield and, second, that the volumes of individual flow units decreased with time. Now since caldera collapse can only take place if a sufficiently large void is created beneath the summit, and because subvolcanic reservoirs must be full prior to the onset of eruption, only during the extrusion of lava or the injection of dikes into the volcano's substructure can such a space be created. Such is in accord with similar conclusions reached by Mouginis-Mark (1981) for Ascraeus Mons. Thus the greater volumes of the earlier sheet and tube-fed flows imply that progressively lesser volumes of magma were required to trigger eruption as time proceeded. The implication here is that the magma flux rate beneath the volcano must have been greater during the earlier stages of its construction, that is, when the main part of the structure was emplaced during the early Amazonian epoch, than during the later stages.

The very large dimensions of Alba flows have no real parallel on the Earth and it is informative to consider the long-term magma supply rate to this huge volcano. On the assumption that all of the present topography of Alba is due to constructional volcanism, the lava pile has a volume of $4.16 \times 10^9 \, km^3$. Now if volcanism occurred over a period of at least 2.3×10^9 a, this gives an overall magma production rate of $2 \times 10^{-3} \, km^3 \, a^{-1}$. Such a figure is significantly larger than present day rates of magma supply to the volcanic pile of Mauna Loa and Kilauea! Calculations also show that several of the larger sheet and tube-fed lavas on the lower flanks must have released two orders of magnitude more thermal energy than the Earth's annual release due to volcanism, and one order of magnitude more thermal energy than the annual conducted heat flow of the Earth (Cattermole, 1989). This surprising conclusion leads inevitably to the implication that eruption of these very large volumes of magma must have constituted an extremely, if not the most, important, source of energy loss from the Martian interior over geologically short periods during the Hesperian and early Amazonian epochs.

7.10 THE SHIELD VOLCANOES OF ELYSIUM

There are three shield-like volcanoes in Elysium and these are significantly different in morphology from those of Tharsis. A fracture belt almost completely encircles the summit of Elysium Mons and outside this ring are numer-

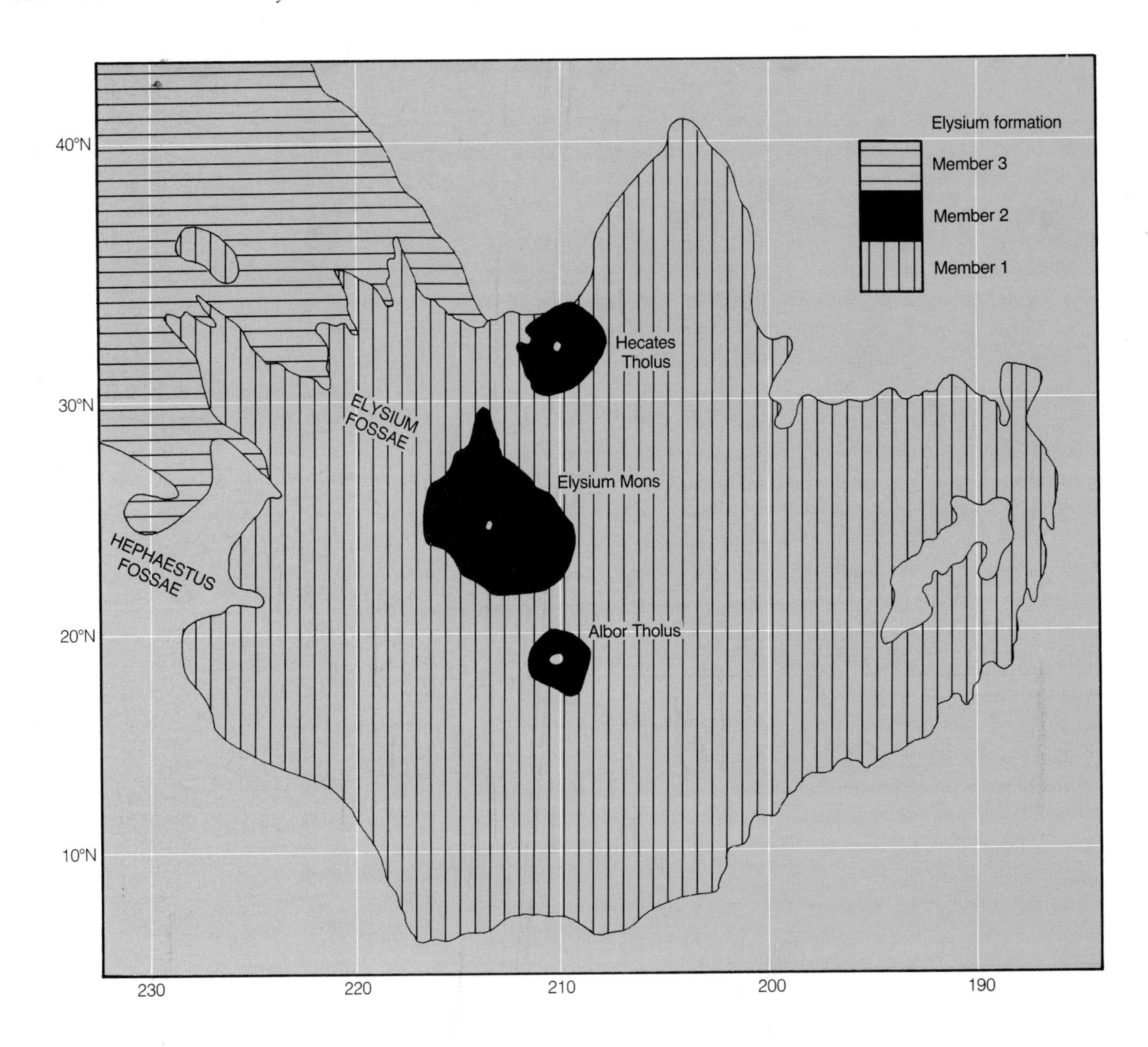

Fig. 7.27 Generalized geological map of the Elysium region of Mars.

ous west-northwest-east-northeast trending troughs, most with flat floors and the general appearance of graben. For reasons not entirely understood, these pass northwestwards into a series of branching channels which extend for several hundreds of kilometres. The lower Amazonian lavas associated with Elysium central volcanoes cover an area of approximately $3 \times 10^6\,km^2$ (Fig. 7.27).

The largest of the Elysian volcano group is Elysium Mons, which has a diameter of 170 km and a single caldera 12 km in diameter. Its summit lies about 9 km above the surrounding plains and it has slopes comparable with those of the large Tharsis volcanoes. The overall geometry is, however, asymmetric since there appears to be a main shield, approximately circular in plan, with the caldera sitting on the crest of a broad ridge that trends northwest-southeast across its crest. Numerous narrow flows outcrop on the main cone

Fig. 7.28 General view of Elysium Mons, showing the 14 km diameter summit caldera, main shield and peripheral radial fractures. Frame width 125 km. Viking orbiter frames 541A44–6.

and there are also large numbers of hummocks up to 5 km across and linear channels (Fig. 7.28).

The other large Elysian structure, Hecates Tholus, is more interesting and has been studied closely by Mouginis-Mark *et al.* (1982). Situated north of Elysium Mons, it rises about 6 km above the adjacent plains and takes the form of a low shield 160 km × 175 km across. At the summit is a nested caldera complex measuring 11.3 km × 9.1 km. Unlike Tharsis shields which largely are constructed from lava flows, Hecates Tholus exposes none. All that is visible is a complex of radial channels not unlike those seen on Ceraunius Tholus (Fig. 7.29). However Mouginis-Mark and his colleagues (1982) showed that the anastamosing courses and dendritic tributary networks were unlike channels associated with volcanic density currents. Furthermore, the absence of channels from the summit region makes it difficult to believe their formation was associated with explosive volcanic activity. As a result they suggest that the channels are fluvial in origin, having been incised into materials less coherent than the lava shield that they mantle.

Impact crater counts indicate that the chanelled surface is actually younger than the youngest region of the Olympus Mons caldera (perhaps as young as 3×10^8 a). The same group therefore proposed that this is a relatively young air-fall deposit they estimate to be about 100 m thick, produced by an eruption cloud with a height approaching 70 km. Because a stable eruption column

Fig. 7.29 *Summit region of Hecates Tholus showing the nested caldera complex, aligned pit rows, channels and generally rather smooth appearance of the circum-summit area. Frame width 60 km. Mariner 9 frame DAS13496298.*

apparently was able to be sustained, the magma volatile content must have been about 1 wt% if the volatile were H_2O and more than 2 wt% if CO_2. Such a volatile component requires that the source magma must have originated at depths of greater than 50–100 km if the volatile was CO_2 and between 4 and 150 km if it was H_2O. While this imposes mantle depths for CO_2-rich magmas, no such restrictions pertain if water was the volatile, and thus the possibility exists of absorption of permafrost or trapped groundwater as the rising magma neared the Martian surface. In order to reach such an altitude, model calculations suggest a volatile-charged Martian magma would need a mass eruption rate of around 107 $kg^{-1} s^{-1}$. Here we have yet another volcano, therefore, in whose growth phreatomagmatic activity has played an important role.

7.11 APOLLINARIS PATERA

Apollinaris Patera is rather isolated and lies southeast of the Elysium group, at 96°S, 186°W. Crater counts indicate an age roughly half that of Alba Patera (Plescia and Saunders, 1979). Its general appearance is that of a broad 400 km diameter lava shield crowned by a 70 km summit caldera which has two different floor levels (Fig. 7.30). The flanks are strongly striated and transected on the west, north and east sides by a prominent scarp. To the south, however, a large fan with its apex at the caldera rim extends for about 350 km and buries the scarp in that sector. On the surface of the fan is an array of what

Fig. 7.30 The isolated central volcano Apollinaris Patera, with its 100 km diameter caldera (north is to the left). The shield surface is truncated by a prominent scarp that is buried on the south side by a massive fan-shaped deposit that spills on to the adjacent lava plains. This could be a lava fan but might also represent channelled pyroclastics. Frame width 370 km. Viking orbiter frame 639A92.

appear to be broad channels, some of which incise the caldera backwall. The rather poor resolution of Viking imagery prevents a detailed analysis and it is not possible to decide whether the fan-like deposit is built from lavas or from channelled pyroclastic deposits.

7.12 CENTRAL VOLCANISM ON MARS

Volcanic activity has been widespread spatially and also over a long span of time. The earliest centralized volcanism generated low-profile paterae, mainly in the upland hemisphere, a particularly prominent group of structures surrounding Hellas. These were active during Late Noachian (Apollinaris Patera) and Early Hesperian times (Hellas group of paterae), and from morphological characteristics, their growth appears to have involved phreatomagmatic eruptions, as well as lava flows. At this early stage, however, the most widespread form of activity involved flood lavas and plains building on a global scale (Fig. 7.31(a)).

The vast patera of Alba Patera was first active during the Early Hesperian epoch, as localized activity became dominant. There was also growth of low-profile volcanoes in Syrtis Major Planum and in the Tempe Fossae region. Activity at Alba continued until the Amazonian, a span of perhaps 1.5 billion years. The more impressive shield volcanoes grew later, along tectonic lines atop broad rises in the Martian lithosphere – those of Tharsis and Elysium. Their development spanned the Early Hesperian to Late Amazonian epochs (Fig. 7.31(b)). The main shield-building activity apparently peaked during the Early-Middle Amazonian, but minor effusive activity continued until the Late Amazonian. Fissure activity accompanied centralized volcanism throughout Martian history.

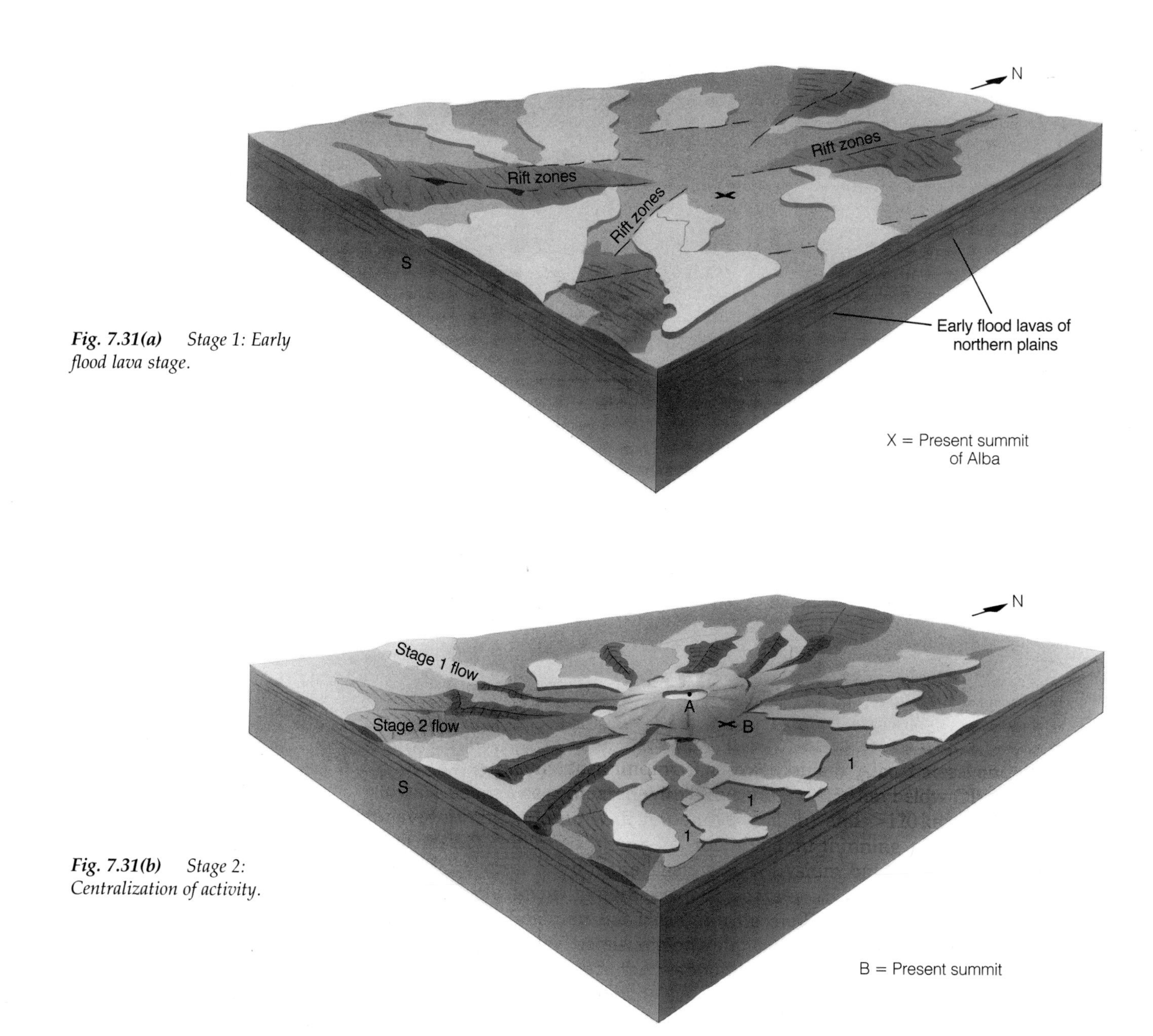

Fig. 7.31(a) *Stage 1: Early flood lava stage.*

Fig. 7.31(b) *Stage 2: Centralization of activity.*

THE PLAINS OF MARS

8

8.1 INTRODUCTION

The large shield volcanoes and canyon systems of Mars are amongst the more spectacular landscape features in the solar system. The widespread plains are not. However, vital clues to the geological history of the planet are locked up in extensive flat-lying deposits. These are the very variable geomorphological features classified as plains. Crater ages suggest that plains span the complete age spectrum (see Table 6.1, page 56), the oldest having developed during the early stage of intense bombardment; the youngest are still forming today. Unfortunately, even after over a decade of study of Viking imagery, geologists are little wiser about the origin of some Martian plains than they were before spacecraft reached the planet.

The greatest expanse of Martian plains occurs north of the line of dichotomy separating the two Martian physiographic hemispheres; these are largely of Late Hesperian and Amazonian age (Plate 5). There are also significant developments of Hesperian-age plains within the impact basins of both Hellas and Argyre, while east of Tharsis are the volcanic plains of the Hesperian-age Tempe Volcanic Province. More plains surround the polar regions.

While the origins of some remain equivocal, this is not true of all. Thus the basement of plains upon which the Tharsis and Elysium volcanoes are built, and which were generated during the late Hesperian epoch, are volcanic units believed to have been erupted from fissures. In other words, they were formed in much the same way as terrestrial flood basalts. These volcanic flow plains cover very wide areas, particularly in the northern hemisphere but also extend southwards around the southern flank of the Tharsis Rise.

As we have seen earlier (Chapter 6), numerous outcrops of ancient (Late Noachian–Early Hesperian) intercrater plains are located within the heavily cratered southern hemisphere. These is much evidence to suggest that volcanism has played an important part in their history, but so also have aeolian and alluvial deposition.

The occurrence of stratified rocks in the sidewalls of canyons in the Valles Marineris attest to a long period of pre-canyon plains deposition in the equatorial regions. While these plains could be built predominantly from volcanic rocks, they might equally well be sediments, or the sequence could represent a combination of the two. Elsewhere the active degradation of upland plains has led to subsequent deposition of young plains immediately north of the line of dichotomy. There is also a growing body of opinion in support of the notion that some mid-latitude northern hemisphere plains may have been laid down in large lakes or even open seas which existed on Mars in the distant past. If this proves to be true, then there are obvious exciting implications for Martian climatic evolution.

Plains north and south of about latitude 30° are particularly difficult to classify and interpret, largely because their albedo changes quite dramatically over quite small distances as well as on the scale of tens of kilometres. This phenomenon may be a function of circumpolar wind activity, which sweeps clear some parts of the plains surface but leaves others coated in aeolian debris. Climatic changes may also be responsible for the cyclic deposition seen in the young laminated deposits exposed near the poles. The regularity of alternating low and high albedo layers exposed in polar canyon walls signifies some form of cyclic deposition, and precessional variations usually have been invoked to account for such cyclicity.

8.2 NOACHIAN AND EARLY HESPERIAN PLAINS

Resurfacing of the Martian southern uplands followed the formation of the ancient cratered crust. A long period of deposition laid down deposits on the plateau surface which subdue or bury the underlying terrain. These **plateau plains** are characterized by smoother surfaces than the heavily cratered areas and have been separated on the basis of albedo patterns into **smooth plains**

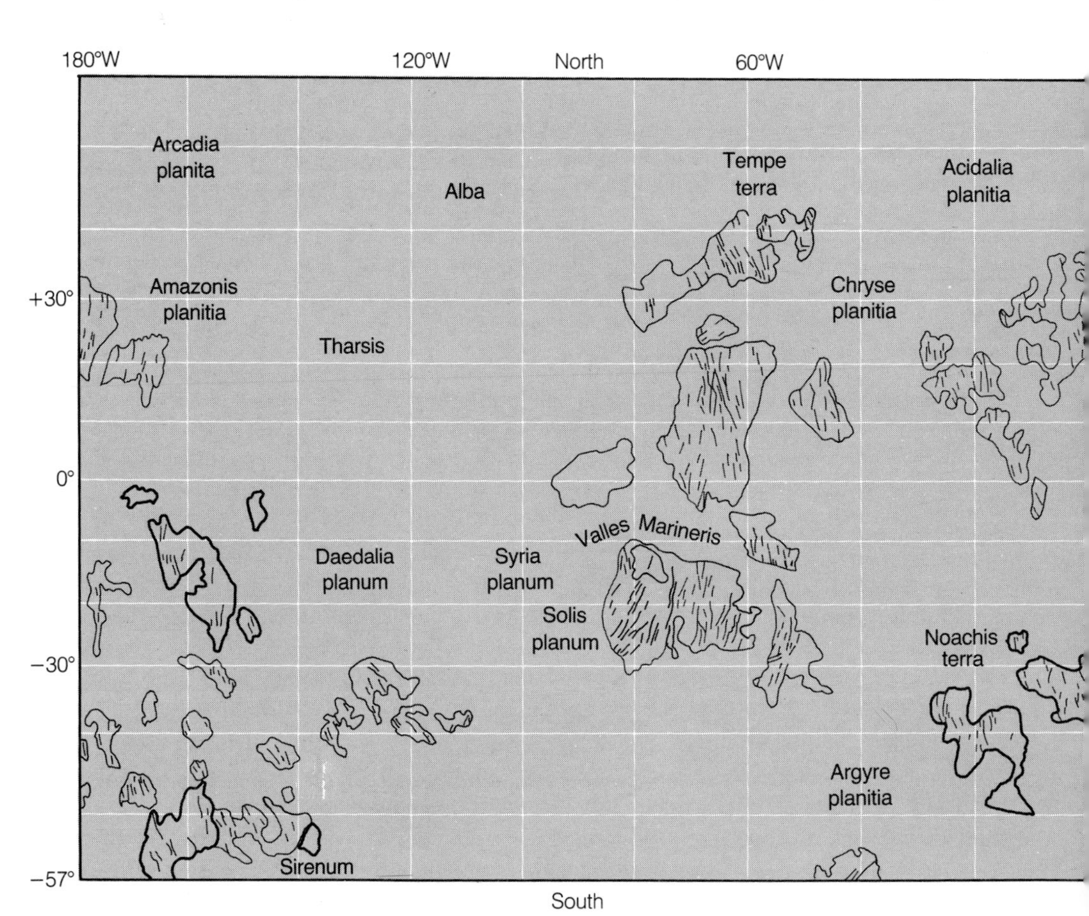

Fig. 8.1 Distribution of ridged plains on Mars. Noachian units are shown in heavy outline, Hesperian-age plains by lighter lines.

and **mottled plains**, the latter forming in response to the deposition of a patina of wind-sorted sediments (Greeley and Guest, 1987). Both types may consist of interbedded lava flows and sedimentary deposits.

Paul Spudis and Ron Greeley (1978) estimate that the aerial extent of these ancient plains is about $2.9 \times 10^7\,km^2$; of this, about 36% is covered by intercrater plains with ridged surfaces (Greeley and Spudis, 1981). Figure 8.1 shows the extent of such ridged plains units on Mars. Noachian-age plains are seen to extend widely in an easterly direction across Noachis Terra, to outcrop in Memnonia and across the southern part of Sirenum Terra. Cratering studies suggest that these Middle Noachian units are among the oldest plains units on Mars (Scott and Tanaka, 1986). Younger, Hesperian-age, ridged plains occur in the western hemisphere in a broad outcrop about 1000 km wide on the eastern flank of the Tharsis Bulge. They cover an area of about $4 \times 10^6\,km^2$. In the eastern hemisphere, the largest expanse of such plains is in Hesperia Planum. Upland depressions, such as Syrtis Major Planum, the western part of Amazonis, and impact basins like Hellas, Argyre and Isidis, also show development of similar plains. Crater ages derived for those in Syrtis Major Planum give 3.6×10^9 years (Hartmann *et al.*, 1981); plains in Hesperia Planum and Lunae Planum are also of Lower Hesperian age.

The ridges characteristic of many of these old plains are similar in many respects to lunar wrinkle ridges, with an asymmetric profile and anastomosing course. While they are widespread on the intercrater plains units, they also

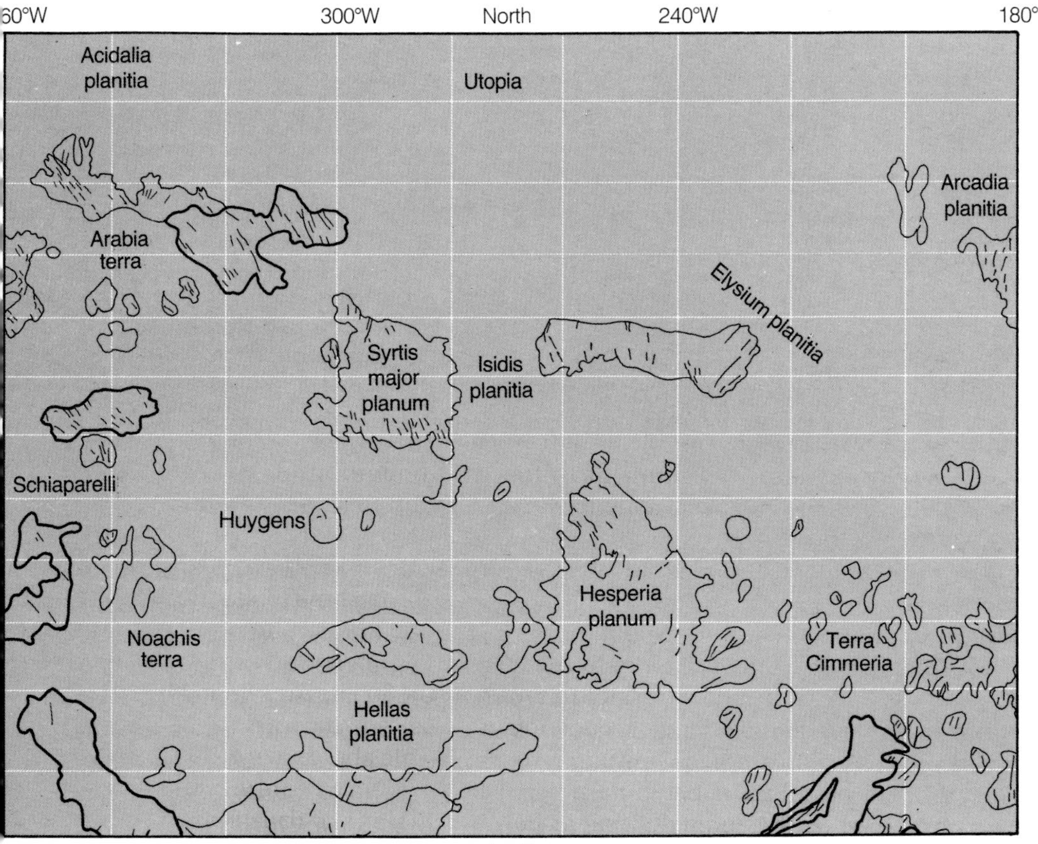

Fig. 8.2 *Ridged plains development on the interior of a 70 km diameter crater. Viking orbiter frame 432S25. Centred at 19.87°S, 180.15°W.*

may be seen traversing the flat floors of large impact craters which appear to have been inundated by fluid lavas (Fig. 8.2). The development of such ridges – invoked by many to be evidence for volcanism – is not in itself necessarily a volcanic process; however, as is well known from lunar experience, such ridges usually form in resilient rocks such as flood lavas. There is thus at least circumstantial evidence for extensive volcanic activity in the southern hemisphere during both Noachian and Early Hesperian times.

More direct evidence for Noachian volcanism may be found in regions like Protonilus Mensae, where there are quite large areas without ridging and with a smoothed geomorphological signature (Fig. 8.3). Aeolian mantling may account for some of the smoothing, but there are numerous other landforms which are difficult to interpret as other than volcanic flows, flow lobes and either exhumed dikes or spatter ramparts (Cattermole, 1990). Such evidence can be augmented by noting, as has Pete Schultz (Schultz, 1977), that there is a close correlation between the occurrence of ridged plains and floor-fractured impact craters, which are widely believed to have been modified by volcanic activity. An early episode of volcanism – prior to 3.9×10^9 a ago – would certainly alleviate the crater extinction problem discussed earlier in the book. Furthermore, as has been observed by Mike Carr (1984), high rates of volcanism might be expected during the period of intense bombardment and brecciation, and of elevated accretional energy dissipation which must have been characteristic of the period around 4×10^9 a ago.

Plate 1 The Chryse plains as seen from Viking 1 lander.

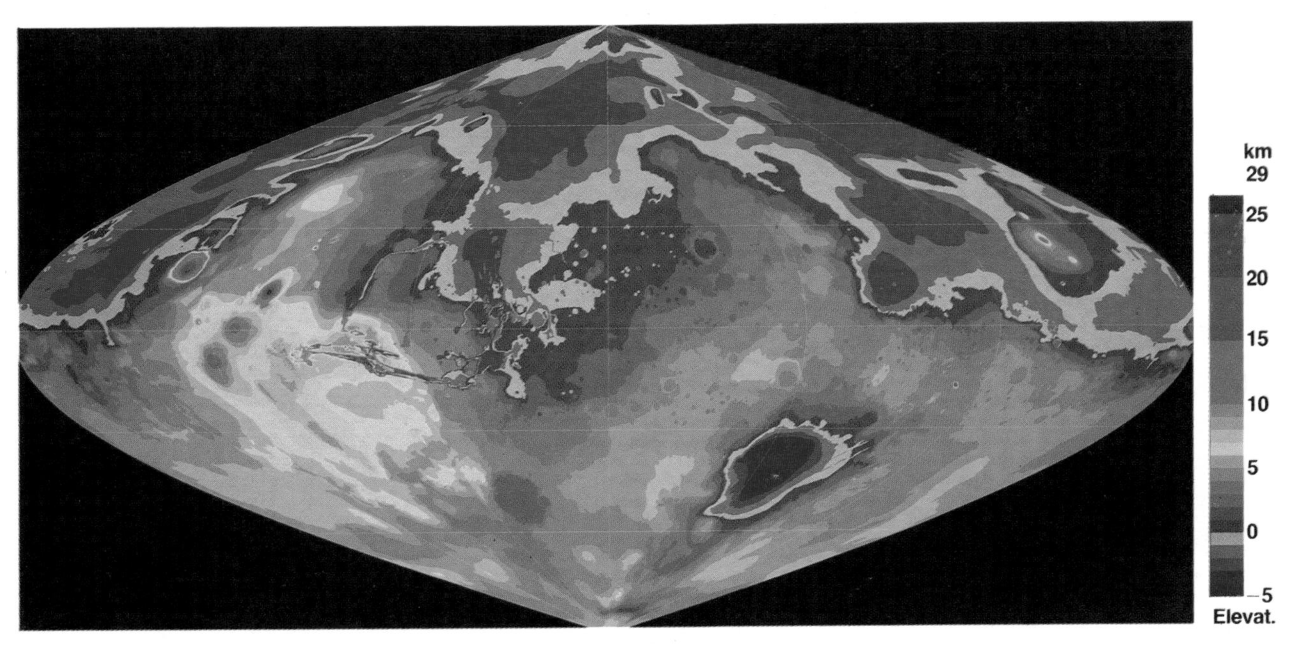

Plate 2 Topographic map of Mars on sinusoidal equal-area projection (courtesy of Sherman Wu, USGS Flagstaff).

Plate 3 Synoptic view of the western hemisphere of Mars (mosaic courtesy of Alfred McEwen, USGS Flagstaff).

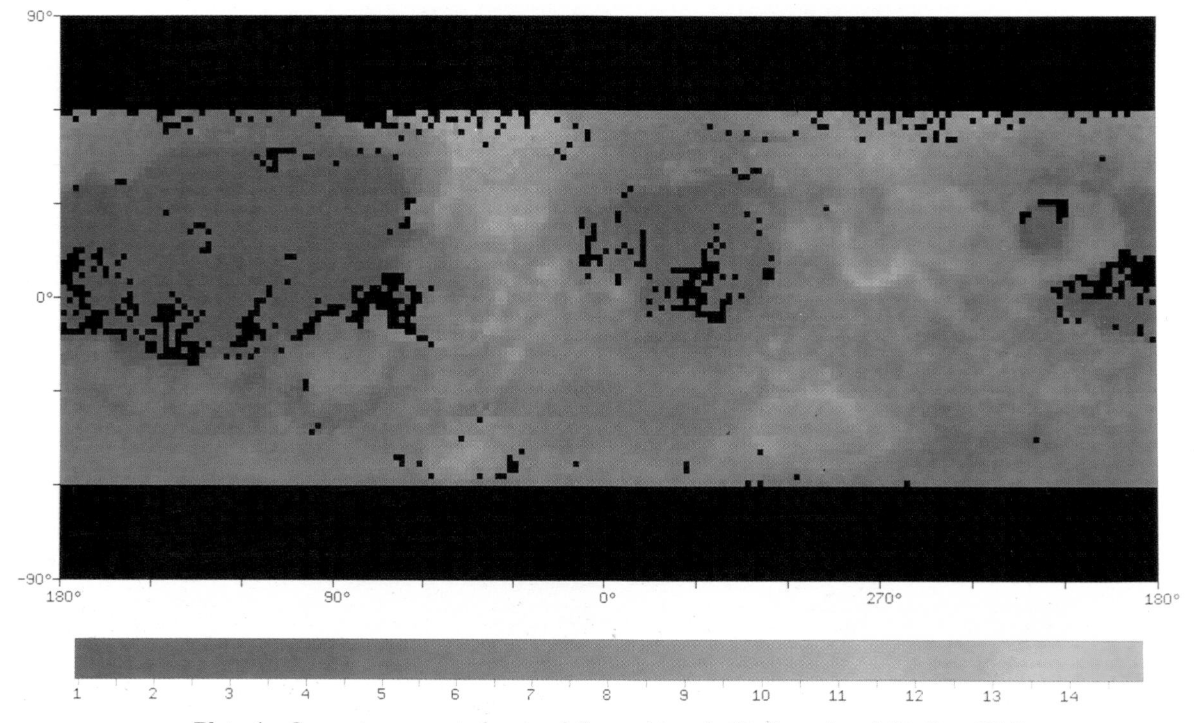

Plate 4 Computer-generated map of thermal inertia (Palluconi and Kieffer, 1981).

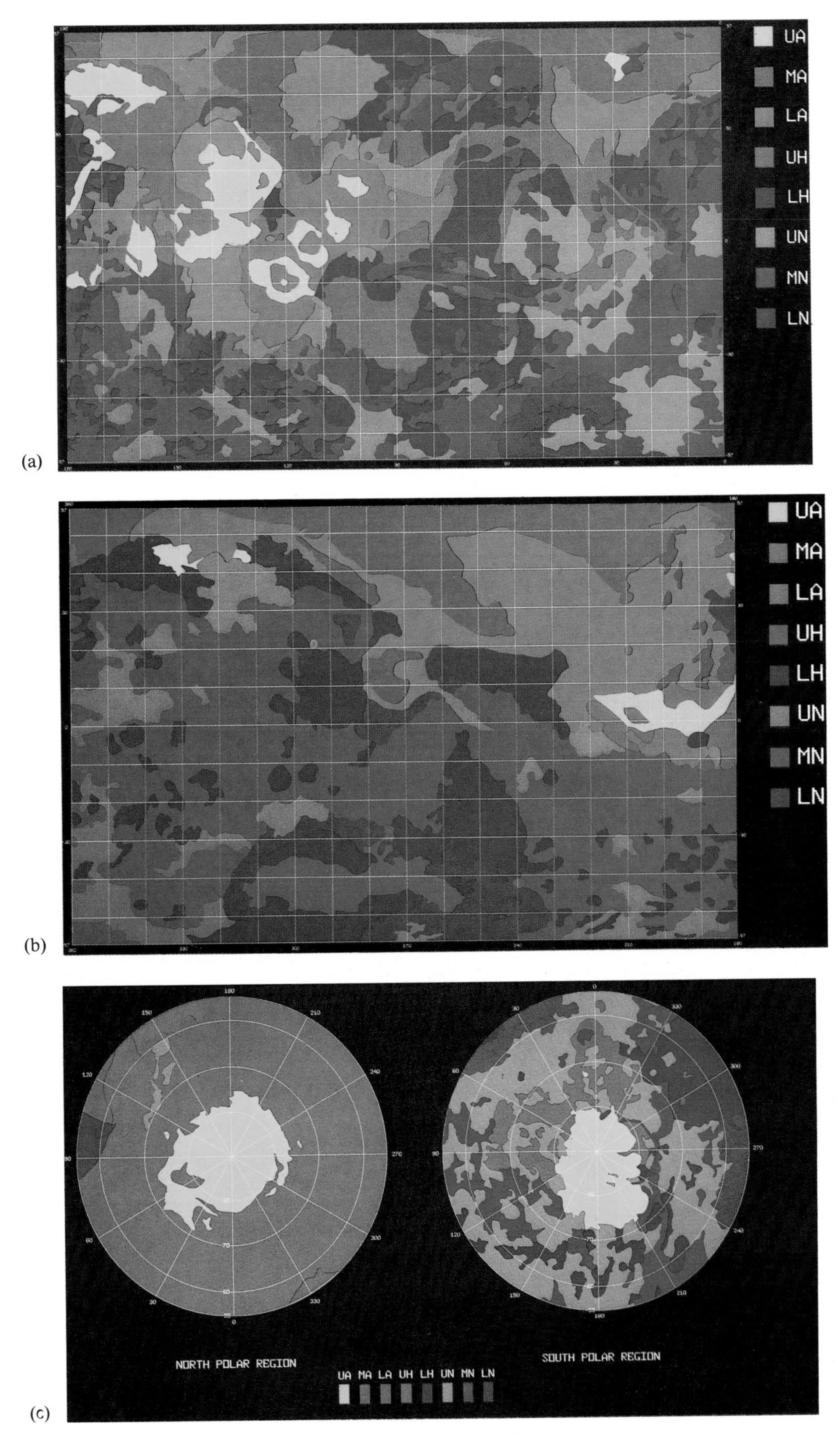

(a)

(b)

(c)

Plate 5 (a,b,c) Stratigraphic maps of Mars (Tanaka, 1986).

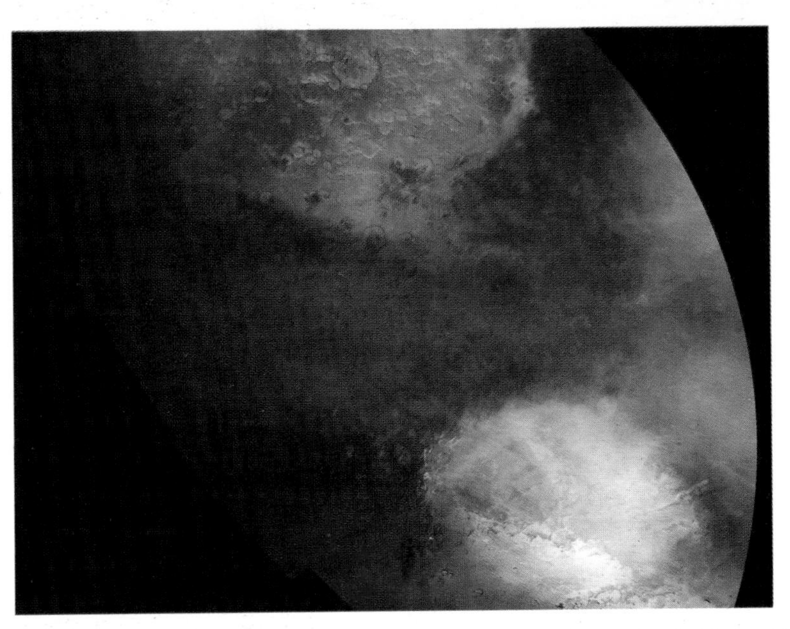

Plate 6 Viking mosaic of the region between latitudes 70°S to 50°N and longitudes 260° to 360°W. The bright white region near the bottom of the image is due to CO_2-frost in the Hellas impact basin which is 2000 km in diameter. Note the two prominent channels incised into the basin's eastern rim massif. The yellow region towards the top of the mosaic is Arabia (USGS Flagstaff mosaic produced by Jody Swann).

Plate 7 The Cerberus hemisphere of Mars. A mosaic compiled from 104 Viking images obtained during early northern summer 1980. The boundary between the upland cratered hemisphere and the northern lowlands crosses the image one-third of the way up from the bottom. A series of poorly indurated sedimentary deposits outcrops just north of the dichotomy; these may be pyroclastic rocks. The dark area left of centre is named Cerberus; to its northwest is the Elysium dome with its superimposed shield volcanoes and graben faults. The white clouds at the top right edge of the mosaic hang over Olympus Mons. To the west of this are extensive drifts of wind-blown materials. The bluish region at the bottom of the picture represents the extent of the seasonal CO_2 polar cap (Image produced by Jody Swann at USGS Flagstaff).

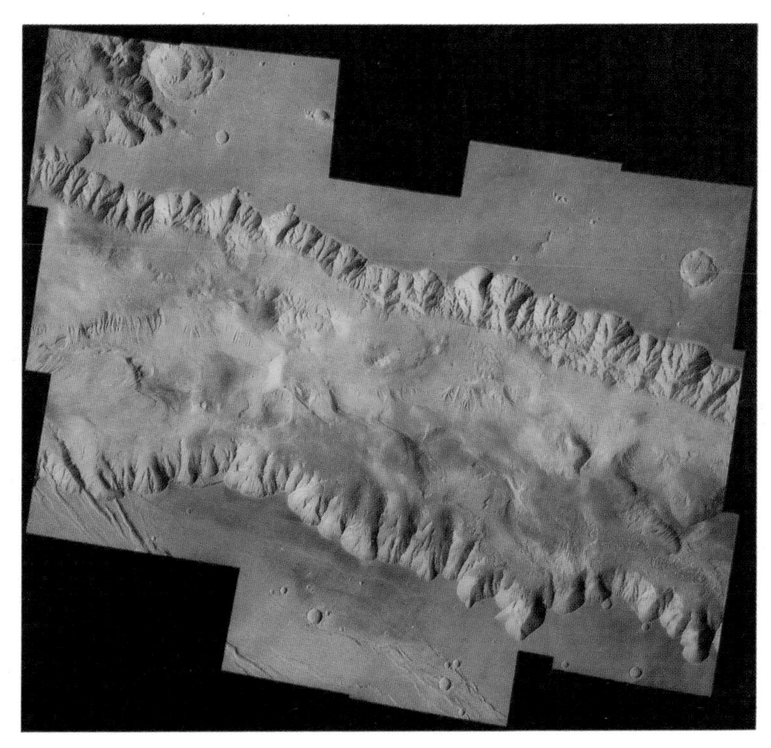

Plate 8 Section of Melas Chasma showing cuspate wall structure, prominent ledges and interior deposits. Note the dark areas which may represent volcanic bedrock swept clear of surficial dust (Mosaic courtesy of Alfred McEwen, USGS Flagstaff).

Plate 9 Mosaic of part of Candor Chasma between latitudes 9° and 30°S and longitudes 69° to 75°W. Layered terrain is widely developed and may have been deposited in a vast ancient lake. The complex form of the canyon floor and walls has been shaped by tectonic, mass wasting, aeolian and probably fluvial activity and volcanism (Mosaic by Alfred McEwen, USGS Flagstaff).

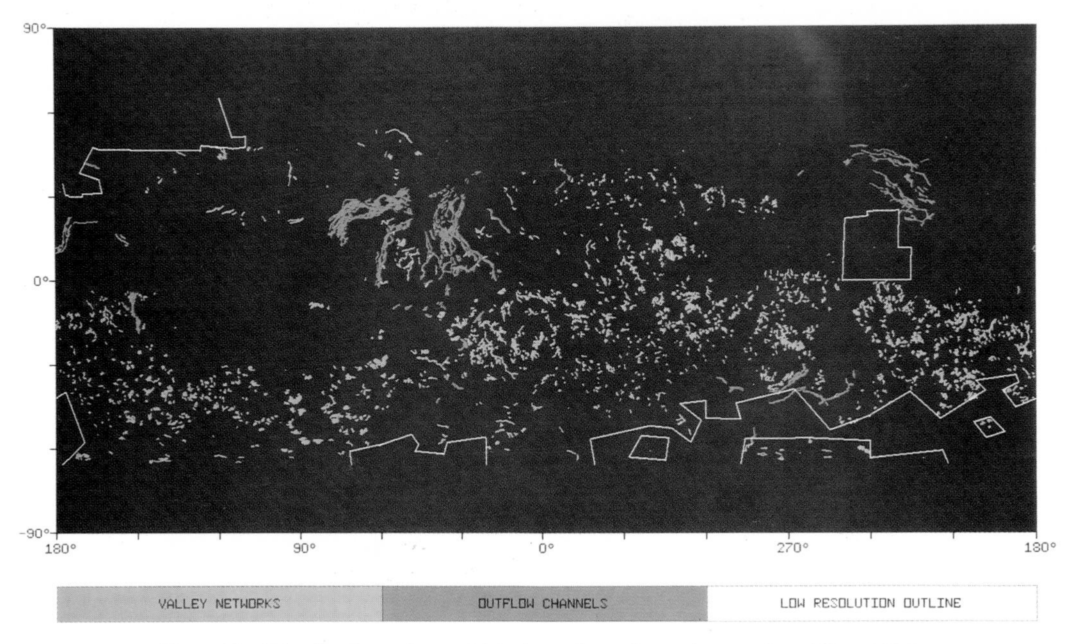

Plate 10 Distribution of runoff and outflow channels on Mars (Carr 1980).

Plate 11 The region of the boundary scarp east of Mangala Valles. Channelling of the ancient cratered plateau preceded the effusion of flood lavas to the north. Note the flow-like ejectamenta associated with the prominent impact crater (USGS Flagstaff Viking mosaic by Alfred McEwen).

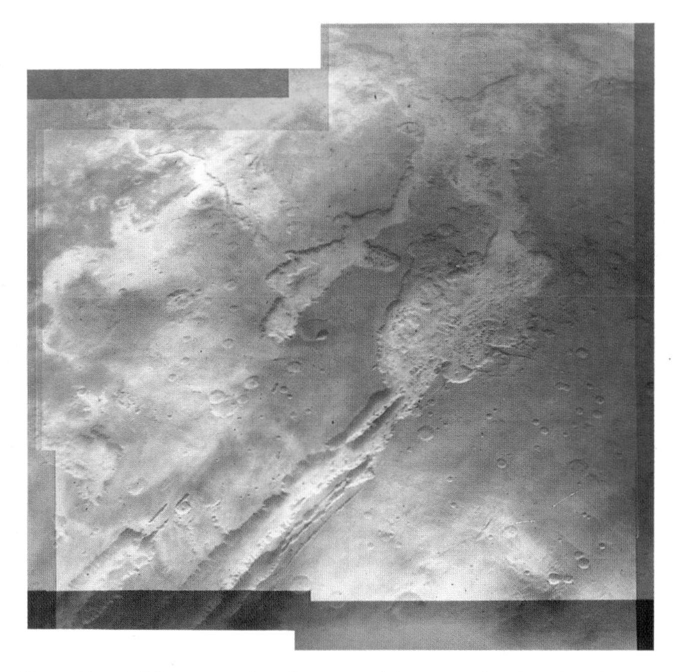

Plate 12 Viking mosaic of equatorial canyonlands, showing chaotic terrain at the eastern end of the Valles Marineris system.

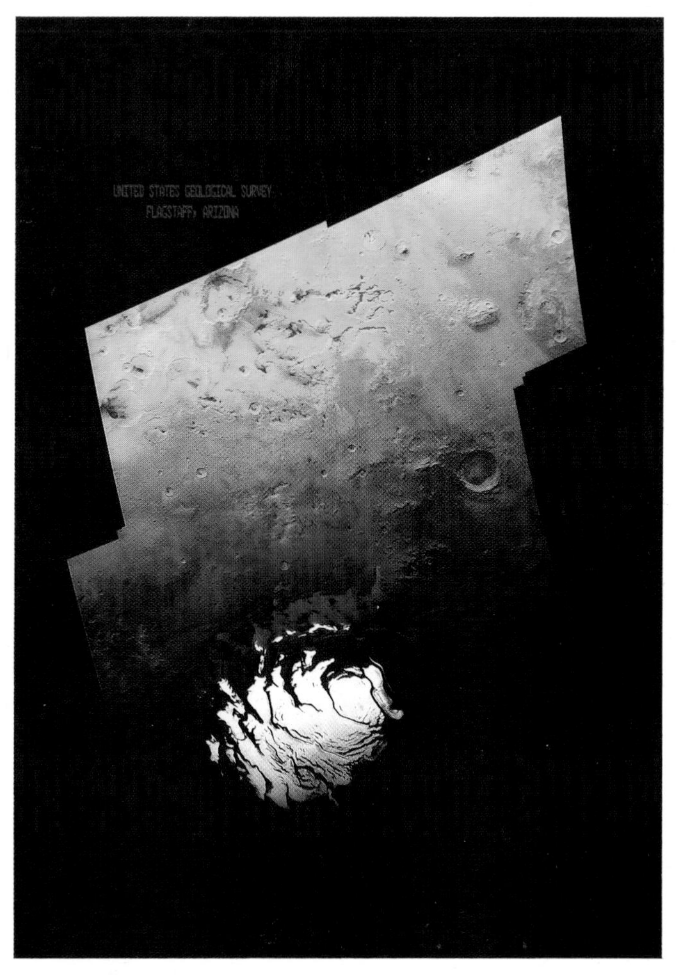

Plate 13 The south polar cap and environs, imaged by Viking during the Martian southern summer 1977. The cap was nearing its final stage of retreat prior to vernal equinox. The circumpolar deposits are often wind-etched with the development of sharp-rimmed pits and escarpments. (Viking orbiter mosaic produced from 24 images by Tammy Becker, USGS Flagstaff.)

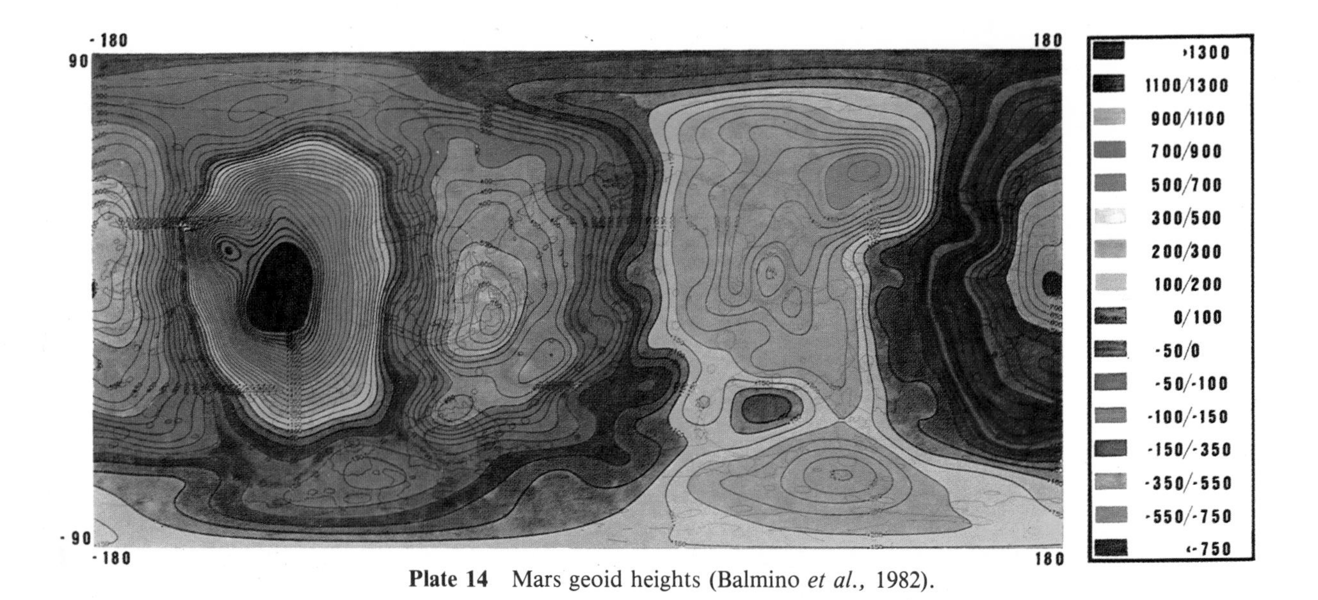

Plate 14 Mars geoid heights (Balmino *et al.*, 1982).

Plate 15 The line of dichotomy as it traverses Amazonis Planitia. The sedimentary deposits to the north and east may consist of pyroclastic rocks, aeolian or palaeopolar deposits. The ancient river courses seen cutting the cratered plateau are associated with the Mangala Valles system (USGS Flagstaff mosaic, produced by Alfred McEwen).

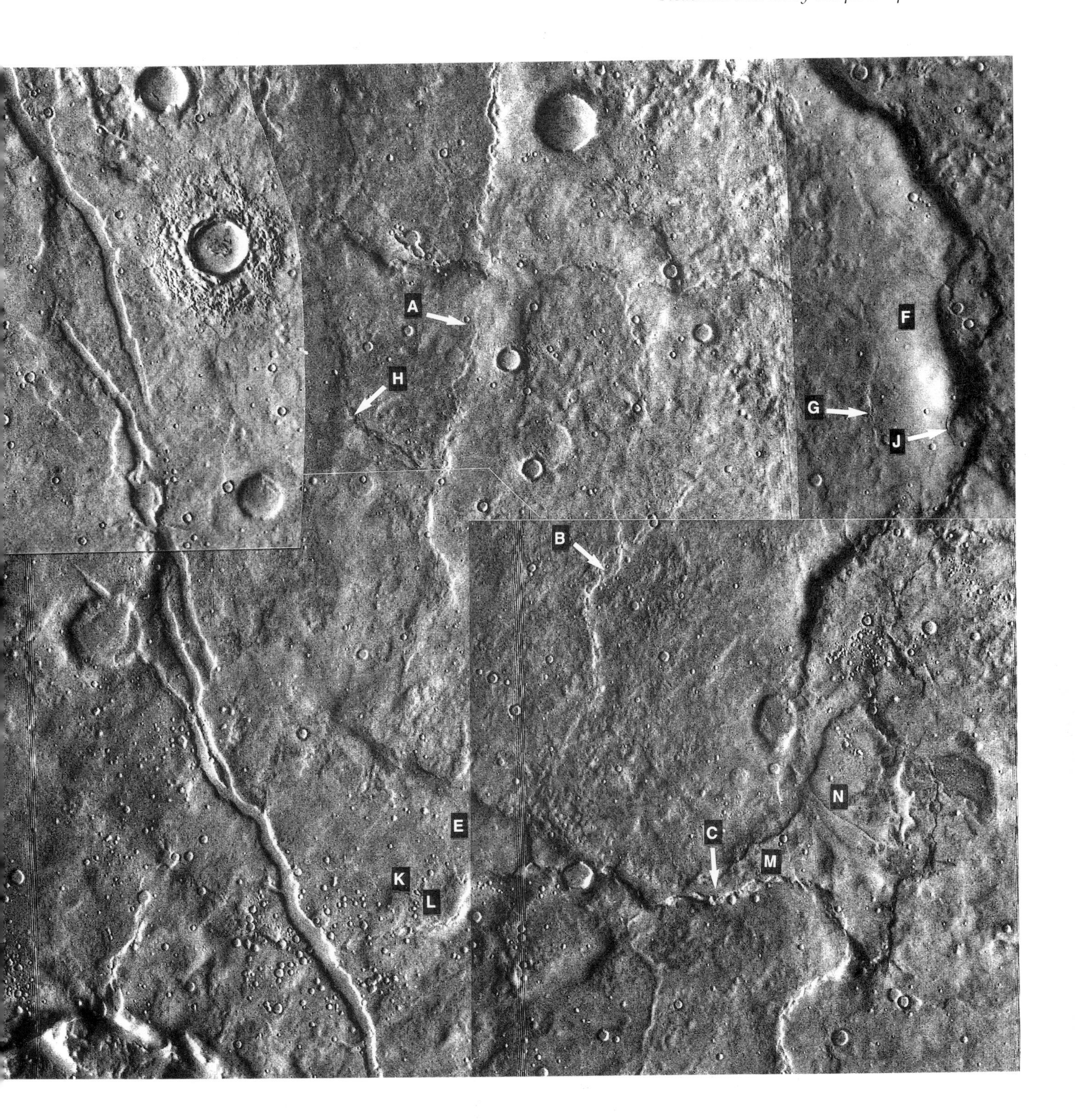

Fig. 8.3 *Noachian-age ridged plains northeast of Arabia. Prominent flow scarps outcrop at A, B and C, with wrinkle ridges at E and F. Possible narrow volcanic flows occur at G and H, while a flow channel exists at J. Linear ridge segments at K–N may be parts of spatter ridges, or exhumed dikes. Viking orbiter frames 641A02–7. Centred at 34°N, 311°W. Frame width 100 km.*

Fig. 8.4 *Wrinkle ridges and volcanic flows, some with tube-channel feeders, near Hadriaca Patera. Viking orbiter frame 413S14. Centred at 29.3°S, 259.5°W. Frame width 170 km.*

Ridges developed within the younger, Hesperian-age plains show more definite evidence of a relationship between regional and local tectonic regimes – ridge segments typically are aligned over large areas. In both Solis Planum and Hesperia Planum, flow fronts and lobate flow terminations accompany the ridges, while on the western side of the latter, there are extensive smooth-facies ridged and lobed plains units associated with the highland volcanoes, Hadriaca and Tyrrhena Paterae (Fig. 8.4). There is also a broad swath of such plains around Hellas. Narrow flows, small cratered domes and what appear to be either exhumed dikes or spatter ridges are clearly discernible on the borders of Isidis, while there are two very prominent low volcanic shields with attendant caldera structures on the surface of Syrtis Major Planum (Fig. 8.5). High-resolution images of the region adjacent to one of the calderas clearly indicate a development of tube-fed volcanic flows. Crater statistics suggest an age of 2.6 $\times$ 10^9 a for flows associated with this volcanic plains complex (Meyer and Grollier, 1977).

Less heavily cratered (Upper Hesperian) ridged plains outcrop in Chryse Planitia and south of Elysium. The landing of Viking 2 in the former locality

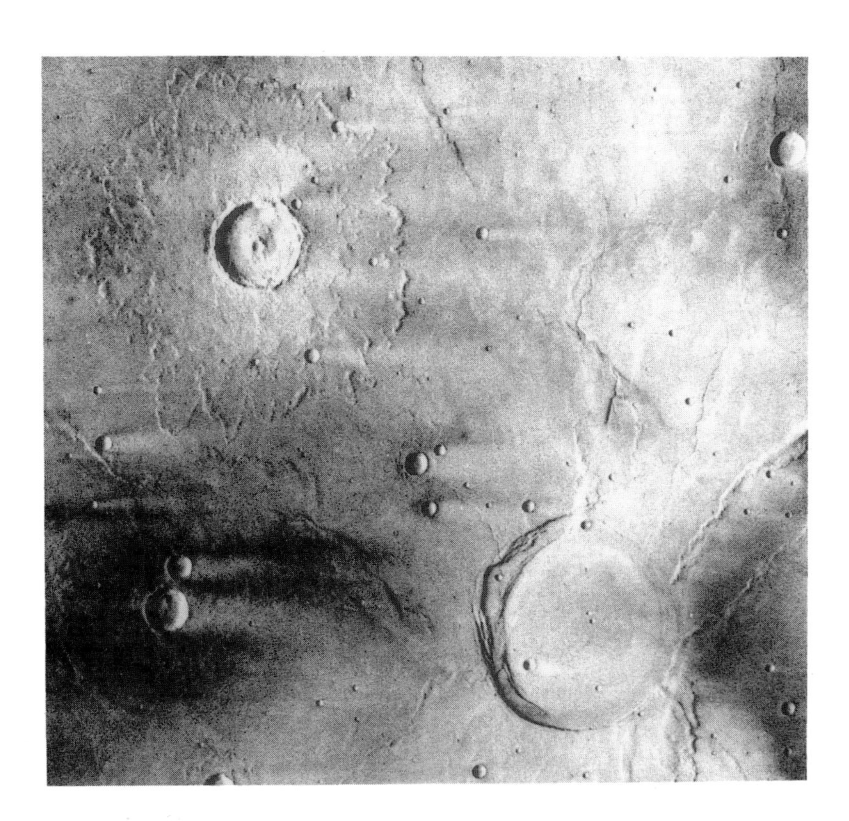

Fig. 8.5 A 50 km diameter volcanic caldera on the ridged plains of Syrtis Major Planum. Viking orbiter frame 372S32, centred at 5.5°N, 289.8°W. Frame width 220 km.

indicated the presence of what appear to be vesicular basaltic blocks on the surface of the plain and confirmed there were basalt weathering products in the surface soils. While this does not prove a volcanic origin for the plain, it is strong circumstantial evidence in support of it and of its surfacing by mafic lavas.

8.2.1 Significance of plains ridges

Studies of similar ridges on the east flank of the Tharsis Bulge by Phillips and Ivins (1979), strongly suggest that they are tectonic in origin. However, the characteristic ridge-arch morphology of ridges on plains units – a feature they share with their lunar counterparts – appears to support the notion that they were formed in competent materials, which appears to rule out ancient impact breccias. Therefore it is more likely that they were produced in low-viscosity lavas, similar perhaps to the lunar maria. The ridge morphology, the extensive distribution of smooth intercrater plains with ridged surfaces, and their association with various kinds of volcanic structures, support the idea of an early phase or phases of flood volcanism which modified the ancient cratered terrain widely, perhaps as early as 3.9×10^9 a ago.

8.2.2 Relationships between channels, plains and volcanism

South of Tharsis, in the regions of Thaumasia Fossae and Aeolis, the numerous valley networks which incise the intercrater plains have been shown to have formed after north-south Tharsis-related faults and northeast-oriented graben (Brackenridge, 1987). The plains themselves are also cut by most faults, but not all; some of the graben do not displace the plains units, suggesting that graben faults may have been forming while plains were still being laid down. Mapping also shows that outflow of low-albedo volcanic flows occurred after both types of faulting had taken place; many of these lavas partially cover tributary channels. Sections visible in some remnants of eroded plains exhibit a light/dark internal stratification which suggests that dark volcanic flows are interbedded with lighter sediments. Valley heads tend to occur at the base of a prominent low-albedo volcanic unit that is sandwiched between a lighter surface unit and an older one beneath; this could be a sill. Brackenridge suggests that the volcanic horizon could have acted as an aquiclude for heated volatiles which escaped as springs along its base. If this was so, then valley development would be explained by hot spring activity associated with the emplacement of a sill into ice-rich fragmental material.

Elsewhere in Aeolis a strong preferred northwest trend can be observed in the numerous channel networks, showing underlying structural control. Furthermore, channel-wall interbedding of light (sedimentary) and dark (volcanic) layers, implies a strong correlation between volcanism and channel development. The more degraded of these intercrater plains are most readily explicable in terms of groundwater release from a frozen subcrust, the former being instigated by mafic early, post-bombardment volcanism.

One of the major problems here, is deciding how such a reservoir of groundwater may have accumulated near the surface of Mars during its early history. Jakosky and Carr (1985) infer from calculations of the pre-Tharsis obliquity of Mars, that enhanced obliquity would have instigated ice-condensation at low latitudes (where it currently is unstable when in contact with the atmosphere). If this assertion is valid, then here is a mechanism for ice deposition in the cratered highlands at an early stage in Martian history.

8.3 HESPERIAN-AGE FLOW PLAINS

The most extensive occurrence of Hesperian-age flow plains is peripheral to the major volcanic provinces of Tharsis and Elysium (Fig. 8.6). There is also a major outcrop of somewhat equivocal volcanic deposits in Malea Planum which may be fluvially modified volcanic flows. The oldest lavas outcrop around Tempe Terra, Memnonia and Ceraunius Fossae and their eruption marked the first of several major volcanic episodes which resurfaced huge areas of the northern lowlands. Extensive flow plains also were erupted from near the crest of the Syria Rise and now cover large areas of Syria and Solis Planae at the western end of Valles Marineris. Particularly extensive lobate flows of Mid to Late Hesperian age are found east and northeast of the base of Olympus Mons shield and also on the western side of Tempe Terra.

The most extensive series of flows emanates from beneath the low pile of Alba Patera, where a sequence of broad sheet flows extends at least 1500 km from its summit. Similar flow plains originated from centres now situated beneath younger volcanoes like Arsia Mons and Uranius Patera, where occasional rimless depressions and discontinuous spatter-type ridges are aligned along what are assumed to be linear source vents or fissures, most of which have been buried by the flows themselves, or younger ones. Flood-type lavas of Late Hesperian to Early Amazonian age also occur around Ceraunius Fossae, where they flood fractures incised into older highland terrain.

Fig. 8.6 Hesperian-age flow plains flooding ancient cratered terrain in southern Tharsis. Viking orbiter frame 056A14, centred at 31.53°S, 130.73°N. Frame width 230 km.

Flows typically are between 60 and 120 m thick, often composite, and have relatively featureless upper surfaces. Many can be traced for hundreds of kilometres and several may coalesce to form broad, overlapping sheets. Individual flows are immense and have volumes in excess of 400 km^3. Flow channels can often be discerned and there are numerous low domes and small depressions, presumably the sources of some flows. By and large, however, the sources of such lavas are obscured by younger flows and it can only be assumed that they issued from fractures or linear vents which were buried by their own products.

8.4 THE MEDUSAE FOSSAE PLAINS

Unusual plains units are located in a region which has the volcanoes Apollinaris Patera, Biblis Patera and Olympus Mons at its apices (Fig. 8.7). In general these deposits have higher relief and albedo than most plains materials, but are less cratered and hilly than the highlands. They outcrop in a broad but discontinuous zone that runs east-west along the lowlands-highlands boundary. Here a series of discontinuous, flat-lying sheets, each about 100 m thick, has smooth, gently undulating or etched surfaces and a total thickness of at least 3 km. Some of the higher sheets show a development of yardangs along plateau edges, which suggests they are not competent lavas but relatively

Fig. 8.7 Smooth-textured high albedo plains units in Amazonis. Some of these are characterized by the development of yardangs at their margins (see enlarged area in Fig. 8.8). Viking orbiter frame 635A83. Centred at 10.33°S, 177.73°W. Frame width 310 km.

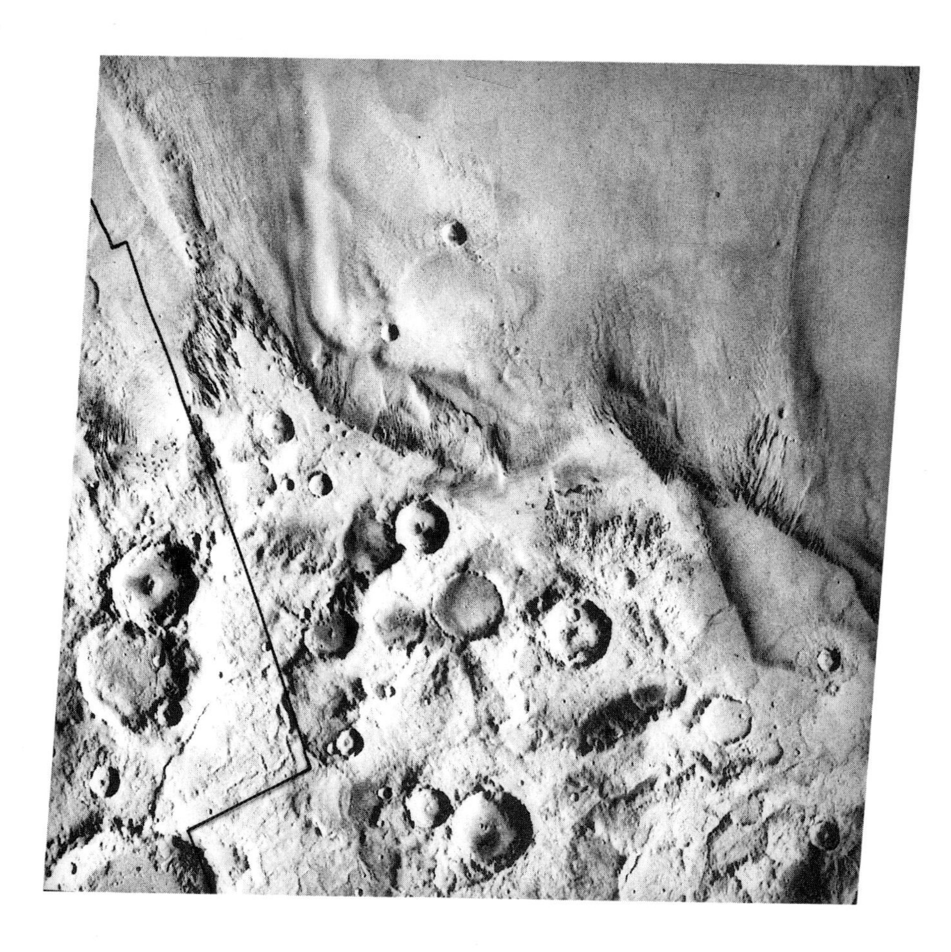

friable rocks (Fig. 8.8). In morphology they bear a striking resemblance to some terrestrial ignimbrites, as was first noted by Mike Malin (1977).

Photogeologic work by Scott and Tanaka (1982) indicates that the deposits cover an area of the about $2.2 \times 10^6\,km^2$ and have an estimated minimum volume of $3.85 \times 10^6\,km^3$. Seven units can be distinguished, while at four locations there is a marked thickening of the sheets, suggesting possible eruptive foci. Early work by Scott (1969) among terrestrial silicic volcanics, allowed him to highlight a number of features common to both:

1. rounded patches of smooth, high-albedo (non-welded) materials that overlie low-albedo jointed (welded) flows;
2. local complimentary joint sets in some (welded) materials; and
3. thick flow sheets of great lateral extent that pursue but subdue the underlying topography.

While Scott and Tanaka's feelings regarding these volcanic plains must remain tentative, the circumstantial evidence for a pyroclastic origin is con-

Fig. 8.8 (opposite) *Higher-resolution mosaic of region identified in Fig. 8.7 above. The yardang development shows a different orientation in different aged units. Note the wind-etched rocks seen on the interiors of the large impact craters towards the bottom of the mosaic. Viking orbiter frames 437S01–8.*

siderable. However, it should be noted that other workers have suggested that they could equally well be a thick sequence of aeolian deposits transported and trapped along the highland-lowland boundary (Lee *et al.*, 1982; Thomas, 1982).

One other interesting feature of this region is the general absence of volcanic domes, which are common in terrestrial silicic igneous provinces. Their absence does not rule out a pyroclastic origin, since theoretical considerations indicate that there should be wider dispersion of pyroclasts on Mars than on Earth for the same mass eruption rate (Head and Wilson, 1981). Thus the absence of domes may be a function of the lower expected relief and an inability of Viking images to resolve very low features.

In the Basin and Range region of the western USA, ignimbrites are closely associated with block faulting where highly volatile silicic magmas were available for energetic eruption. The high concentration of northwest-southeast faults and graben within the Amazonis-Aeolis region, for instance Medusae Fossae, shows that there was indeed extension in the Martian crust at this time and it is possible that such faulting could have produced roof failure in a large magma chamber located beneath the deposits, producing extensive ash flows. Calculations by King and Riehle (1974) suggest that if Martian ash flows were generated, they would remain in a fluidized condition for between three and nine times the period typical of the Earth, giving scope for very extensive flat-lying sheets which could travel great distances from their source.

8.5 TEMPE TERRA PLAINS PROVINCE

While there is a general lack of specific evidence regarding the source vents of the very extensive Hesperian flow-plains lavas, this is not the case with those in the Tempe volcanic province. This interesting volcanic complex developed within the uplands which form a northwestward continuation of Lunae Planum, on the opposite side of Kasei Vallis. The Hesperian-age volcanic sequence occupies an area of $3.4 \times 10^6\,km^2$ and comprises three distinct kinds of terrain:

1. rugged hilly terrain,
2. faulted terrain, and
3. smoother uplands which are an extension of the Lunae Planum plateau (Scott, 1982).

The nature and relationships between the various units are shown in Fig. 8.9; blocks of the Noachian cratered plateau, mantled in younger lavas and (?) pyroclastics are seen to be embayed by younger Hesperian volcanic plains, sometimes heavily mantled with aeolian debris. Distinct narrow lava flows can be seen on some of these plains units. Debris aprons hug the plateau-plains escarpment and accumulate in enclosed depressions where the uplands are being actively degraded.

On the surface of the plateau is a suprising variety of smaller volcanic constructs, several major volcano-tectonic structures with little or no relief, and steeper-sided volcanic mountains, which may be dissected shields volcanoes. Details of several of these structures have been given by Underwood and Trask (1978), Scott and Carr (1978), Wise (1979), Plescia (1980) and Scott (1982). The major ring structures are somewhat similar to Alba Patera, though with even less vertical relief. The largest is about 250 km across.

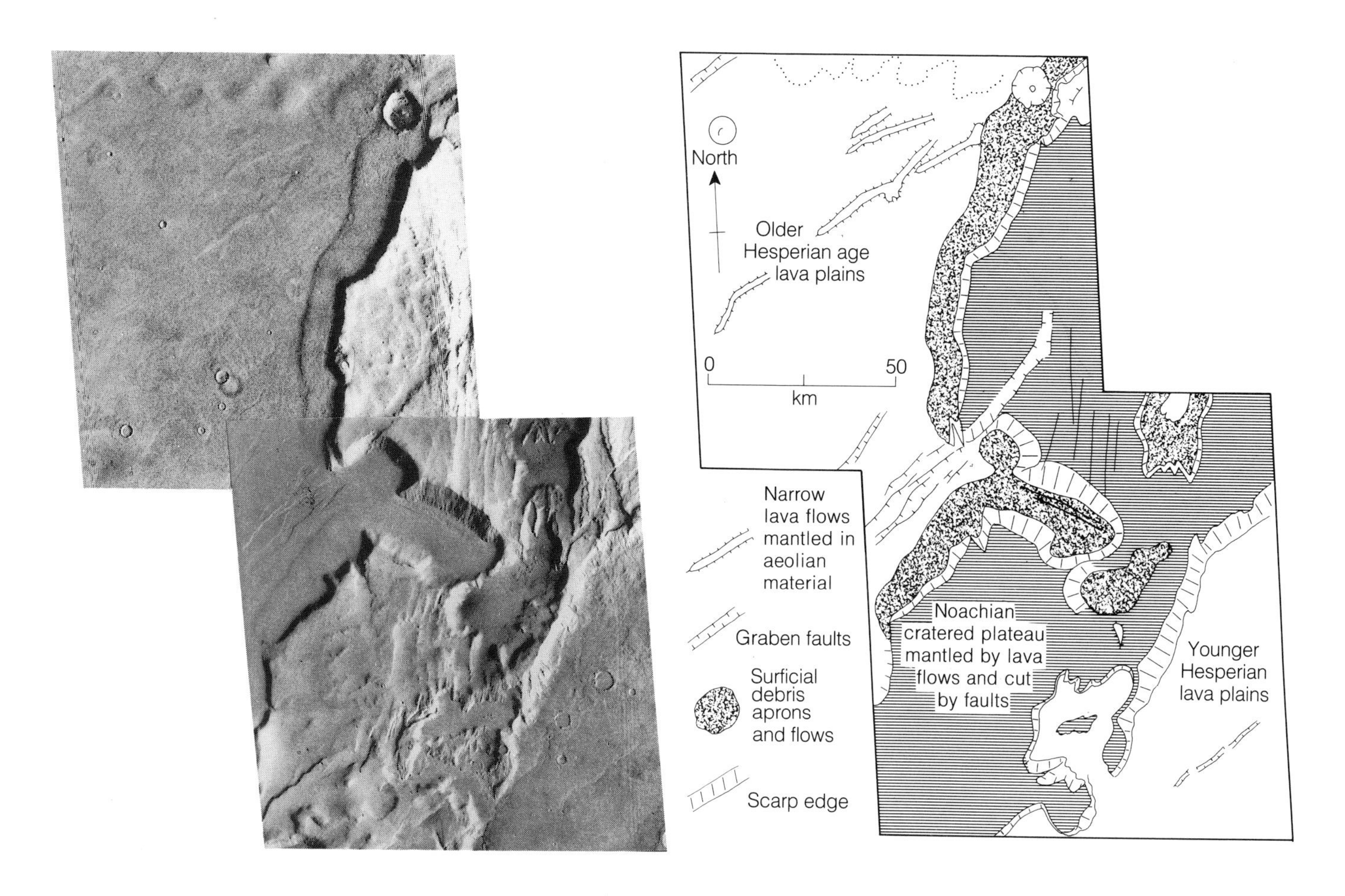

Fig. 8.9 Western margin of plains units in Tempe Terra. Viking orbiter frames 255S50–1. Centred at 46°N, 84°W. Mosaic length 220 km. The accompanying map indicates the geological features of this region.

The numerous smaller-scale volcanic features, in particular the substantial number of low shields which have been built on the resurfaced parts of the fractured plateau, bear a striking resemblance to those which developed during the plains-type volcanism of the Snake River Plains, Idaho, which produced widespread volcanic plains from source vents arranged along rifts. Thus small, low shields have well-defined summit pits which often are aligned in the direction of the southwest-northeast fracturing. There are also many elongate vents and thin sheet flows, some of which appear to have originated in fissures (Fig. 8.10). Plescia (1981) estimates that low shields account for about 75% of the constructional landforms present. Their obvious development along rift faults, as well as the presence of fissure-fed sheet flows and elongate depressions, reveals, as it does in the Snake River Plain, a close genetic relationship between volcanism and crustal extension. What appear to be absent from the Tempe plains are tube-fed flows, so common a feature in the Snake River province; however, it is possible that such flows were emplaced but that their roofs have not suffered collapse.

In addition to the low shields, there are several steeper-sided constructs with diameters in the range 5–10 km. These are probably composite cones and have no discernible lava flows associated with them. Such features are not found in the Snake River Plains and their occurrence in Tempe may imply a more extended range of volcanic style, involving more viscous, silicic magmas than are common in the Snake River province. Supporting evidence for such a

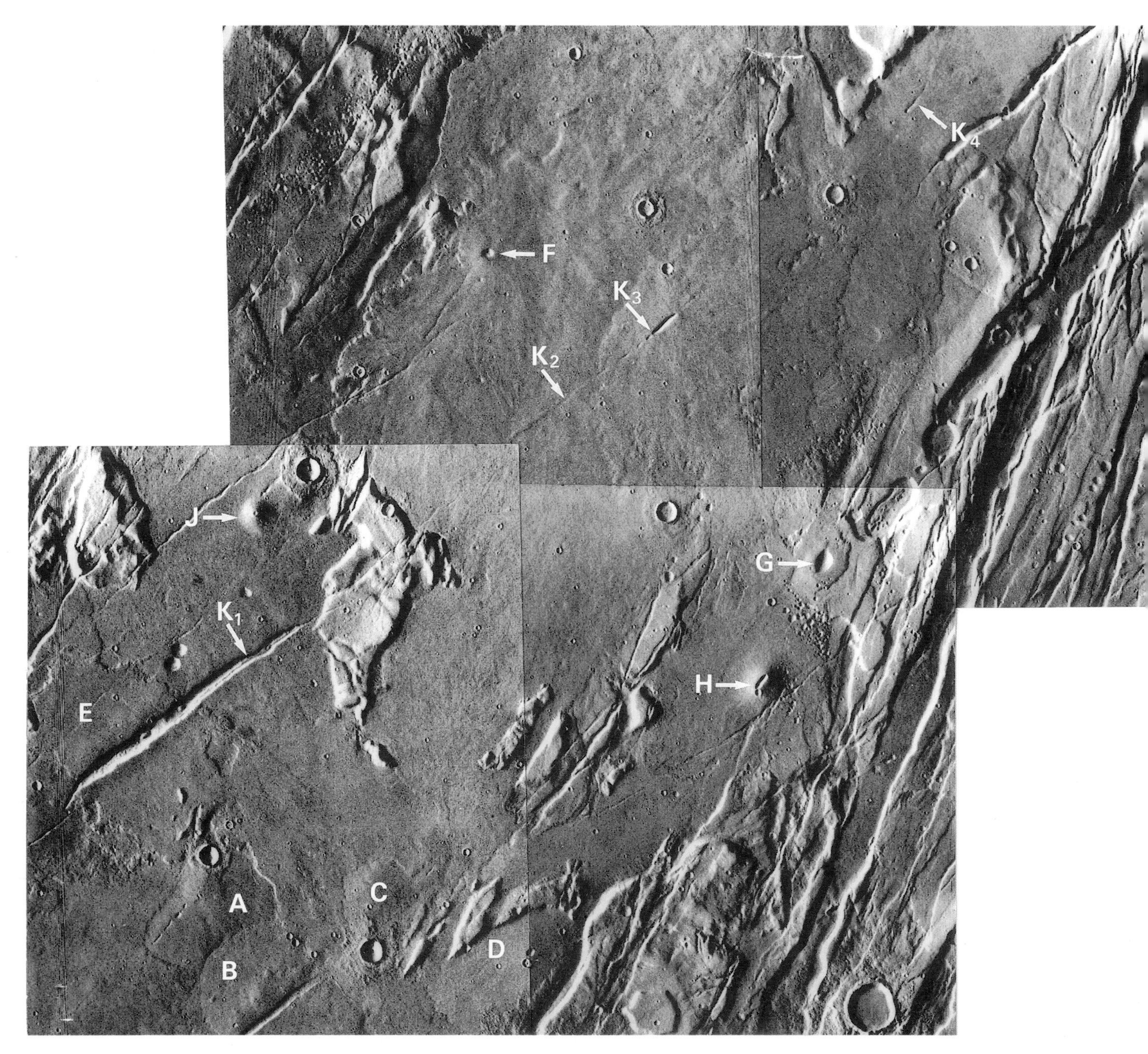

Fig. 8.10 *Plains-style volcanic province of Tempe Terra. Note the low shields (A–D), of which A has a row of aligned axial depressions, and D a single ovoid vent. Similar vents occur atop low domes at F and G, and on steeper-sided cones at H and J. A sheet flow outcrops at E and is seen to predate the SW–NE faults. Also a volcano-tectonic feature traverses the region in a SW–NE direction, consisting of a graben fault (K_1) and elongate depressions (K_{2-4}). Viking orbiter frames 627A26, 27, 29, 41. Mosaic width 270 km.*

hypothesis is provided by a large dissected patera structure located in the northeast of the plateau which, by analogy with highland paterae eleswhere on Mars, may have developed during a phase of explosive volcanism generating ash-flow and air-fall pyroclastic deposits.

8.6 VOLCANIC PLAINS OF AMAZONIAN AGE

Most of the younger flood lavas are associated with the massive shield volcanoes of both Tharsis and Elysium. For this reason they have been discussed already in Chapter 7. However it must be noted here that while they are closely related to central volcanoes, they extend outwards for distances quite unheard of in the terrestrial environment. For instance, Olympus Mons flows extend for more than 400 km from the base of the shield, while Arsia Mons lavas can be traced back more than 600 km to their source.

8.7 PLAINS DEPOSITS OF HELLAS

The Hellas impact basin shows a widespread development of plains units, some of which extend well beyond the rim (Fig. 8.11; Plate 5). The basin itself was produced in Noachian times, when the rim and adjacent cratered plateau materials were uplifted. Subsequently, a long period of modification took place, during which time erosion, channel formation and deposition occurred. The basin was then flooded by fluid basalt-like lavas which were deformed subsequently to produce ridged plains. Other volcanic plains units were generated from major paterae, such as Hadriaca and Tyrrhena Paterae; in general, their deposits are external to the basin's inner ring. Subsequently, the interior of Hellas was mantled by a variety of materials which were later modified to give what Greeley and Guest (1987) term 'a dissected floor unit'. It is not possible to constrain exactly what these Late Hesperian age deposits are, but they are believed to comprise a mixture of aeolian, fluvial and volcanic materials. The production of the sediments from the rim materials was most likely accomplished by a mixture of fluvial run-off and groundwater sapping.

At a later stage, during Amazonian times, there was significant fluvial activity on the basin's eastern rim, with the activity of major channels like Dao and Harmakhis Valles. These spread out sediments widely on the eastern floor which were later modified by deflation and collapse. The geomorphological signature of these younger materials suggests that both fluvial and periglacial activity may have played a part in their production (Crown *et al.*, 1990). Within this plains mantle there is a lineated floor unit and a channelled plains rim unit which may be only slightly modified areas of the original sedimentary mantle. The latter has subdued topography (Fig. 8.12) and low remnant mesas and narrow channels, while the former exhibits straight and curvilinear lineaments within smooth plains material and may have formed by tectonic modification of mantling material.

Other Hellas plains have a reticulate pattern of ridges, rugged on the kilometre scale, while others are punctuated by knobs a few kilometres across. Extensive ridged plains extend beyond the southern rim and include a number of low-relief presumed volcanic paterae, including Amphitrites and Peneus Paterae and several unnamed ring structures with radiating ridge patterns (Fig. 8.13).

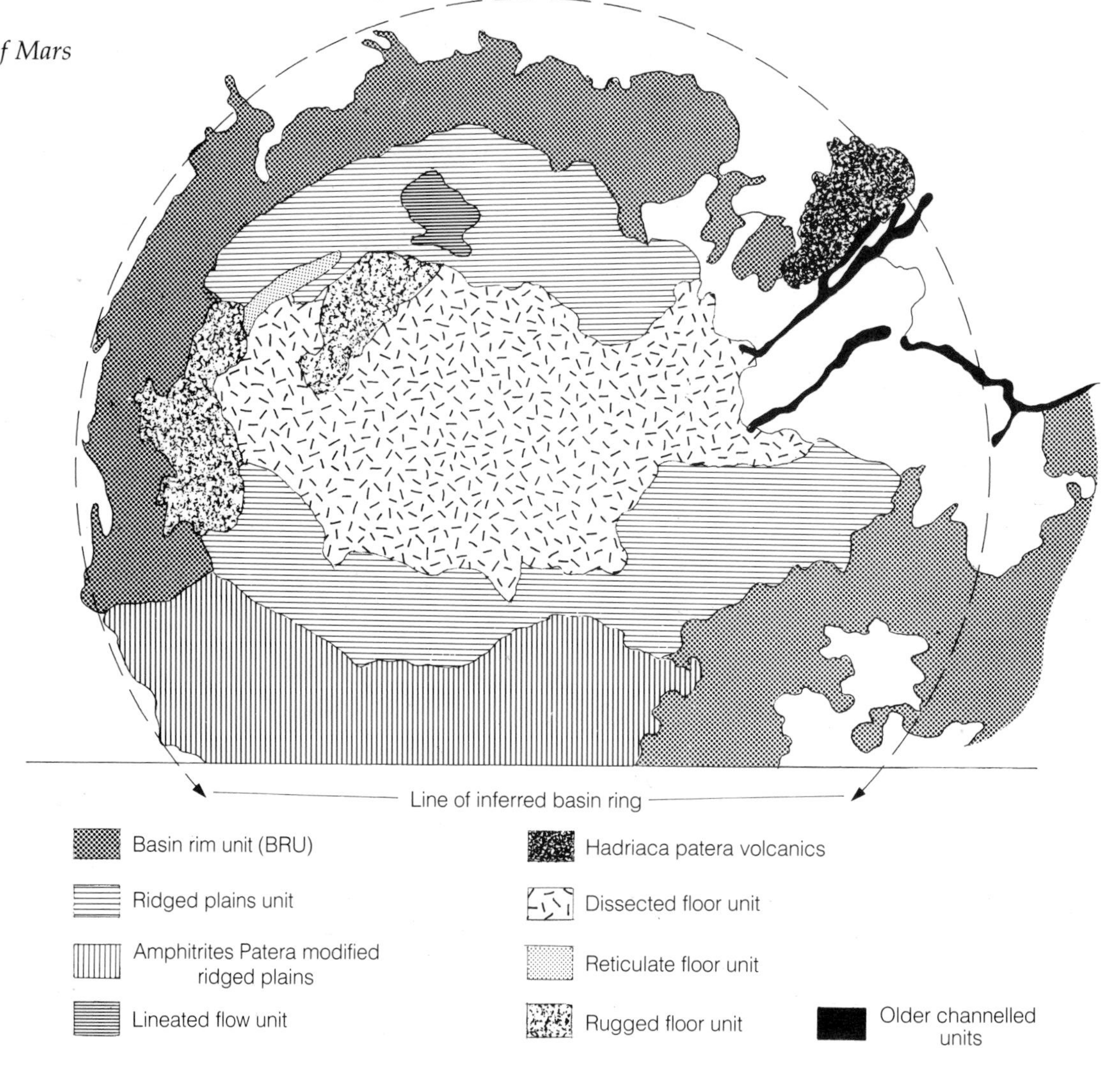

Fig. 8.11 Geological sketch map of units in the Hellas region.

Fig. 8.12 High-resolution image of subdued channel-floor unit on the northern rim of the Hellas basin. Viking orbiter frame 789A26. Centred at 26.21°S, 290.42°W. Frame width 10.5 km.

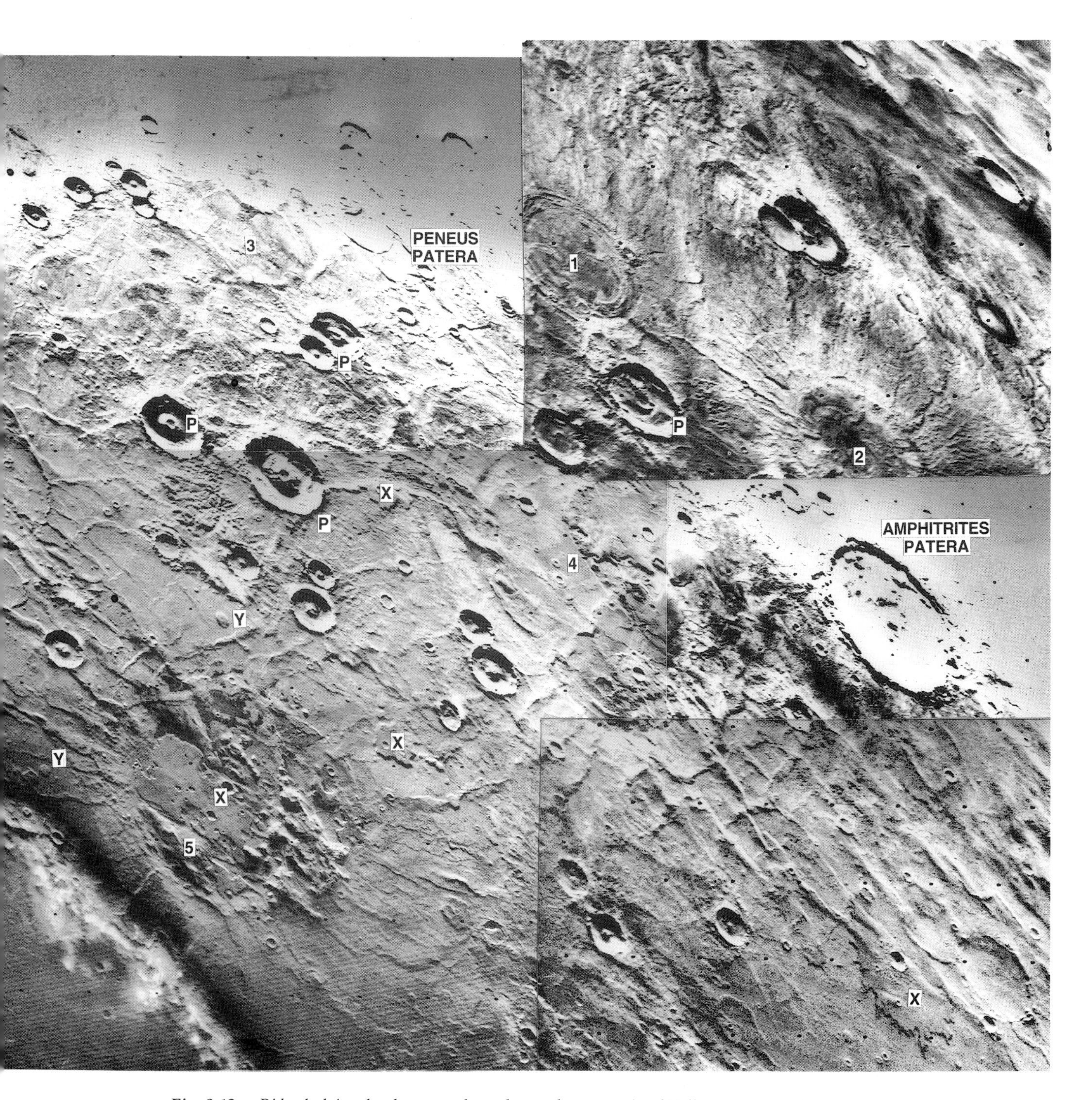

Fig. 8.13 *Ridged plains development along the southern margin of Hellas. Ancient volcanic paterae include: 1. Peneus Patera, 2. Amphitrites Patera and (3–5) three equivocal ring structures with associated radial striae. Note the craters with central depressions on their summit massifs (P) and the stripped and cratered plains unit left as mesas (X). Viking orbiter frames 056B31–5. Centred at 63°S, 305°W. Frame width approximately 2000 km.*

Fig. 8.14 Knobby plains development in the northern plains, north of the volcanic shield of Hecates Tholus. Viking orbiter frames 086A36–48.

8.8 THE NORTHERN PLAINS

8.8.1 Introduction

Emplacement of the very extensive northern plains was accomplished during the Late Hesperian epoch. This apparently was achieved by the laying down of a complex of lava flows, aeolian and alluvial sediments (Plate 7). Resurfacing and modification of parts of the northern lowlands continued through the Amazonian epoch. Sedimentation and volcanism produced diverse plains deposits which can be divided roughly into **smooth plains** and **knobby plains**. The former, in places, clearly are composed of aeolian deposits, while some knobby plains appear to have been produced by erosion of older units (Fig. 8.14). (This resurfacing episode was entirely separate, of course, from the much earlier events, still an enigma, which were responsible for resurfacing and lowering the planet's northern hemisphere.)

8.8.2 Mass-wasting deposits of fretted channels

Flat-lying surficial deposits resulting from mass wasting are located both within and beyond the mouths of fretted channels in the Deuteronilus Mensae region and east from here. These overlie older channelled terrain and have a very low incidence of impact craters, indicating a relatively young age. Plains formed from this material may be up to 600 km wide and are marked by low-albedo, sinuous intertwining albedo patterns which become increasingly more mottled westward. They appear to be fluvial deposits with channels marked by sand bars and relict islands; mottled regions may represent deposition from the ponded terminus of an ancient fluvial system. The fact that such channels are restricted to the upland-lowland interface, suggests that only here was there a sufficiently large down-channel slope to allow for transportation of the sedimentary debris downstream.

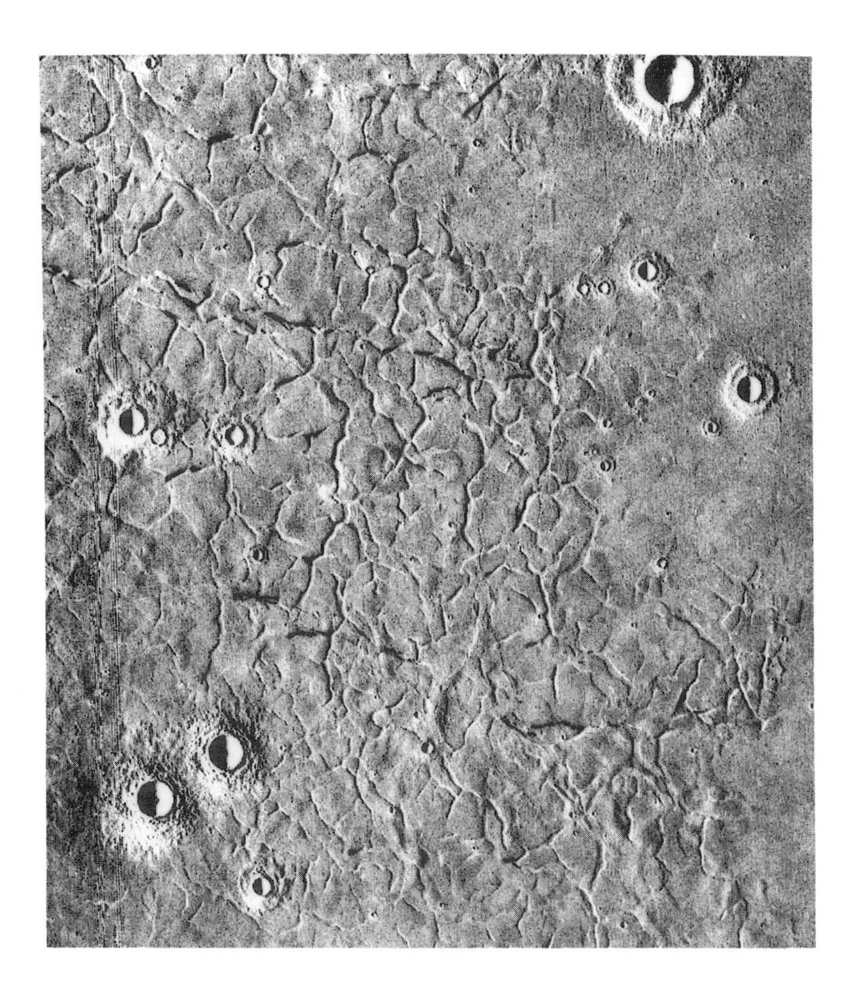

Fig. 8.15 *Fractured terrain in the northern lowlands. Viking orbiter image 32A18.*

8.8.3 Northern lowland plains in mid-latitudes

One of the more distinctive geomorphological imprints seen on the lowland plains is a polygonal fracture pattern which divides the plains into a series of giant blocks anything from a couple of kilometres to 10 km across (Fig. 8.15). In two short papers McGill (1985a,b) proposed that such patterns had developed in relatively thin layers of sedimentary rocks, where contraction and compaction had occurred over the underlying topography. In this way some curvilinear fracture patterns would represent outlines of impact craters buried beneath the sediment cover. A more recent study by Lucchitta *et al.* (1986), confirmed this suggestion. There are several reasons why the idea seems plausible: first, polygonally fractured deposits tend to be located in low-lying areas that apparently received an influx of sediments; second, the fracture patterns are concentrated near embayments in the upland hemisphere where major outflow channels debouch (Fig. 8.16); and third, crater ages indicate the channels and fractured plains to have similar ages.

The giant polygons, although they are at least an order of magnitude larger than any terrestrial analogue, have been interpreted by McGill (1985a,b) and by Lucchitta *et al.* (1986) to be giant desiccation or compaction features in frozen

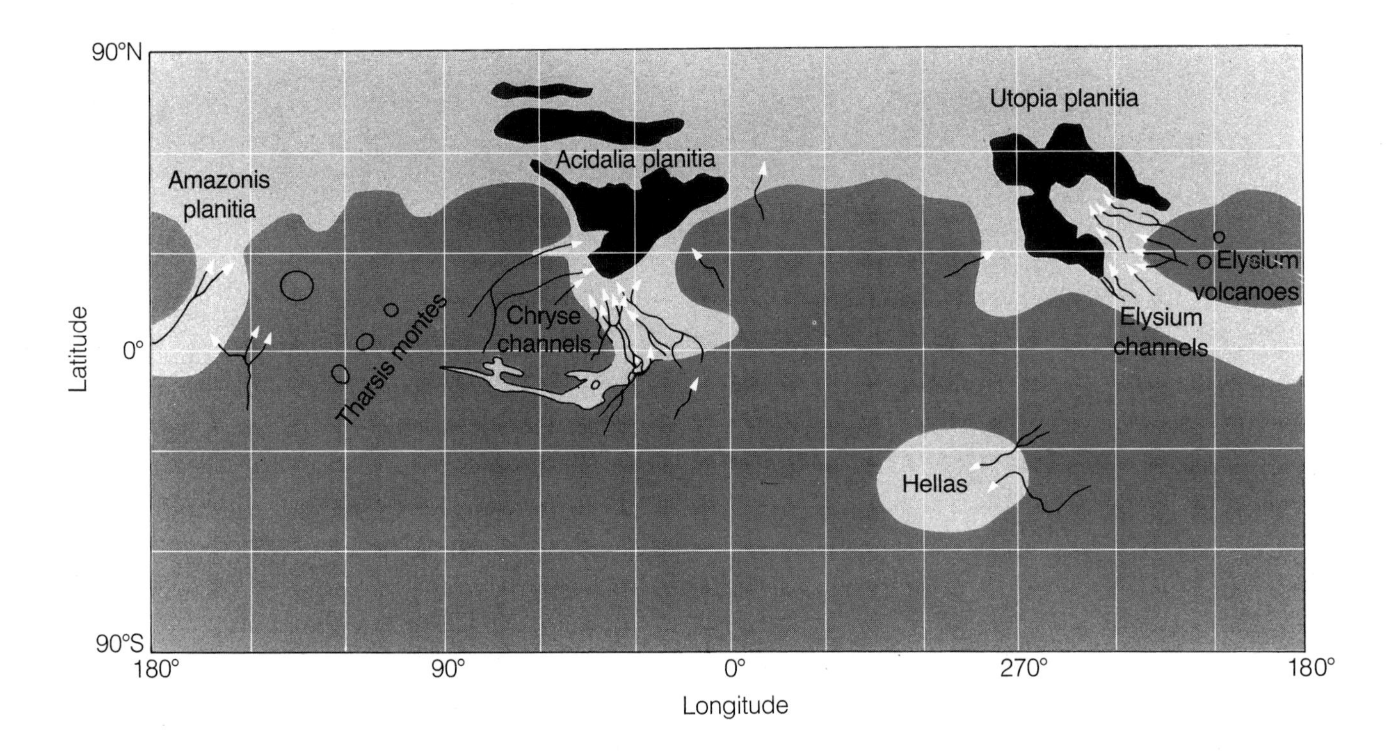

Fig. 8.16 *Map of polygonally fractured terrain in the northern plains of Mars (after Lucchitta et al., 1986).*

outflow channel sediments. However, Pechmann (1980) prefers to consider them as tensional fracture patterns developed in the plains by regional doming. What is clear, however, is that the pattern of cracking implies uniform tension in all directions within the plane of the surface. Closely similar patterns (in form, not scale) occur on Earth where horizontal extensional stresses develop because of cooling or desiccation.

Interestingly, on their recent geological map of the western equatorial region of Mars, Scott and Tanaka (1986) assign an Upper Hesperian age both to the Chryse outflow channels (which debouch into Acidalia Planitia) and the polygonally fractured plains in Acidalia, suggesting some causal relationship. Greeley and Guest (1987) assign a similar age to the fractured plains of Elysium and Utopia Planitia. However, the prominent Elysium channels appear to be slightly younger (Early Amazonian), and connected with the Elysium shield volcanism. This anomaly may be more apparent than real, since the Elysium channel deposits simply may be obscuring an older channel system which was directly responsible for the Hesperian plains. In general terms, age relationships tend to support the plains origin outlined by Lucchitta and her colleagues.

Other distinctive geomorphological features occur on the plains within a few tens of kilometres of the lowland-upland boundary. Like much of the polygonally fractured ground, these are located near the mouths of the Chryse channels that debouch into Acidalia Planitia and below the highland scarp in western Utopia. One particularly interesting set of features comprise sinuous ridges between 0.5 and 1 km wide set in elongate depressions (Fig. 8.17). Lucchitta *et al.* (1986) observe that they bear a striking resemblance to the ridges which are generated near the mouths of Antarctic ice streams and on ice shelves, particularly where shoals are developed. For this reason they suggest that the Martian uplands were bordered by frozen materials which had mech-

Fig. 8.17 Sinuous ridges in Utopia Planitia. The cratered uplands are located to the left of the image. Viking orbiter frame 572A03. Centred at 35.19°N, 274.79°W. Frame width 165km.

anical properties similar to those of modern Antarctic ice shelves. If their suggestion is accepted, then the northern lowland materials represent outflow channel sediments that were laid down in a partially frozen ocean. However, Parker *et al.* (1986) compared the ridges to terrestrial lacustrine or shallow marine coastal spits and barriers, which imply a predominantly liquid environment in which waves are responsible for the features seen.

Recent work by de Hon (1987) has shown that in many locations close to the mouths of major outflow channels (e.g. Maja Vallis, Ladon Vallis) there are relatively featureless plains units which appear to represent lacustrine deposits. Temporary ponding in channels would have been expected under Martian conditions, with the result that local sedimentation both within channels and near their mouths would have taken place. Flood-plain sediments are widespread downstream of channel mouths (Fig. 8.18), while lacustrine deposits have been identified in such areas as Chryse Planitia, Terra Sirenum and Lunae Planum. Possible deltaic sediments are revealed on Viking images in Amazonis and Elysium Planitia.

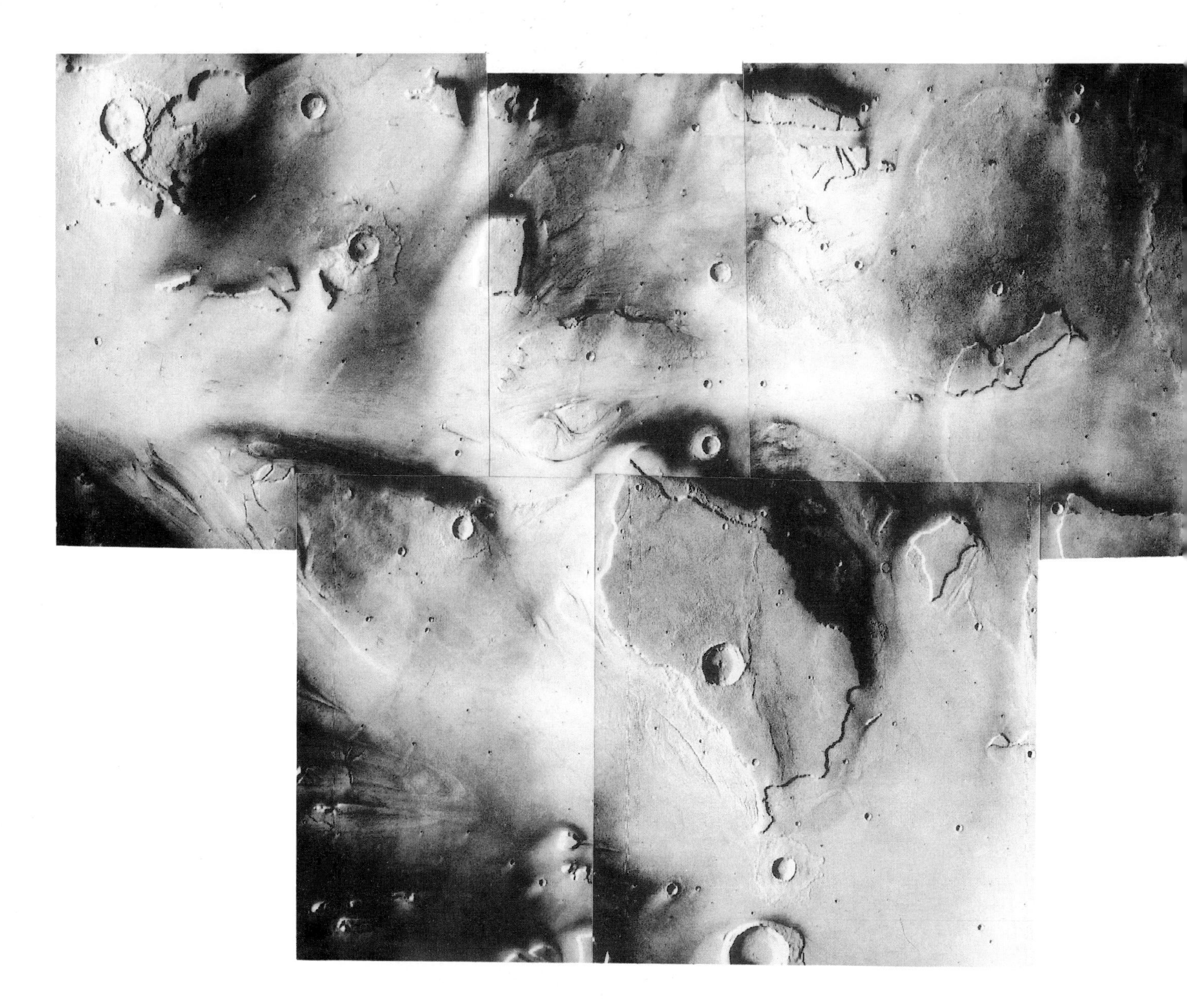

Fig. 8.18 The mouth of the Kasei Vallis channel system. Between the relict mesas cut through lightly cratered and ridged plains, narrow channel floors show strong striations in the direction of former flow. To the east (right) of the constricted sections, flat-topped and smooth surfaced units may represent lacustrine and deltaic sediments. Viking orbiter frames 558A13–17. Centred at 30°N, 51°W.

On the basis of all this evidence, there is now a growing awareness among Mars specialists that extensive lakes, deltas and seas may once have existed on Mars. Currently, several groups are endeavouring to identify critical landforms which may assist in answering the very important question of whether or not there was a widespread northern hemisphere ocean on the planet and, if so, whether this was an open sea or a largely frozen one. Definitive answers to these vital questions unfortunately may have to await the return of higher resolution imagery from future Mars missions.

8.8.4 High-latitude plains

The character of plains units in higher latitudes (above 35°–40°N) is very different from those elsewhere. These plains are poorly understood and appear

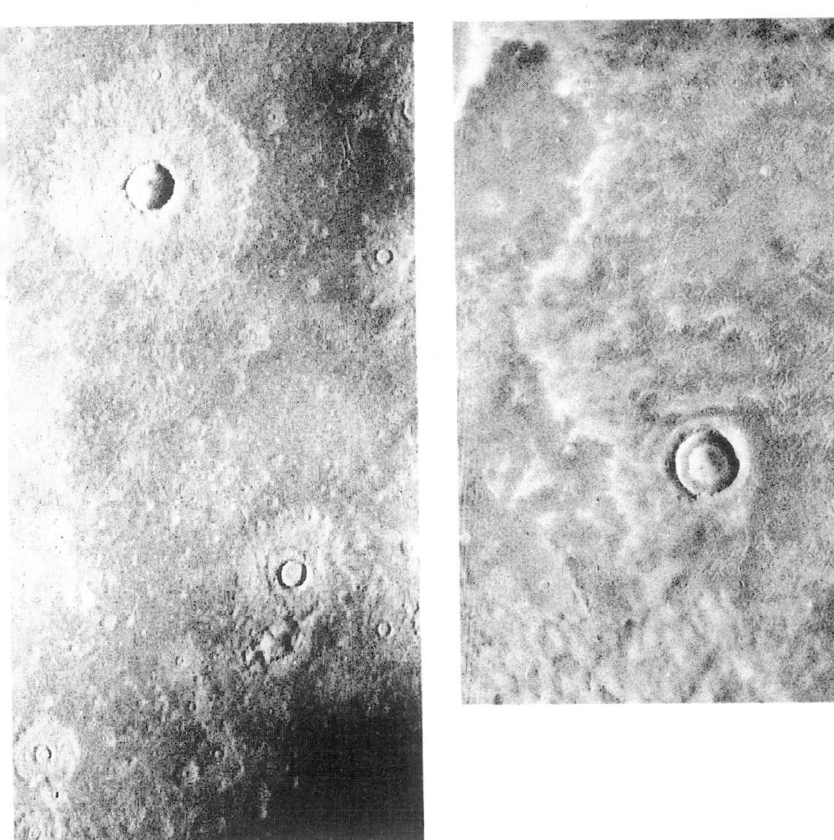

Fig. 8.19 (a) *Softened topography northwest of Alba Patera. A channelized lava flow and numerous impact craters have been blanketed in (presumed) aeolian materials. Viking orbiter frames 007B9–11. Frame is 120 km wide.* (b) *Mottled plains. Viking orbiter frame 670B23.* (c) *Knobby plains in Arcadia. Note the heavily mantled flow lobe on the left of the frame. Viking orbiter frame 319S49.*

to be blanketed in aeolian debris which softens the topography over vast areas. Little or no direct evidence is found for primary volcanic deposits, although such may well be present beneath the sedimentary cover (Fig. 8.19a). Landforms developed within these high-latitude plains include surface mottling, ridges, grooves and knobs. Impact craters at these latitudes tend to have encircling pedestals of raised ejecta and, where aeolian burial is deeper, are entirely blanketed in sediment, giving them the appearance of volcanic domes (Fig. 8.19a). Mottled plains are particularly enigmatic, and almost encircle the planet between latitudes 50° and 70°N. Mottling is largely due to the relatively high albedo of impact crater ejecta compared with the intercrater regions (Fig. 8.19b). All of the landforms which outcrop within these plains appear to be older than those on adjacent plains units. Some plains at these latitudes, while showing mantled volcanic features, also display larger numbers of small knobs (Fig. 8.19c).

Such studies as have been completed suggest that while there has been considerable volcanic input into these Hesperian-age plains – there was major volcanism in the Tharsis and Elysium regions at this time – there has also been much modification. Periglacial processes, tectonism and sediment compaction may all have been involved. Because densities of impact craters >5 km diameter are surprisingly uniform for these extensive plains units, it would appear that, however they were formed, they were produced in a relatively short time.

The extensive circumpolar plains deposits will be described in Chapter 11.

8.9 PLAINS AND MAJOR RESURFACING EPISODES IN MARTIAN HISTORY

One of the principal effects of resurfacing events is that they 'depopulate' a surface in terms of visible impact craters. Neukum and Hiller (1981) developed a technique whereby they could identify resurfacing events by breaking up the cumulative size–frequency crater curves into distinct branches, by noting where crater-depopulation episodes generated a departure from the standard crater curve. They determined the crater retention ages for both the older (underlying) and younger (overlying) surfaces. On this basis, if there had been a global resurfacing event at a specific time in Martian history, the surface corresponding to the end of the event should have the same crater-retention age everywhere, independent of any previous geological history.

By studying crater-retention ages of various plains surfaces on the planet, Frey *et al.* (1989) established that the most widespread common-age resurfacing event to have affected the planet has a crater-retention age $N(1)$ – cumulative number of craters larger than 1 km per $1 \times 10^6\,km^2$ – of 25 000. Units with this (Lower Hesperian) age are widespread in Lunae Planum, in the adjacent cratered plateau, in the ridged plains of Tempe Terra, in the cratered landscapes of the lowland-upland boundary zone and south of Valles Marineris. Study of plains units further west, towards Tharsis, gives a younger age $N(1) =$ 14 000. Model chronologies linking retention ages with absolute ages, by correlation with lunar data, give an age of around 3.8 Ga for the Lunae Planum event (Hartmann *et al.*, 1981).

An earlier major resurfacing episode took place at $N(1) = 85\,000$; units with this age outcrop in Tempe Terra, much of eastern Mars and in knobby terrain in Elysium. Frey *et al.* (1989) suggest that this represents the generation of the cratered plateau materials (Late Noachian). Older again are surfaces with $N(1) = 250\,000$ which cover the heavily cratered parts of Tempe Terra, large regions south of the dichotomy in eastern Mars and in the oldest part of the knobby terrain in Elysium. Even greater ages (600 000–800 000) are found in Xanthe Terra, Arabia and Memnonia.

Thicknesses of the Lunae Planum resurfacing-age materials appear to vary widely (Frey *et al.*, 1989). In the region of the dichotomy (transition zone between upland cratered and lowland materials) in eastern Mars, west of Isidis, the thickness ranges between 225 and 325 m; however, east of Isidis this falls to between 115 and 185 m. In contrast, around 100 m is appropriate to cratered units of the same age south of the line of dichotomy. Then again, ridged plains of Lunae Planum resurfacing age are about 315 m thick in Tempe Terra and 270 m thick in Lunae Planum, but only 105 m thick in parts of Coprates. If the thicknesses of ridged plains resurfacing materials of post-Lunae Planum age are added, an aggregate maximum thickness of around 500 m is established.

9 THE EQUATORIAL CANYONS

As the Mariner 9 spacecraft orbited Mars, it revealed not only the shield volcanoes of Tharsis and the planet's polar caps, but an enormous equatorial canyon system. Appropriately it was named Valles Marineris, after the probe which discovered it. Unbeknown to astronomers, it was this enormous feature which unwittingly had been depicted on many early drawings as one of the notorious 'canals'.

9.1 GENERAL FEATURES OF VALLES MARINERIS

Compared to the famous Grand Canyon of Arizona, Valles Marineris is over four times deeper, six times wider and at least ten times longer. It is a global-scale feature rather than a local one, and is more closely akin to terrestrial structures like the East African Rift than to the Grand Canyon. Neither is it a single canyon, more a maze of sometimes interconnecting ones which straddle the Martian equator between longitudes 30° and 110°W. With a total length of approximately 4500 km – that is, nearly a quarter of the circumference of Mars – its western extremity is located in the high plains near the summit of the Tharsis Bulge and it extends eastwards towards an enormous region of collapsed ground (**chaotic terrain**) located between Chryse Planitia and Margaritifer Sinus (Plate 2). The deepest sections plunge at least 7 km below Mars datum, while individual canyons attain widths of 200 km.

9.2 DETAILS OF CANYON PHYSIOGRAPHY

Valles Marineris can best be considered in three sections (Fig. 9.1): in the west, near the focus of a plethora of radiating fractures that centre on the Tharsis Bulge, is a region incised by relatively short and often interconnecting canyons called Noctis Labyrinthus. They extend over an area of at least 400 000 km^2.

Commencing about 600 km east of the summit of Arsia Mons, this maze of canyons and fractures bounds Syria Planum on its northern side. The much-fractured ground into which the labyrinth is incised is embayed and sometimes overrun by the Upper Hesperian volcanic flows of the Syria Planum formation; it is therefore older than this. The fracturing continues along the western side of Syria Planum and continues towards Solis Planum as a series of NNW-SSE graben called Claritas Fossae. To the north, fractures trend just east of north, but are soon obscured by the volcanic flows associated with Tharsis Montes.

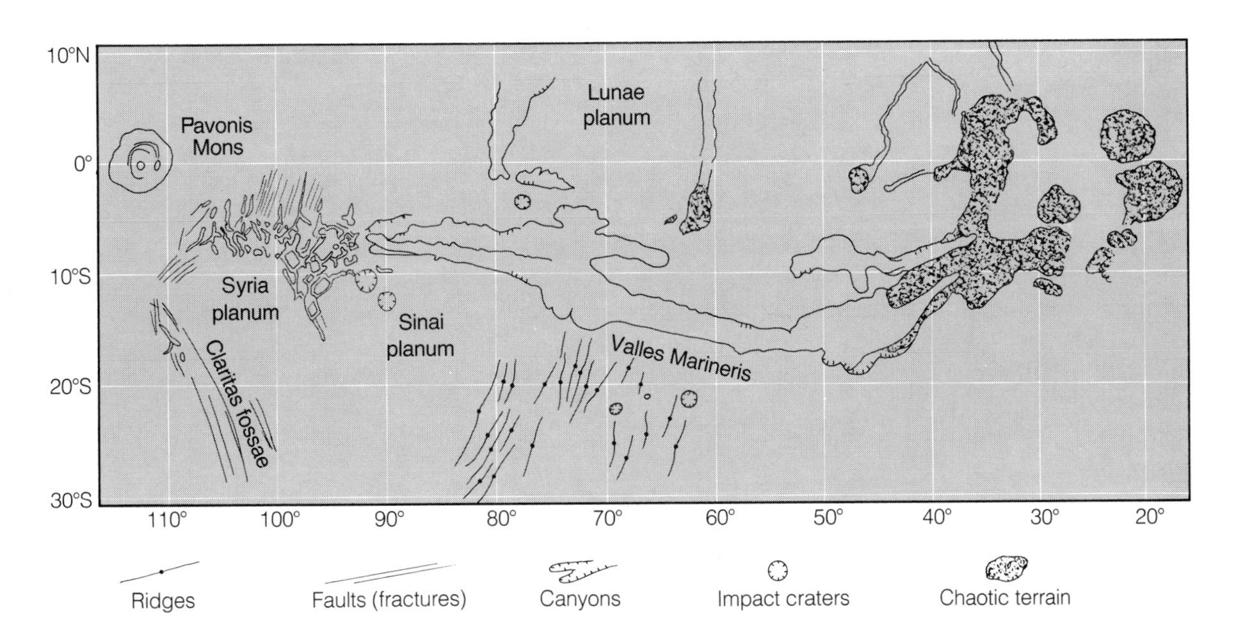

Fig. 9.1 Generalized map of Valles Marineris canyon system.

This intricate labyrinth is unlike the remainder of Valles Marineris. Maze-like in its overall appearance, it comprises a series of intersecting canyons whose position and orientation was controlled by sets of intersecting fractures. The latter almost certainly developed in response to the doming which accompanied the growth of Tharsis, and which elevated this region to a height of 10 km above datum. In places the faults which controlled canyon development can be observed intersecting the walls, usually at angles close to vertical.

Individual canyons tend to be narrow and rather short; because the controlling faults define an intersecting network, the canyons now break up the high plateau surface into a region of blocky remnants (Fig. 9.2). Towards the west the labyrinth is composed of rows of coalesced pits, individual depressions often having different floor levels. This suggests that canyon growth was accounted for largely by subsidence as opposed to longitudinal removal of debris by fluvial runoff or ice erosion. Towards the east, pits tend more often to coalesce and ultimately join with the more or less contiguous canyons of the main section.

Noctis Labyrinthus gives way eastwards to the central section of the system. This encompasses a series of huge parallel-sided canyons that extend over a zone 2400 km long and, in places, 700 km wide. The general parallelism of the canyon walls and pervasive W-E to WNW-ESE trend of the system as a whole, indicates they developed along major extensional (rift) faults.

The central 2400 km long section is multiple throughout its entirety. Within its confines a variety of landforms have developed: flat-floored graben faults, graben with subsidence pits, chains of coalescent depressions and, finally, parallel-walled canyons. A transitional sequence can be observed, there being a general tendency towards the latter as the eastern end is approached.

Immediately east of Noctis Labyrinthus two major canyons bound Sinai Planum on the north. These are Ius Chasma to the south and Tithonium Chasma to the north. The Hesperian-age plateau surface on both sides of each is incised by numerous graben and pit rows which share the same trend (Fig.

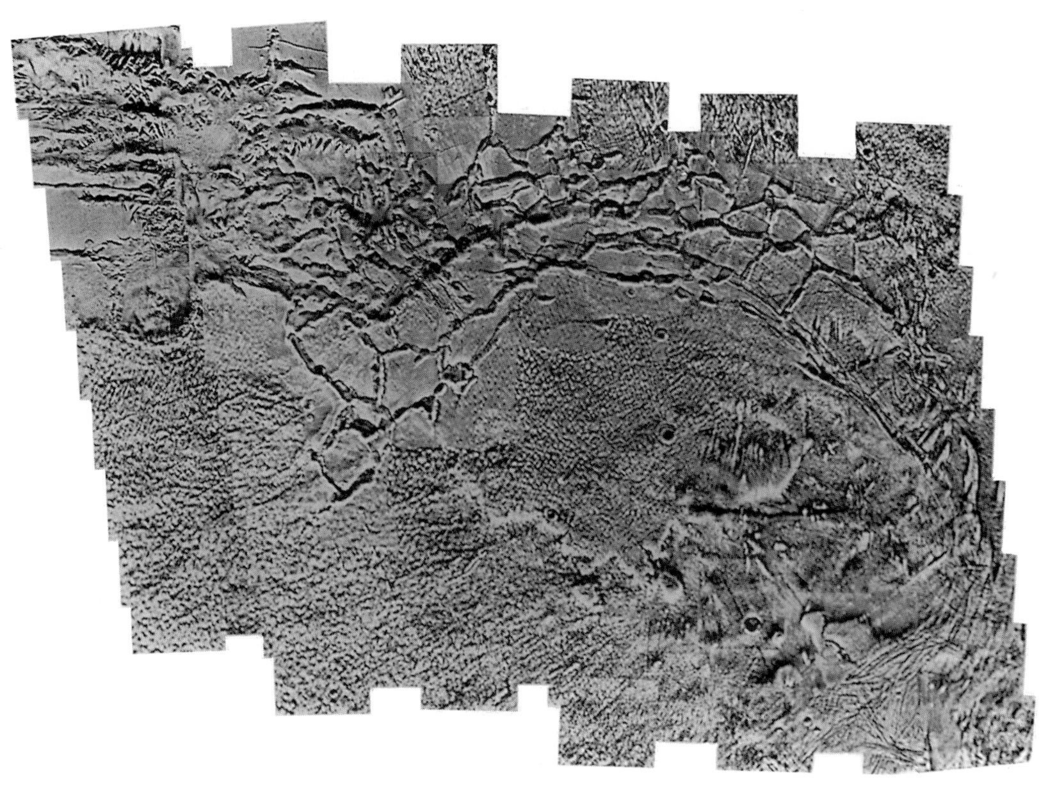

Fig. 9.2 *Noctis Labyrinthus, a maze of interconnecting short canyons developed at the western end of Valles Marineris. Viking orbiter mosaic.*

9.3). Sections of these are often offset by transverse faults, some of which themselves are graben.

The more northerly of these two canyons forms a single trough and eventually narrows towards the east, until at about 78°W it is separated from the broader central section by a line of large depressions. Its more southerly counterpart comprises two troughs separated by a ridge; towards the east, each of these becomes wider and eventually the two merge into the broad Melas Chasma.

Melas Chasma is shown in Plate 6 and forms a part of the most continuous section of the system, since it also connects eastwards with Coprates Chasma, which in turn passes into Eos Chasma, the latter merging with the chaotic terrain that develops from Valles Marineris at its eastern end. Between 65°W and 80°W Valles Marineris attains its greatest width (Fig. 9.3). Along this section Melas, Candor and Ophir Chasmata run in parallel, producing a broad depression 600 km wide and up to 7 km deep. The Candor and Ophir Chasmata are closed at both ends, as is the separate and more northerly depression, Hebes Chasma, which borders Lunae Planum on its south side. On the plateau surface beyond their walls, the same trend is continued by lines of partially coalescent pits – further evidence of a strong underlying structural control on canyon formation. All three canyons have a rather spatulate appearance and large areas of their interiors are covered with mass-wasted debris.

Separating Ophir, Candor and Melas Chasmata are intervening ridges which may be partially breached; the intercanyon rocks appear to be remnants of the

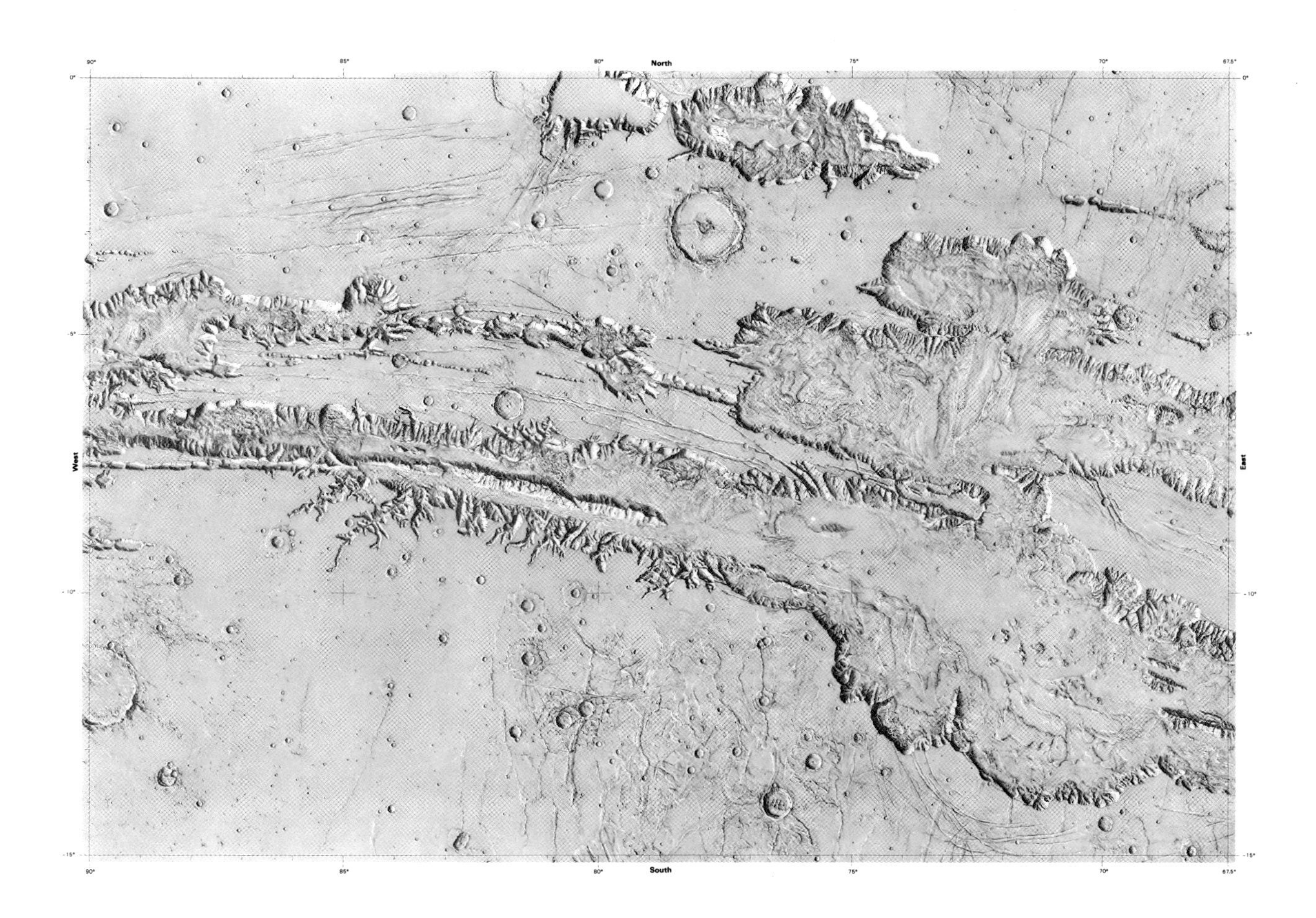

Fig. 9.3 *Shaded relief map of the western section of Valles Marineris, from Tithonium Chasma to western Coprates Chasma. Note the graben and pit rows, the ridged plateau surface to the north and south, and the short side canyons.*

ancient plateau surface. A particularly prominent gap in the south wall of Candor Chasma has allowed debris to flow southwards across the canyon divide and on to the floor of Melas Chasma (Plate 7). The divide between Candor and Ophir Chasmata is similarly built from the older plateau rocks but stacked up against it and partly burying it are younger, finely gullied deposits with a relatively smooth upper surface. A mesa-like remnant of similar rocks, located closely west of the former, exposes distinct layering on its western flank. These stratified sequences appear to rest discordantly upon the underlying bedrock and have been asigned an Amazonian age by Scott and Tanaka (1986).

Melas Chasma passes eastwards into the well-defined, parallel-sided Coprates Chasma which comprises several parallel troughs, crater rows and graben. This extends between 68°W and 52°W, a distance of about 1000 km. Eastwards, the nature of the canyons changes. The parallel-sided canyons give way to less well-defined depressions with increasingly hummocky floors. Eos and Capri Chasmata have distinctly scalloped walls, blocky floors and merge eastwards into chaotic terrain that extends ENE towards Aureum Chaos (Plate 8). North of them and separated by a remnant of Lower Hesperian plateau sequence 200 km wide, lies Ganges Chasma on whose floor are several spectacular landslides. This likewise merges eastwards into a large area of chaos.

9.3 GENERAL FORM OF CANYON WALLS

The steep canyon walls tend to be relatively smooth but deeply gullied, with numerous rounded embayments into the plateau surface. The shoulders between individual gullies are sharply defined, branching and fluted; they resemble terrestrial escarpments formed largely by dry mass wasting under both desert and glacial climates (Lucchitta, 1981a). In many places the shoulders are truncated by low fault scarps. Where this occurs, as commonly it does in Coprates Chasma, triangular facets develop. This faulting must have followed gully development, suggesting that tectonic activity accompanied both canyon formation and modification. Talus slopes form large aprons at the foot of the walls; these have slopes of around 30°.

Stratigraphic considerations suggest that the bedrocks behind the equatorial escarpments are, in the upper regions, resistant volcanic flows and, lower down, less well-consolidated impact breccias. The entrapment of ground water or ground ice within the relatively porous breccias may have played an important part in the generation of landslides.

9.4 WALL RETREAT AND LANDSLIDE DEPOSITS

Several huge embayments give many sections of the canyons' walls a markedly scalloped appearance. This is particularly true of the northern walls of Ius and Tithonium Chasmata, and also parts of Ganges Chasma (Plate 6). Such embayments are due to wall failure. The scars often have a somewhat rectilinear form – another indicator that existing structural controls dictated canyon geometry. Where sections of the steep walls have failed, massive landslides have left vast alcoves and spread debris over the canyon floors (Fig. 9.4). One particular slide in Ius Chasma is 100 km wide and extends from one wall to the other. Several other slides are actually banked up against the opposite rampart.

Slide morphology is variable. The overall form appears to be dependent upon whether or not slippage was constrained by existing topography (Lucchitta, 1979). Slides which were largely unconfined usually have large slump blocks at their heads and vast grooved aprons (Fig. 9.4). On the other hand, where the flow of debris was confined between canyon walls, such as in narrow troughs, debris aprons usually have developed transverse ridges, particularly where the slide materials became piled up against obstacles near the toe (Fig. 9.5).

9.4.1 *Dyanamics and genesis of landslides*

It is possible to gain some understanding of how fast these huge slides moved. Thus, by measuring the dimensions of obstacles overridden by certain slides, Lucchitta (1978) estimated velocities of between 100 and 140 km s^{-1}. Such high velocities imply that the material must have had a very low coefficient of friction. This could be accounted for by assuming the flow to have been lubricated by a cushion of air, as is suggested to have occurred in certain terrestrial slides (Shreve, 1966). However, this seems unlikely to have occurred on Mars and greater support has been tendered for the idea that either water or ice may have helped both trigger and fluidize the slippage. Certainly, the depth of canyon walls is greater than the depth to the base of a permafrost

Fig. 9.4 *Massive landslides in Ganges Chasma. That below the north wall has an upper blocky layer (probably disrupted cap rock), and a striated lower layer. Viking mosaic P16952. Mosaic width 65 km.*

layer (Fanale, 1976). Thus, water-saturated materials with low shear strengths could well have existed behind canyon walls, making them susceptible to failure.

The longitudinal striations seen on some Martian debris aprons are very similar to those observed in parts of Alaska, where slides have developed above glacier ice (Lucchitta, 1981b). The best known of Martian striated landslides are those which occur in Ganges Chasma (Fig. 9.4). These have emerged from a 2 km high backwall and have moved at least 60 km from their source. While Shreve (1966) explains the features of the Alaskan Sherman Landslide as being due to air cushioning, it seems that such a catalyst is not necessary to produce the features of striated slides in Valles Marineris. Rather, the frictional resistance of the debris would have been low in any case, due to the enormous

Fig. 9.5 Section of Ius Chasma, showing the north wall (in shadow) and ribbed landslide banked up against opposite wall. On the plateau surface to the south of the canyon, parallel graben and lines of depressions can be seen. Viking orbiter frame 057A45. Frame width 130 km. Centred at 4.46°S, 85.09°W.

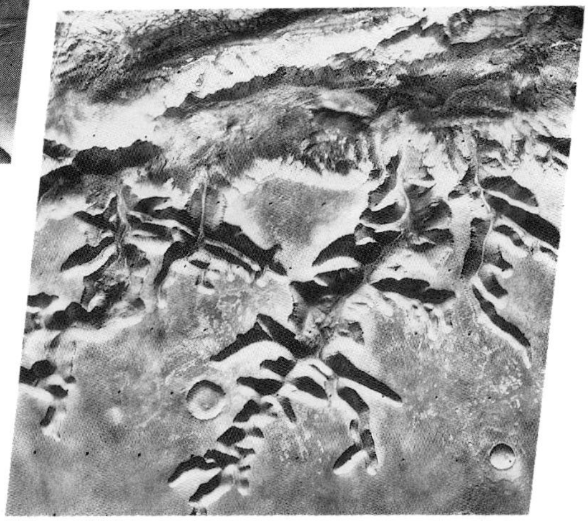

Fig. 9.6 Tributary canyons developed in cratered plateau to the south of Ius Chasma. Viking orbiter frame 059A20. Frame width 128 km. Centred at 7.82°S, 84.46°W.

energy and velocity attained by the great mass of material during its descent of several kilometres.

Slides may therefore have originated as massive mudflows with large slumped blocks at their heads. Precisely what triggered collapse is speculative but seismic tremors associated with faulting and/or volcanism, could have liberated water trapped in aquifers behind a mantle of ice that coated canyon walls.

9.4.2 *Age of landslides*

Many of the landslides appear to be of roughly similar age (Lucchitta, 1979). They were produced after the major episode of faulting that generated escarpments of great relief in the equatorial regions. Indeed, most slides were emplaced after the escarpments had been gullied and dissected by tributary canyons. This episode seems to have coincided approximately with major volcanism associated with Tharsis Montes.

9.5 WALL RETREAT – TRIBUTARY CANYON DEVELOPMENT

While scalloping is common on north canyon walls, the development of tributary canyons is characteristic of several sections of south walls. This is particularly evident on the south side of Ius Chasma, where the canyon backwall is deeply dissected (Fig. 9.6). The canyons tend to be relatively short, have V-shaped cross sections and rather rounded headward terminations; also, they have developed along orthogonal fractures.

Sharp (1973) has suggested that side canyons were generated by a sapping process involving groundwater seepage – probably as a result of the sublimation of buried ground ice – which would have weakened the side walls. The preferential development of such canyons on the south side of Ius Chasma was likened by Sharp and Malin (1975) to the situation at the Grand Canyon, where a regional southerly dip – which controlled the southward migration of groundwater down the bedding planes – has meant that there is a better development of tributary canyons up-dip on the north rim. Should this in any way be analagous, then the situation at Ius Chasma, by implication, would be the reverse of this.

9.6 INTERIOR DEPOSITS

Within the walls of the canyon system are widespread sedimentary deposits which are younger than both the plateau surface and the canyon floors. These include finely layered deposits and fine dust or sand.

9.6.1 Laminated deposits

Between the walls of the extensive canyon system there is sparse direct evidence for the activity of running water. However, layered (often rhythmically layered) deposits outcrop in numerous places, particularly within the central section of Valles Marineris (Section 9.2). There is ample evidence to suggest that these were deposited after canyon formation, and are not simply remnants of the plateau into which the canyons were incised.

The laminated units outcrop widely in Ophir, Candor, Melas and Hebes Chasmata (Plate 7). The thickest sequences give rise to flat-topped mesas which, at first sight, resemble the old plateau surface. However, they do not show the fracturing or pitting so typical of this and they have a distinctly smoother appearance. The mesa sides also are finely fluted, largely uncratered and quite different from canyon walls (Plate 9).

Layering is present on quite a fine scale – certainly down to the resolution limit of the available imagery. At the junction of Ophir and Candor Chasmata, the horizontally bedded deposits discordantly overlie older rocks. It is clear that they formed after the canyons had developed, and it is also apparent that they themselves have subsequently been eroded. McCauley (1978) suggests that since such finely layered deposits imply deposition under quiescent conditions, they could well be lake sediments. While this has to be speculative, it is quite conceivable that lakes could form either by slow release of trapped groundwater by seepage of some kind, or catastrophically. McCauley favours a catastrophic release, largely because the canyons beyond Coprates merge into chaotic terrain, believed to have an origin in rapid fluid release.

Lakes could, however, only exist if the water could be stabilized by a thick ice cover. Indeed, any potential Martian lake would quickly form such an ice cover (Wallace and Sagan, 1979). Its lifetime would then be determined by the balance between seepage of groundwater into the lake and sublimation from the ice surface. This poses a problem: how could cyclic deposits be laid down under a permanent cover of ice? Admittedly, subsurface erosion does occur on Earth, particularly where loess is forming. Nevertheless, to account for the observed Martian deposits would necessarily mean such a process (variously referred to as 'tunnelling', 'piping' and 'sink-hole erosion') would have had to have taken place on a very grand scale indeed! Evidently more work needs to be done on this particular topic before a more positive statement can be made.

9.6.2 Young dust deposits

The laminated deposits, which are generally rather thin, may reach a thickness of 3 km in Candor Chasma. The only younger sediments that can be identified are thin deposits of dust. These may simply represent dust fallout from the atmosphere or be wind drifted. Either way, a veneer of this material coats most mesa tops within the canyon system.

The lighter-coloured material that outcrops on mesa tops is spectrally reddish and has very low thermal inertia (Lucchitta, 1987). This could conceivably represent palagonite tuff. Where the covering is relatively thin, streaky light and dark deposits outcrop; these are spectrally blue and could represent volcanic materials (Plate 8). If this is so, they could be amongst the youngest volcanic rocks on the planet.

9.7 FORMATION OF THE CANYON SYSTEM

The question of the formation of Valles Marineris remains something of an enigma, although new work is helping to clarify matters all the time. The notion that the canyon system is simply analagous to the Earth's East African Rift appears untenable, since simple extensional tectonics cannot account for the observed configuration of canyons and the pattern of faulting (Wise *et al.*, 1979). However, large-scale faulting followed by scarp recession, can be identified as major contributary factors in canyon development.

Canyon inception can be traced to the initial opening of deep troughs parallel to families of shallow graben which developed radial to the Tharsis Bulge, possibly in Late Noachian but more probably in Early Hesperian times. The dating of this first phase of canyon-forming activity can be established by noting that none of the Late Noachian/Early Hesperian-age plateau lavas spilled over into the canyons, implying they were erupted prior to tectonic disruption (Lucchitta *et al.*, 1989). The fracturing widely disrupted the plateau surfaces of Lunae and Sinai Planae, fracturing the lava caprock and exposing the underlying brecciated megaregolith.

That faulting dominated canyon development is clear from the exceptionally straight wall segments in central Valles Marineris. However, exactly how the canyons attained their great depth (up to 7 km) is not entirely clear. Doubtless the negative gravity anomaly known to exist over the central section holds some clue in this respect but its significance currently remains an enigma. Deepening and lateral growth may have proceeded by a combination of fracturing, crustal spreading, collapse following the withdrawal of magma to lateral sites of effusion, or removal of underlying bedrock by solution or sapping.

The actual volume of removed material has been estimated at $10^6\,km^3$ (Sharp, 1973) but there is little direct evidence to suggest that fluvial processes could have achieved this. True, the fluvial landforms that emerge from the regions of chaos into which the eastern canyons merge suggest that removal of large volumes of debris may have occurred there. However, it is difficult to envisage that fluvial activity has removed much debris from the western sections. One possibility is that as subsidence and wall retreat progressed, downfaulting continued to lower the canyon floors, and interior deposits continued to accumulate. To what extent fluvial, glacial and aeolian processes moved or removed such internal materials remains an unknown factor. The presence of extensive laminated deposits opens up the possibility that lakes once existed within the confines of the canyon walls. Draining of these hypothetical lakes eastwards may have given rise to the Simud and Tiu Valles channel systems. Younger materials also accumulated and these now lie unconformably upon the laminated sequence. Some have rather low albedo and may represent volcanic accumulations which emanated from local vents.

It has been suggested that the higher albedo layers in the laminated canyon sequence could represent carbonates. Interestingly, one very recent suggestion (Spencer *et al.*, 1989) concerns the role of acidic groundwater in removing carbonate rocks from the Martian subcrust in the Valles Marineris region. As we have seen, there is unequivocal evidence for a substantial inventory of volatiles on Mars. Calculations suggest that there may have been as much as 10 bar of CO_2 outgassed during the planet's early history; much of this would likely have become fixed as carbonate rock in the crust. Local accumulations of such rocks may have become unusually thick where suitable sedimentary basins occurred; one could been present beneath the present canyon region.

Now the Viking results have indicated around 3 wt% of sulphur in the surface dust at the Lander sites. This could have come from an originally sulphur-enriched lithosphere, in which case SO_2 could have been vented from

Fig. 9.7 *Diagram showing possible formation of lines of depressions by solution of subsurface carbonate rocks (after Spencer et al., 1989).*

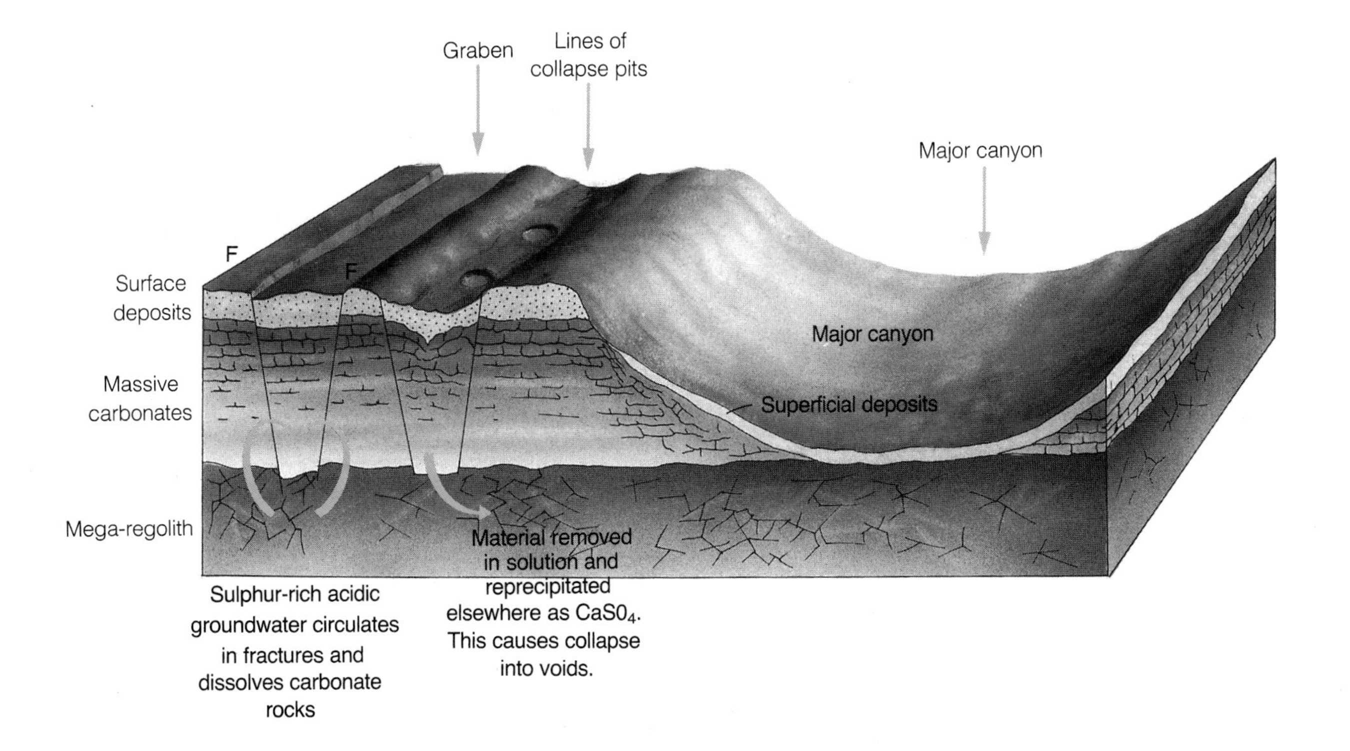

volcanoes during eruptive periods, sending up H_2SO_4 aerosols; alternatively, weathering of sulphur-rich igneous rocks may have occurred; either way, such activities would give rise to highly acidic, sulphate-rich groundwaters. With concentrations of the order of 10^{-3} by volume, these would vigorously attack carbonate rocks, liberating CO_2 and H_2O; if concentrations were higher, $CaSO_4$ would precipitate. The latter could be carried in suspension, thus significantly increasing the solubility.

On this basis, Spencer *et al.* envisage sulphur-rich groundwater circulating in fracture zones and slowly dissolving carbonates concentrated below the surface (Fig. 9.7). Dissolved material would then be removed in solution and redeposited elsewhere, probably as $CaSO_4$. Removal would lead to collapse at the surface, and could provide a plausible means of generating the lines of collapse depressions which are so characteristic a feature of some sections of Valles Marineris, and which may mark the initial phase of canyon growth.

This is an interesting possibility and, not surpisingly, several groups are now studying the possible development of karst topography on the Red Planet. Mars continues to stimulate geologists and geomorphologists to explore new avenues of research, based on terrestrial experience.

MARTIAN CHANNELS AND CHAOTIC TERRAIN

10

10.1 INTRODUCTION

Several aspects of the origin of the equatorial canyons remain something of a puzzle to planetary geologists. However, it is clear that is some connection between canyon formation and the extensive collapsed regions which have been called chaotic terrain. These chaotic regions, largely concentrated in the equatorial regions between 10°N and 20°S, also seem to have spawned several major channel systems. For instance, at the eastern end of Valles Marineris, Hydraotes Chaos passes northward into Simud Vallis, while the lengthy Ares Vallis channel emerges from Iani Chaos (Fig. 10.1). Such channels pose some very important questions about Martian palaeoclimate.

The channels which emerge from areas of chaotic terrain have been termed **outflow channels** (Sharp and Malin, 1975); they tend to lack tributaries, emerge fully developed from chaotic regions and become wider and shallower downstream. Other channel landforms (**fretted channels**) are concentrated along the line of dichotomy; these have flat floors which are characterized by abundant debris aprons and flows. A third type, **runoff channels**, which have well-developed dendritic tributary networks, are largely confined to the ancient cratered uplands. Global channel distribution is shown in Plate 10.

The principal question that arises from the existence of any channels, is whether or not they were generated by running water, for if available evidence is found to indicate a fluvial origin, then the connotations for the planet's climatic evolution are substantial. At the present time liquid water is unstable at the surface; the current temperature range of −143°C to +13°C, and the surface pressure of between 5 and 10 mbar, dictates that liquid water will quickly either evaporate or freeze on Mars, depending upon the latitude, altitude, time of day and season. In view of this difficulty, several investigators have pursued alternatives to water as an erosive agent; such alternatives include wind, ice, lava and faulting. While some of these may have played a part in channel formation, the concensus among the planetary community is that running water probably played the dominant role in generating at least two of the channel types.

10.2 VALLEY NETWORKS IN THE ANCIENT CRATERED TERRAIN

Of all the valley systems visible on Mars, the dendritic networks which characterize the ancient cratered hemisphere are most like terrestrial river systems

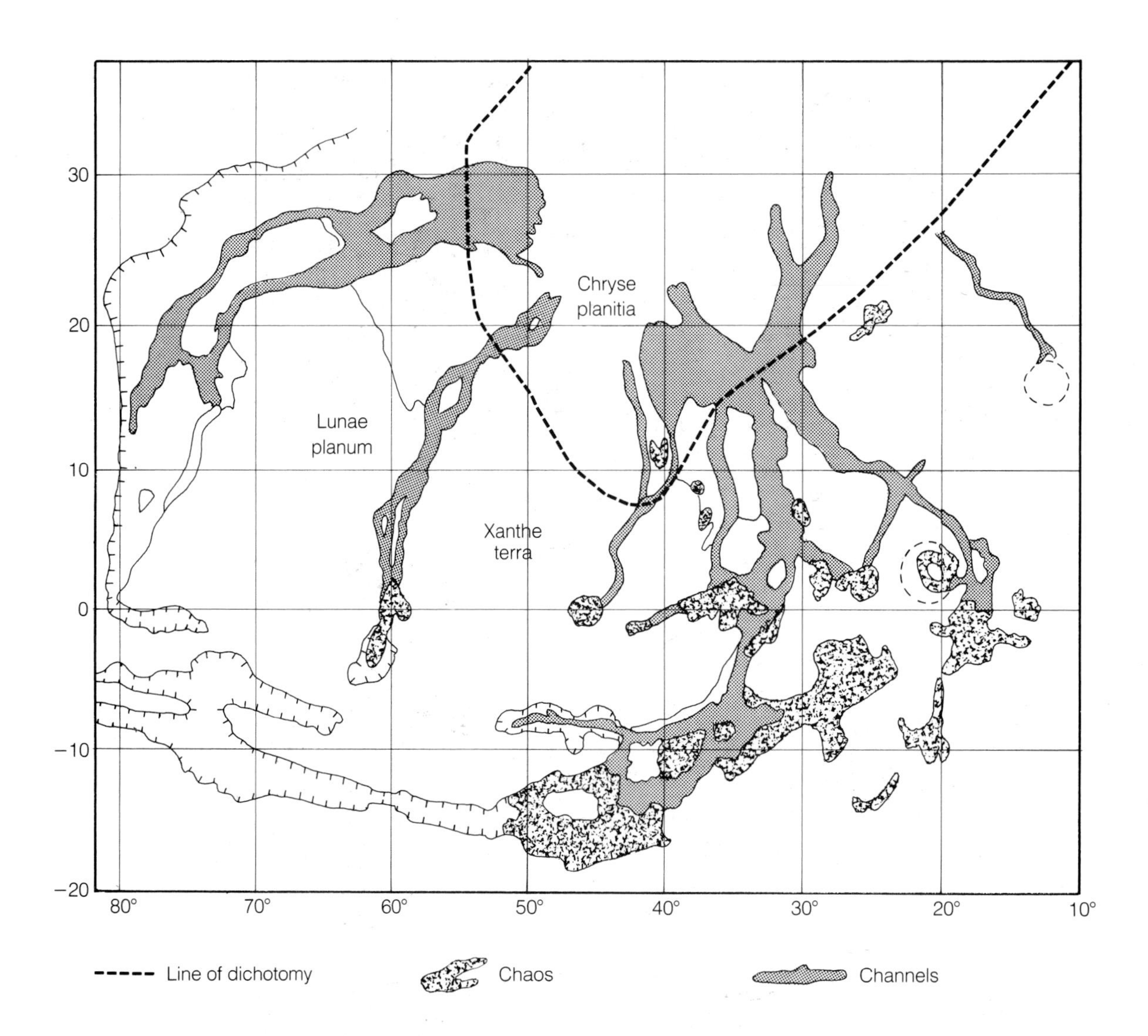

Fig. 10.1 *Map to show the relationship between canyons, chaotic terrain and channels in the equatorial regions of Mars.*

(Fig. 10.2). These were first recognized from Mariner 9 images. Typically, these have dense tributary systems, are roughly V-shaped in cross-section and broaden downstream. They have been termed runoff channels (Sharp and Malin, 1975). While most are relatively small, a small number of more major systems, such as Nirgal and Ma'adim Valles (Fig. 10.3), share some of their features but tend to have much blunter and less well-developed tributary networks. Baker (1982), noting the great age of both of these larger systems, suggests that perhaps these started as runoff channels but were slowly modified by wall retreat in their lower courses. The smaller valleys are between 1 and 2 km wide and may form systems several hundreds of kilometres in length. With the exception of some tributary networks associated with Ius Chasma and some with the volcano, Alba Patera, nearly all are located in the heavily cratered regions between 30°N and 40°S. This can hardly by a coincidence and seems to imply that they themselves are ancient (Carr and Clow, 1981). Indeed, crater counts suggest that most of the dendritic valley networks were formed between 4.0 and 3.8 Ga ago, during the period of heavy bombardment.

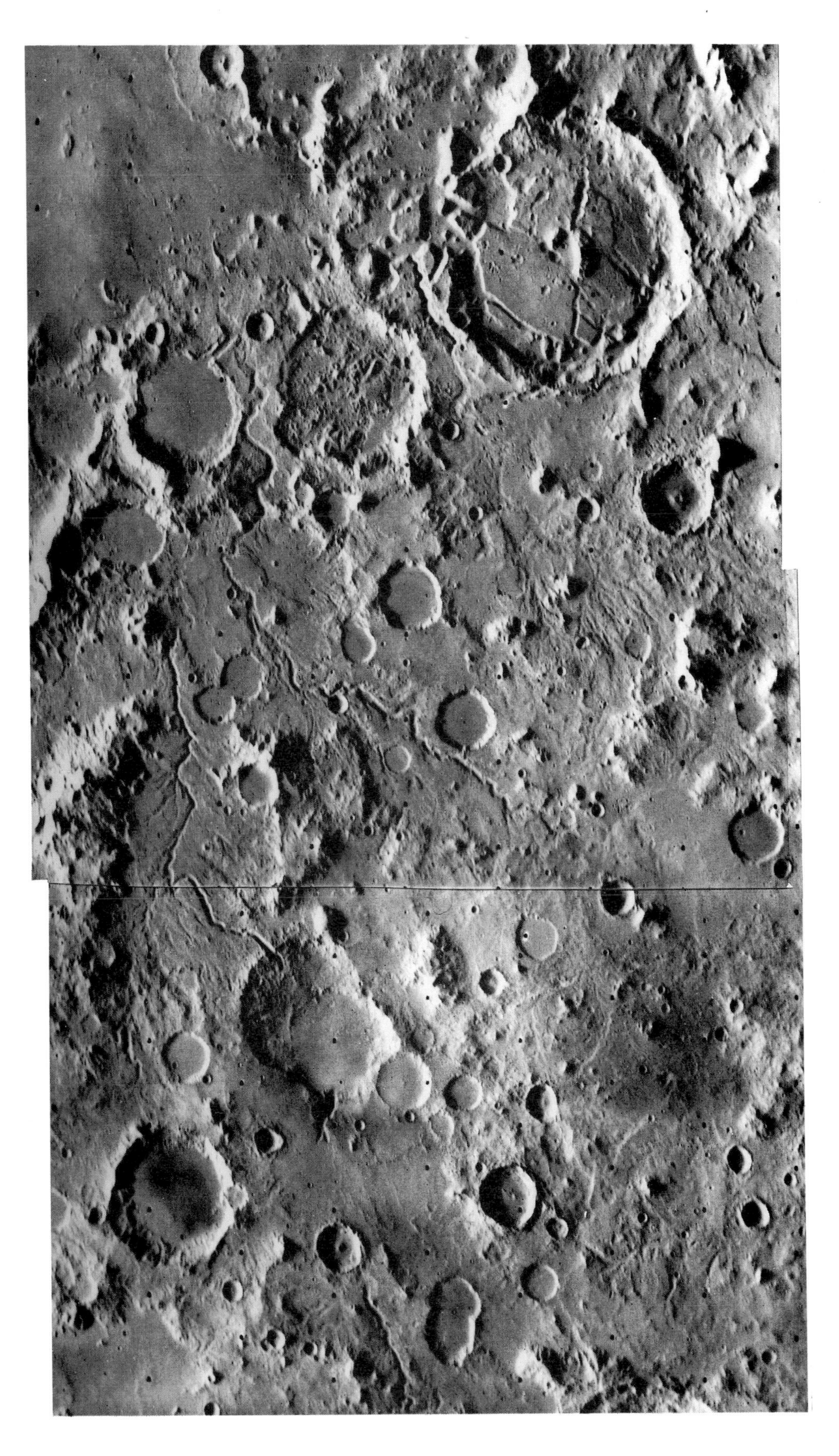

Fig. 10.2 *Runoff channel networks in the ancient cratered plateau southeast of Isidis. Viking orbiter frames 067B65–7. Centred at 0.50°S, 270°W. Frame measures 250 km × 430 km.*

Fig. 10.3 *Downstream course of the ancient channel Ma'adim Vallis. Viking orbiter frame 639A91. Centred at 18.63°S, 124.74°W. Frame measures 510 km × 960 km.*

10.2.1 Form of the valley networks

The drainage patterns delineated by the valley networks are either rectangular or parallel, showing a considerable measure of structural control (Mars Channel Working Group, 1983). In the former case, the patterns have developed in areas of fracturing. The parallel patterns, on the other hand, often result from drainage off the flanks of large impact craters and have relatively small drainage areas compared with the rectangular ones.

When studied in detail, the networks are more complex than at first sight. The most widespread valleys have scalloped or runnelled walls and rather indistinct walls. Indeed, their walls generally have been degraded to rather a low slope angle (Fig. 10.4). Individual large networks frequently are broken up by areas of cratered terrain, giving them a rather poorly integrated appearance. Baker and Partridge (1986) have termed these degraded valleys. Of the two types of valley recognized, on the cratered plateau, these effect the greater degree of plateau dissection.

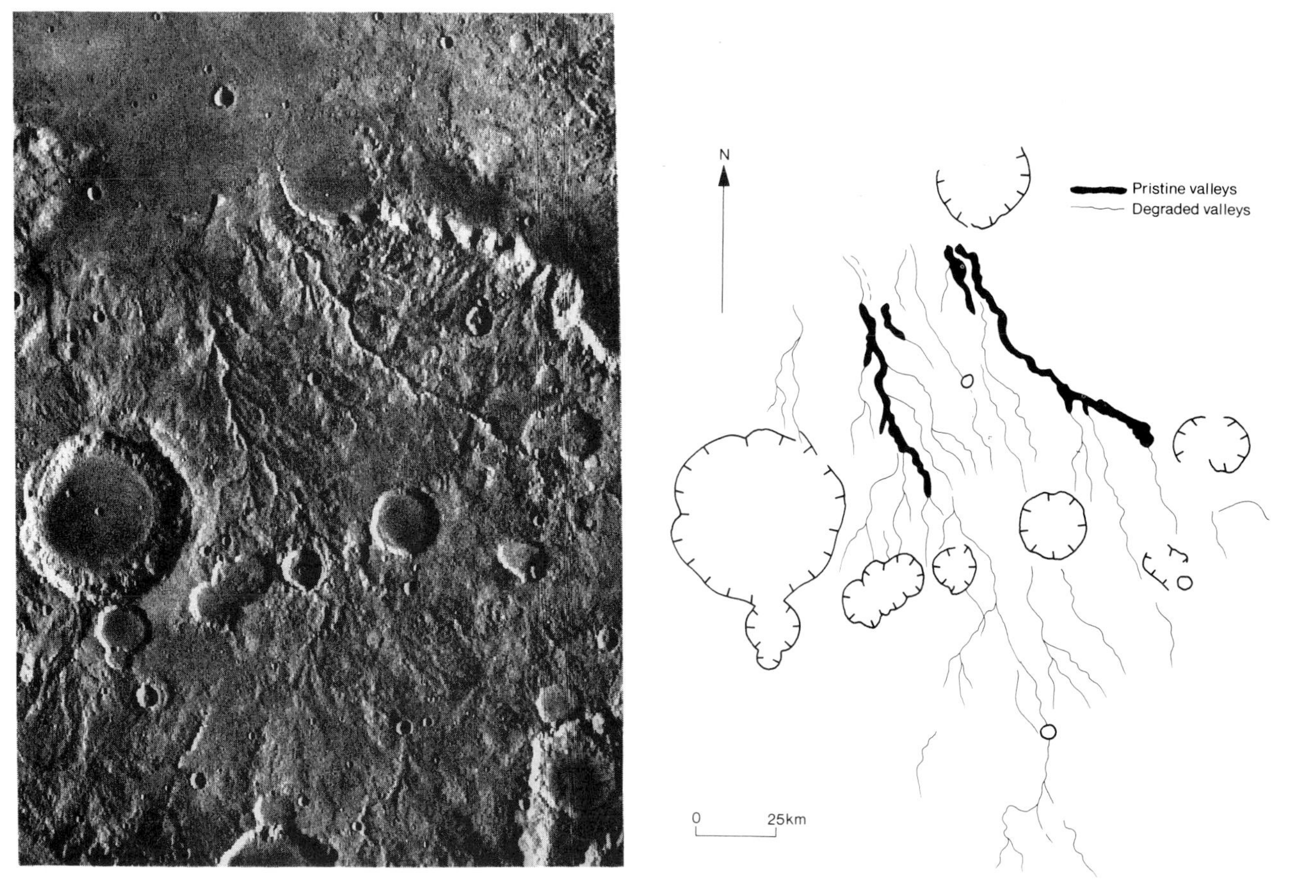

Fig. 10.4 (a) Pristine and degraded channel development in Terra Tyrrhena. Viking orbiter frame 525A02. Centred at 12.24°S, 279.90°W. Frame measures 260 km × 190 km; (b) key map for (a).

More local in their development than the above are deeply incised valleys with steep or even vertical walls and U- or box-shaped cross-sections (Fig. 10.4). These have been called pristine valleys by Baker and Partridge (1986), and while such valleys are more local in their development than degraded ones, they may still attain lengths of 600 km. Compared to the degraded valley systems, pristine valleys generally have rather poorly developed tributary networks and often show amphitheatre-like headwalls. Furthermore, the plateau surface in the interfluves is relatively undissected. Pieri (1980) suggests that such characteristics indicate that groundwater sapping may have been important in their growth.

The broad pattern observed is of systems that consist of both degraded and pristine segments. The pristine segments tend to lie in downstream locations, while degraded ones occupy upstream positions. Although the former extend into heavily cratered terrain, they tend to have been incised more widely into intercrater areas – a natural consequence of their occupation of downstream locations. Degraded valleys also may be observed in downstream positions but where they do, they frequently are obscured either by lava sheet or sedimentary units.

10.2.2 Formation of the networks

A number of features indicate that the degraded networks were incised before the pristine ones. Thus, where degraded tributaries enter pristine valleys, they usually do so as hanging valleys; then again, knick points mark positions where pristine valleys end upstream within degraded valleys, giving rise to a cirque-like head. Baker and Partridge (1986) attempted to derive crater ages for the two types and, acknowledging that large errors were inherent in their method, obtained ages of around 4 Ga for the degraded networks and 3.8–3.9 Ga for the pristine valleys.

The geomorphological features of the networks suggest the following scenario: during the period of major bombardment, valley networks were cut into the rugged and primitive cratered crust. The parallel drainage patterns of many principal valleys indicate that their development was strongly dictated by the positions of large impact craters which provided local gradients down which drainage occurred (Fig. 10.5). A later period of resurfacing produced the intercrater plains, which locally buried sections of the existing channels. Subsequently, after the main phase of bombardment was complete, a second stage of valley formation was initiated. This second generation of (pristine) valleys gradually encroached up into the headward regions of the older networks, occasionally extending into the heavily cratered regions of the plateau.

Even though it is difficult to map the exact extent of the degraded valleys (due to their low slope angles), their very wide extent and high density of tributary development have obvious implications for Martian palaeoclimate. Their close similarity with fluvial systems developed on Earth leaves little doubt that they were produced by fluvial action. None of the alternatives so far proposed, can sensibly account for their production.

The more restricted distribution and less well-developed tributary systems of the younger valleys, have led to several workers attributing their development to groundwater sapping, rather than to fluvial run-off (Pieri, 1980; Baker and Partridge, 1986). If this is so, then the implications are that an active hydrological cycle must have been in operation, which could have recharged the high water tables that would have been necessary to sustain spring discharge at valley heads. There seems little getting away from the conclusion that a period of sustained erosion must have occurred to incise the networks in the first place. It is difficult to imagine any other means for provision of a rechargable hydrological system without surface runoff. Under present climatic conditions it is difficult to envisage how runoff can be returned to subsurface reservoirs, since the ground is permanently frozen, although Clifford (1981) has suggested a cycle involving basal melting of polar ice. The most attractive explanation, therefore, is that when the valleys were cut, climatic conditions were more temperate than they are today, allowing the groundwater system to be replenished from precipitation (Carr, 1984). It then becomes possible to envisage valleys being cut by a combination of fluvial action and groundwater sapping.

10.3 OUTFLOW CHANNELS

These channels are particularly spectacular and have provoked much discussion amongst planetary geologists. With the exception of those on the east

Fig. 10.5 *Channel networks in Terra Tyrrhena, showing parallel drainage patterns and initiation of valleys on outer flanks of large impact craters. Viking orbiter frames 625A08–10. Mosaic length approximately 500 km.*

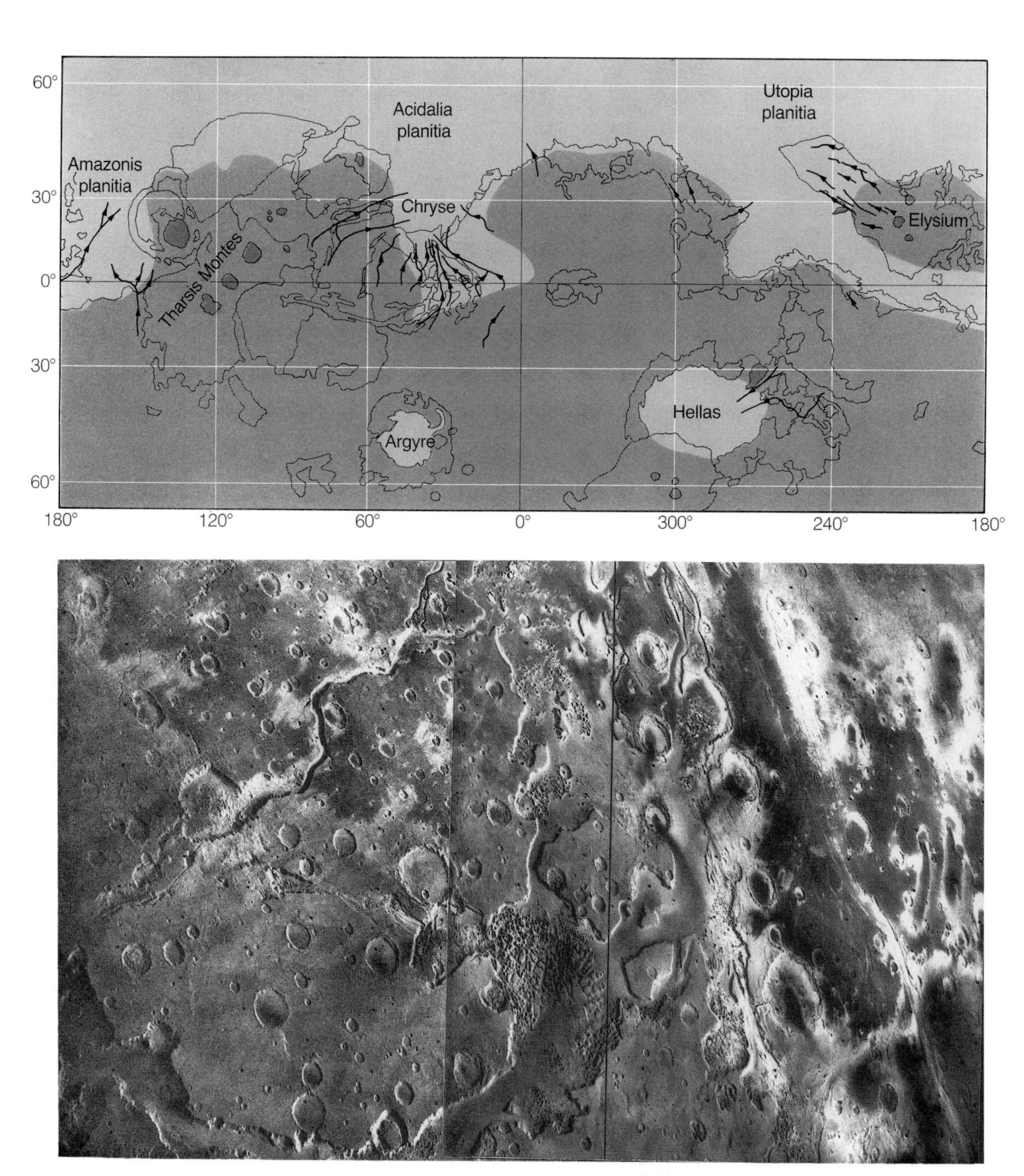

Fig. 10.6 (top) *Distribution of principal outflow channel systems on Mars. The dashed line represents the upland-lowland boundary; (above) outflow channels at the east end of Valles Marineris (left to right: Shalbatana; Simnd; Tiu and Ares Valles). Viking orbiter frames 006A01–3.*

rim of Hellas and on the western flanks of the Elysium volcanoes, they emerge from the cratered plateau, usually from areas of collapsed chaotic terrain, and terminate after transgressing the boundary with the northern lowlands (Plate 10). Several particularly large channels arise to the east and north of Valles Marineris and converge on Chryse Planitia, others debouch on to Amazonis (Fig. 10.6). The lower reaches of one of the latter, Mangala Vallis, are well seen in Plate 11. Their common association with the line of dichotomy must be more than coincidental; thus any formational theories must take this into account.

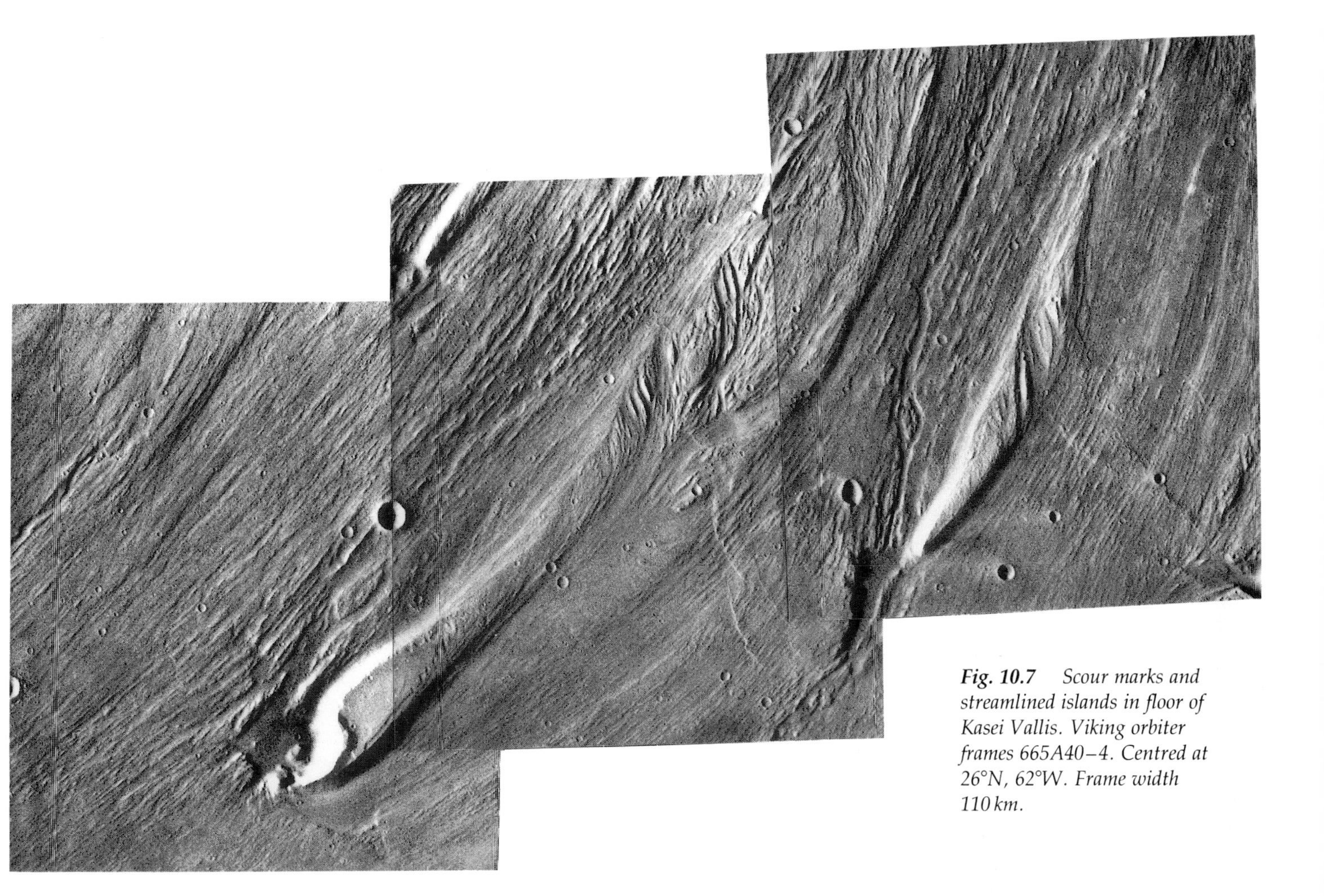

Fig. 10.7 Scour marks and streamlined islands in floor of Kasei Vallis. Viking orbiter frames 665A40–4. Centred at 26°N, 62°W. Frame width 110 km.

10.3.1 Morphology of outflow channels

Outflow channels typically arise full-blown from chaotic regions and are widest and deepest at their upstream ends. Scour marks, while typical of channel floors, are not confined to them, indicating that flow often extended beyond the main channel limits. Scouring is particularly evident where obstacles such as impact craters have caused flow to diverge around them, where ridges, subsequently breached by the flow, have confined flow, and where channels are forced to turn corners and change direction (Fig. 10.7).

Typical of such channels are Tiu and Ares Valles which emerge from the large areas of chaotic terrain at the eastern end of Valles Marineris (see Fig. 10.1). Tiu Vallis is over 600 km long and in places over 25 km wide; Ares Vallis is at least 1200 km in length and 100 km in width. Both originate in collapsed chaotic ground and flow northward, debouching on to the southern floor of the Chryse basin. Neither channel has any significant tributaries. On the floor of each is a series of interlacing channels and streamlined erosional remnants similar to those found in terrestrial braided streams. The teardrop-shaped mesas, often topped by impact craters, frequently exhibit stratification in their

Fig. 10.8 *Tear-drop islands and stratified deposits on floor of Ares Vallis. Viking orbiter frames 004A50–2. Centred at 20.6°N, 31.2°W. Frame width 60 km.*

scarp faces, indicating either that erosional stripping has revealed bedrock or that terracing has been produced by deposition of sediment during channel-forming activity (Fig. 10.8).

Two of the most extensive systems of channels emerge from north of the central section of Valles Marineris; the more westerly arises on the west side of Lunae Planum in the broad Echus Chasma and gives rise to the Kasei Vallis channel; the more easterly emerges from Juventae Chasma, on the opposite flank of Lunae Planum, and forms Maja, Vedra and Bahram Valles (Fig. 10.9). All of them debouch on to Chryse Planitia and both Kasei and Maja Valles exceed 1200 km in length. Sections of Kasei Vallis are over 200 km wide and, like Maja Vallis, show all the features typical of their kind. Where they enter Chryse Planitia, the flow lines diverge across the plains and give rise to a broad complex of shallow channels, streamlined islands and longitudinal grooves (See Fig. 8.18).

10.3.2 Ages of the channels

Unlike the runoff channels, which show a narrow range of ages, the outflow channels appear to have been formed over a lengthy period (Masursky *et al.*, 1977). For instance, some large discontinuous channels (e.g. Ladon Vallis) which are emplaced in the cratered uplands, predate the decline in impact rates which marked the end of the Great Bombardment, around 3.9 Ga ago. On the other hand, Mangala Vallis, which debouches into Amazonis Planitia, is very sparsely cratered and has been assigned an Upper Amazonian age (Scott and Tanaka, 1986). Most of the large circum-Chryse channels appear to have an Upper Hesperian age, that is, a time of formation about 2.5 Ga BP.

During a study of the plains and channels in the Lunae Planum-Chryse Planitia region, Theilig and Greeley (1979) were able to define at least four phases of channel development, the first two of which affected the ancient

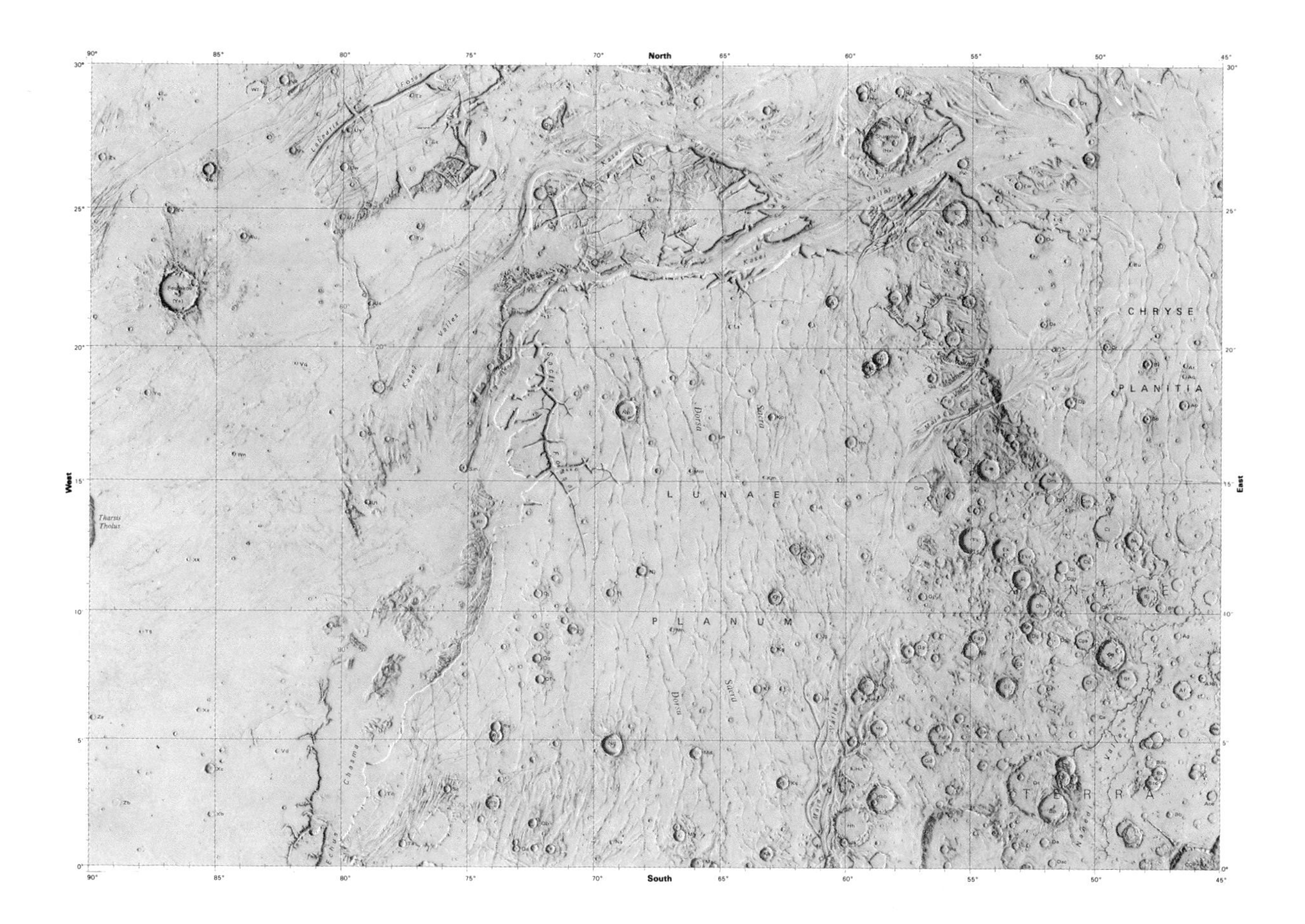

Fig. 10.9 *Shaded relief map of channel systems incised into Lunae Planum and debouching into Chryse Planitia. USGS 1:5M Map I-1511.*

cratered terrain and which may have resulted in the formation of a groundwater system that provided a source of ground ice. These both predated the formation of the extensive volcanic plains of Lunae Planum and Chryse Planitia. The latter were then incised by further channels: Vedra, Maumee and Mahram Valles formed first and laid down a sedimentary unit on the western slope of Chryse; subsequently this was eroded by Maja Vallis, which originated in the chaos around Juventae Chasma.

10.3.3 *Origin of outflow channels*

The Martian outflow channels have been likened to the channelled 'scablands' of E. Washington, USA by a number of authors (Baker and Milton, 1974; Baker and Kochel, 1978; Baker, 1982). These represent some of the most widespread terrestrial catastrophic flood features and are the only ones which approach the dimensions of Martian outflow channels. In a comparison between the scablands and the Martian outflow channels, Baker (1982) and Baker and Milton (1974) point out that teardrop islands, longitudinal grooves, angular channel-floor depressions and terraced margins are common to both.

The scablands were produced when, during the Late Pleistocene, one of the containing walls of Lake Missoula – an ice dam – was breached, causing a catastrophic flood that swept over the surrounding region. The lake, of enormous proportions, had covered large parts of Idaho and W. Montana.

Estimates suggest that peak discharge rates of $10^7\,m^3\,s^{-1}$ may have been attained during flooding (Baker, 1982). It was these very high rates that accomplished the erosion, in just a few days, of the scabland channels which cover an area of almost 100 000 km^2.

If such an origin is accepted – and the consensus view is that it should be – then the problem arises of how such a massive release of water on to the Martian surface could come about. Nummedal and Prior (1981) suggested that the channels had their origins in debris flows which had their source in regions of chaotic terrain. They envisage the channels as having been carved by fast-moving slurries of waterlogged near-surface material, similar in many respects to the subaqueous debris chutes developed along the Mississippi delta front. Lucchitta (1982), on the other hand, proposed a method of ice erosion, the ice being fed by Artesian springs. In pointing out the similarities between the outflow morphology and that of present-day ice streams and glaciers on Earth, she argues that these much better match the scale of the Martian channels. The ice itself could have accumulated in the source regions either from atmospheric precipitation, wind-drifted frost and dust deposits, or from the ground as springs. The latter would seem to have been the major source, since the source regions (chaotic terrain) show signs of major disturbance following the removal of substantial quantities of material.

McCauley *et al.* (1972) and Masursky *et al.* (1977) proposed that the Martian floods may have been similar to Icelandic jokulhlaups, a kind of glacial burst. These are generated when basal melting of an ice sheet is accomplished by volcanic heating. The only channelled region to which this might apply is the west flank of Elysium; however, there is no evidence for volcanic eruptions in the chaotic terrain, and this idea seems rather unlikely for the major outflow systems.

Mike Carr (1979) suggested that flooding may have been the result of the breakout of water under high confining pressures in aquifers buried beneath a thick permafrost layer. Such water, if mobilized, would tend to percolate slowly towards topographic lows, whereupon a hydrostatic head would gradually build. Because of the wide variation in relief on Mars, this could enable the water to reach the surface under very high pressure, thus causing disruption (to form the regions of chaos) and channelling (to produce the outflow channels). Such a process is not difficult to imagine operating on the eastern slope of the Tharsis Bulge and could account for channel formation in a region devoid of volcanic heat sources, such as the source regions of the Lunae Planum and Chryse channels.

There are sufficient differences between individual outflow channel systems to assume that more than one process could have played a role in their formation. Thus, as Carr suggests (1984), the catastrophic release of groundwater may have resulted in the liquefaction of an aquifer and the arrival at the surface of a waterlogged slurry. Initially, this may have formed a flood but, subsequently, as the discharge rate fell away, the slurry might have frozen to become a dirty glacier. Then, at the close of fluvial and glacial activity, the wind may have modified any landforms produced earlier. Any channel could, therefore, bear the geomorphological imprint of a number of quite different processes.

Fig. 10.10 Fretted terrain development at Protonilus Mensae. Viking orbiter frames 268S25–30. Width of mosaic 305 km.

10.4 FRETTED CHANNELS

In various places the ancient cratered plateau is so drastically modified as to be almost unrecognizable. Where such modification is very intense, the plateau has been eroded into a series of relict mesas and knobs, into areas of massive collapse or is heavily dissected by flat-floored channels. The latter style of modification is strongly developed along the line of dichotomy between latitudes 30°N and 45°N and longitudes 280°W and 350°W. In a paper published as long ago as 1973, Sharp termed this kind of region 'fretted terrain'. It is particularly well developed in the regions of Nilosyrtis and Deuteronilus Mensae, where high-standing remnants of the cratered plateau are separated either by arcuate alcoves floored by lightly cratered deposits or flat-floored channels with strongly lineated surfaces (Fig. 10.10). The two components may be separated by escarpments as high as 2 km. Further from the boundary, only tiny isolated remnants of the ancient plateau remain, giving rise to what McCauley *et al.* (1972) called knobby terrain.

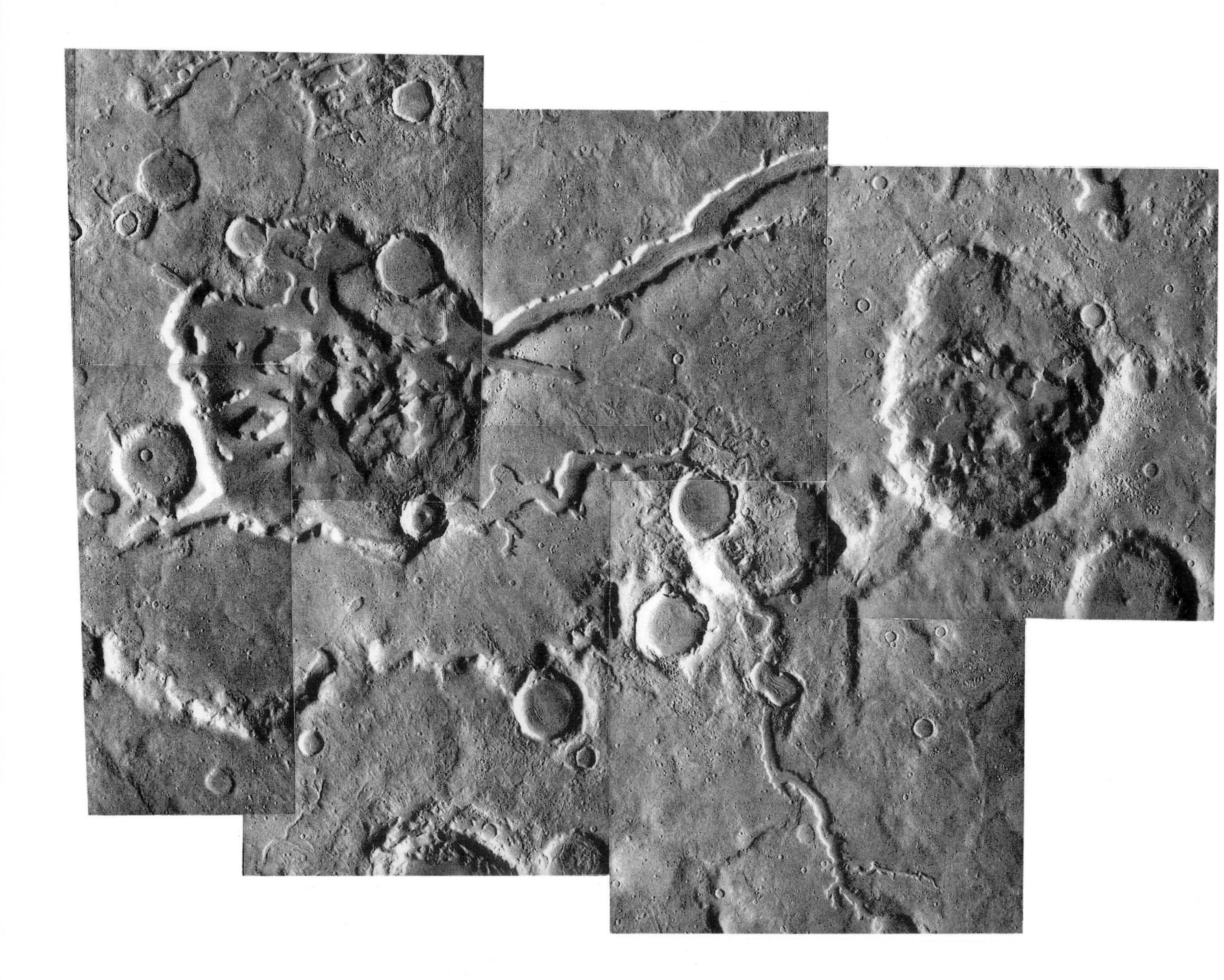

Fig. 10.11 *Fretted channel development, showing emergence from large degraded impact crater and lineated debris flows on channel floors. Protonilus Mensae region. Viking orbiter frames 268S7–11. Width of mosaic 300 km.*

The flat-floored and somewhat sinuous fretted channels often penetrate deeply into the upland hemisphere. The channels, anything up to 40 km across, frequently merge with large impact craters which usually are highly modified and degraded (Fig. 10.11). Almost everywhere talus aprons may be seen at the base of escarpments or isolated mesas. In the latter case these may extend for up to 20 km across the surrounding plains. The surfaces of such aprons are finely striated, the striations being arranged normal to the scarp walls. Where the debris is confined, as it is between channel walls, the striae are longitudinal; they diverge and converge around obstacles and, where two valleys meet, the striations merge, rather like those in modern terrestrial glaciers. The striae evidently represent flow lines. The very small number of impact craters observed, indicates that they are very young features indeed.

The observed relationships seem to imply that the cratered plateau has been – and probably still is being – eroded by a process of scarp retreat, the eroded

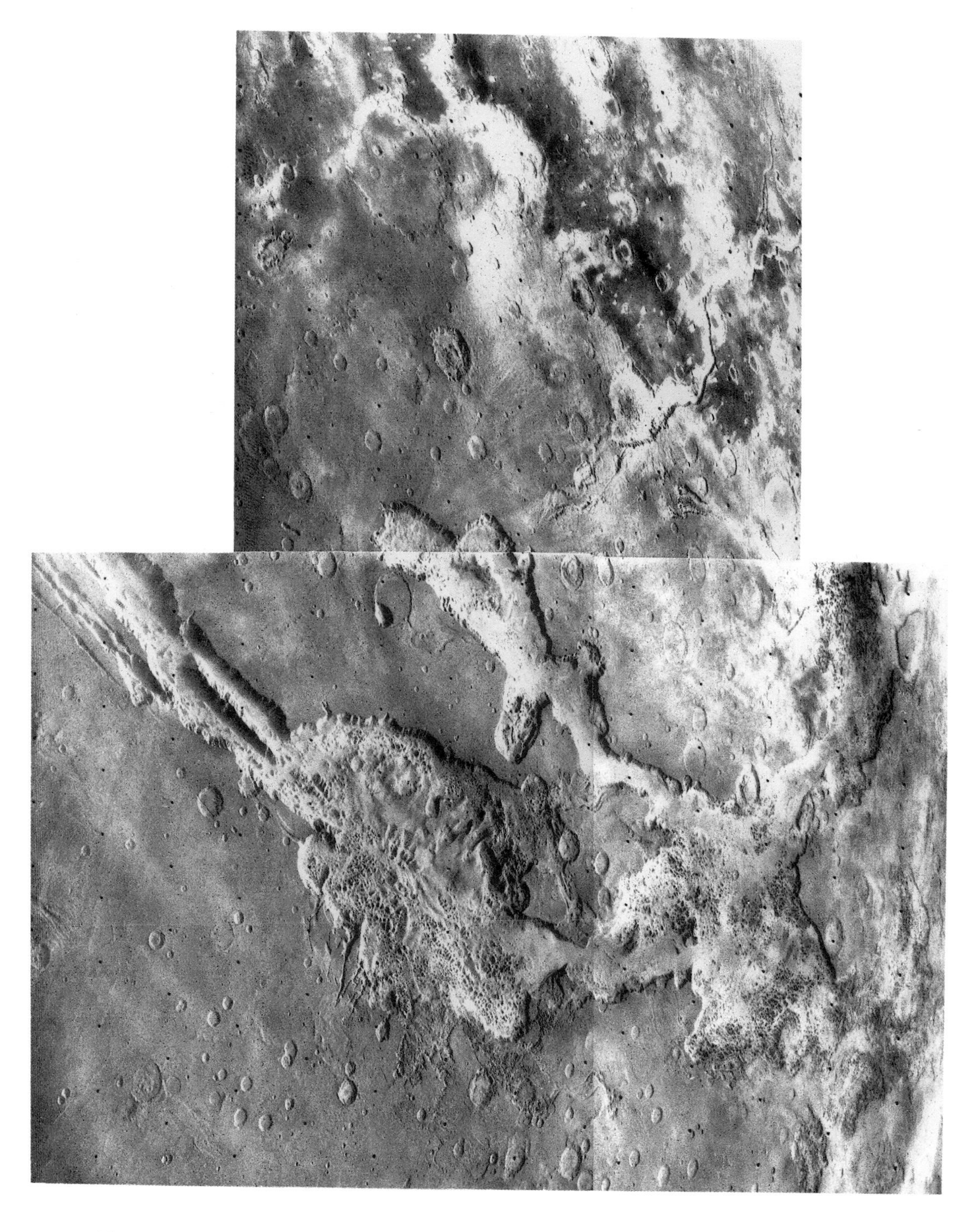

material being slowly moved towards lower ground by gravity, in other words, by mass wasting. Squyres (1978) showed that the most prominent debris flows outcropped in two 25°-wide latitude bands, centred on 40°N and 40°S. Currently, in these zones ice is precipitating out from the atmosphere during midwinter and must be accumulating on the ground. He postulates that this ice mixes with debris being eroded from the scarp faces, and forms an ice–

Fig. 10.12 *Chaotic terrain development near Hecates Tholus. Viking orbiter frames 086A32–9.*

rock mixture that shares many of the features of a terrestrial rock glacier. Downslope flow is accomplished largely by slow movement of the interstitial ice, the debris glacier slowly creeping towards the northern lowlands. The continual removal of the debris from the scarp faces allows erosion to continue, slowly and inexorably adding to the degradation of the cratered plateau.

10.5 CHAOTIC TERRAIN

Numerous references to the extensive regions of chaotic terrain have been made. The largest are located in the equatorial regions between latitudes 10°N and 20°S, and longitudes 10°W and 50°W. These regions of complex blocky remnants, valleys and fractures, may be over 100 km across and occur mainly in the equatorial zone (Fig. 10.1). The most prominent areas occur at the eastern end of Valles Marineris and along the western side of Lunae Planum (Plate 12). Other occurences are found south of Apollinaris Patera and north of Hecates Tholus, where the chaos has developed largely upon the rings of old impact basins.

Chaotic regions tend either to be rather irregular in shape or approximately circular; in the latter case their shape was evidently controlled by large impact craters. The subsided terrain is always outlined by inward-facing escarpments. Individual blocks tend to be rather angular, irregular in shape and usually preserve the plateau's upper surface, albeit in a tilted position (Fig. 10.12).

The occurrence of chaotic terrain at the upstream ends of major outflow channel systems and the beginnings of many of the equatorial canyons has already been noted and implies a genetic connection which cannot be overlooked. Most of the theories seeking to explain its formation invoke the melting of ground ice and subsequent collapse of the ground (Sharp, 1973; Masursky *et al.*, 1977). In view of the connection between chaos, the outflow channels and canyons described in the preceding sections, this seems the most likely explanation and is the one most widely favoured at present. In other words, chaotic terrain forms an integral part of the planet's complex hydrologic cycle.

THE POLAR REGIONS, WIND AND VOLATILE ACTIVITY

11

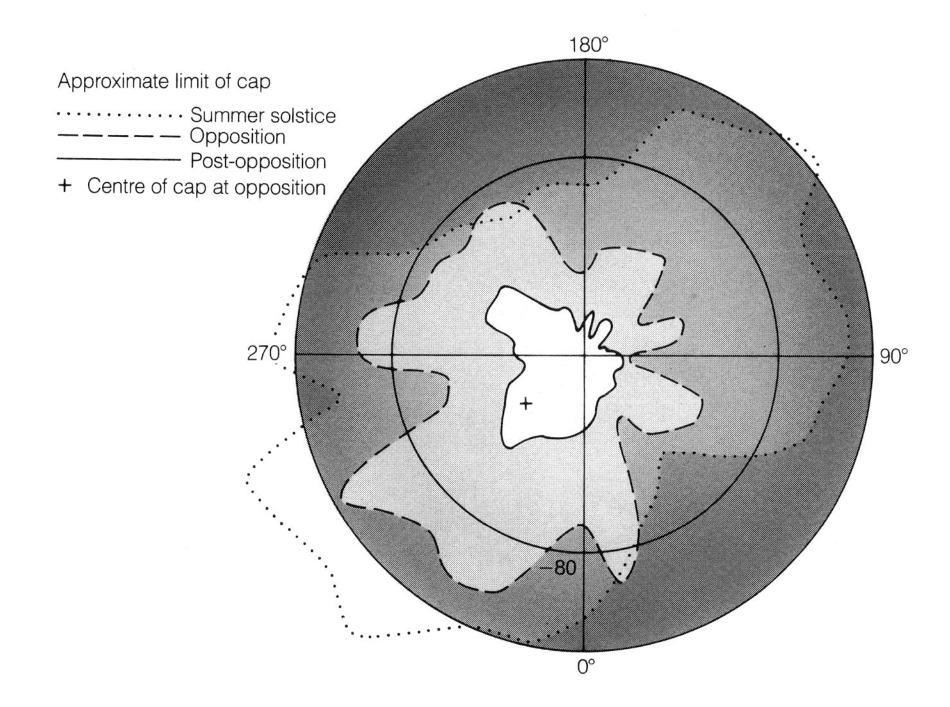

Fig. 11.1 Seasonal changes in the appearance of the northern polar cap, from Earth-based astronomical observations (BAA Memoir – apparition of 1941).

11.1 INTRODUCTION

Mars' polar caps are sufficiently large to be visible from Earth, even with quite modest telescopes. For many years prior to spacecraft exploration, observers noted seasonal changes in the size and shape of the caps, together with albedo changes in the circum-cap regions (Fig. 11.1). These are a function of the Martian seasons which, because the planet's axial inclination is closely similar to the Earth's, lead to alternate melting and refreezing of the ices present, rather as happens on our own world.

They are of particular interest for two reasons: first, they are major reservoirs of volatiles, and, second, they are regions of present-day geological activity. The cyclic movement of volatiles into and out of the caps is a part of the planet's hydrologic cycle which has become much better understood since the Viking data have been analysed.

Surrounding the polar ice, and indeed passing beneath it, are extensive layered deposits. These are incised by sinuous valleys and extend equatorwards to about latitude 80°. Circumscribing these is a broad zone of dunes which bears witness to the influence of wind (Plate 13). Great importance attaches to Martian winds since not only are they responsible for the movement of sedimentary materials, but also of volatiles which may become entrained in the dusty atmosphere. As will be seen later, there is a clear link between the seasonal cycle of carbon dioxide and the corresponding ones for dust and water.

11.2 NATURE OF THE POLAR CAPS

Like those of Earth, the Martian polar caps wax and wane with the seasons. At the present time southern summers are shorter but warmer than those in the north, while winters are longer and colder. This reflects the fact that Mars, as does the Earth, reaches perihelion during southern summer, with the result that the climate experienced by the southern high latitudes is much more extreme than that of the north. Because of this, the south cap shows the greater variation in size, at its maximum extending as far as 50°S, 15° closer to the equator than its northern counterpart.

Each cap consists of two components: a **seasonal** and a permanent or **residual** one. The seasonal caps consist largely of carbon dioxide which condenses out of the thin atmosphere during autumn and then dissipates in spring. As it condenses, clouds of carbon dioxide accumulate, hovering over the poles and obscuring the process of cap growth. By studying the seasonal variations in atmospheric pressure recorded by the Viking spacecraft, Hess *et al.* (1979) were able to estimate that the seasonal cap is only a few tens of centimetres thick.

As each hemisphere experiences its spring its cap retreats, leaving temporary outliers of ice behind. These do not last long, generally disappearing within a matter of days. Eventually a residual cap is left, this being incised by numerous deep valleys that gradually emerge from the frost cover (Fig. 11.2). Their orientation gives to each residual cap a distinctive swirl pattern. The northern residual cap is about 1000 km across but the southern shrinks to a mere 350 km.

11.2.1 Composition of the residual caps

The northern residual cap is almost certainly due to water ice. Temperatures above it have been measured at −68 °C, which is well above the frost point of CO_2 but close to the frost point of water in an atmosphere holding a small amount of precipitable H_2O. The southern residual cap is a mixture of water ice and CO_2 ice, and much lower temperatures were recorded (around −113 °C); little if any water was detected.

11.2.2 The role of wind and dust storms in cap behaviour

The differences between the two residual caps is intimately linked with the movement of dust and atmospheric activities. Dust storms are a feature of the Martian seasonal cycle, thus Viking recorded over 35 during 1977; two of these

Fig. 11.2 *Viking orbiter mosaic of the north polar region, showing the residual ice, laminated deposits and circumpolar dune fields.*

gradually developed into global events. The global storms coincide with the retreat of the southern polar cap, near perihelion. Because of the eccentricity of the planet's orbit, insolation is 40% greater then than at aphelion and leads to increased wind activity; however, it is doubtful that it could raise enough dust to generate the massive storms observed. More important than this effect is that near perihelion a large temperature gradient exists between the newly exposed circumpolar surface and the residual cap. Where considerable topographic features exist, this gradient is believed to be sufficient to generate winds fierce enough to raise large amounts of dust and inject them into the atmosphere (Kieffer and Palluconi, 1979).

Another dust-raising phenomenon is to be found in tidal circulation on the global scale. The wind strength largely depends upon the degree of atmospheric heating, which in turn is influenced by the amount of dust in the air. If the air is heavily dust laden, winds will tend to become more intense until they are sufficiently strong to raise more dust. This regenerative process commences on a local scale, when tidal winds and disturbances arising from local topography produce small-scale dust storms which are believed to be fairly widespread. Interesting in this connection is the recognition of 'dust devils' on a number of Viking images. Support for this idea also comes from the generally pinkish hue of the Martian sky as imaged by the Viking lander, suggesting that the air may be more or less permanently dust laden.

Returning to the differences between the polar caps: the northern seasonal cap grows during perihelion, when large amounts of dust are held in the atmosphere; it is likely therefore that considerable dust is entrained in the ice that is precipitated. Thus, during northern spring and summer, dirty CO_2 ice is exposed to solar radiation through a relatively transparent atmosphere, with the result that all of the CO_2 dissipates. This leaves only a water-ice residual cap. In contrast, when the southern seasonal cap forms, the air is relatively clear. The clean CO_2-ice cap is then protected from the sun's rays by a dirt-laden atmosphere, with the result that much of the CO_2 ice remains in the residual cap.

This is the present-day behaviour of the polar caps. The pattern must change with time, due to the precessional cycle; thus, 25 000 years from now, the southern pole should have the water-ice residual cap.

11.2.3 Atmospheric dust and the seasonal caps

James Pollack *et al.* (1979) have suggested that the fate of atmospheric dust may be closely linked with the formation of the seasonal caps. Their basic premise is that airborne dust particles act as nuclei for condensation of water ice, and that, as either hemisphere approaches its autumn season, the suspended particles get an additional coating of frozen carbon dioxide. This CO_2 coating ensures that the particles are heavy enough to precipitate out of the air, contributing to the growing volume of the seasonal polar caps. However, Jakosky and Martin (1987) point out that Viking IRTM data show that the temperature above the cap often exceeds the frost point of CO_2. This observation is difficult to reconcile with the ideas of Pollack *et al.*; consequently the process of polar ice deposition remains one of active research.

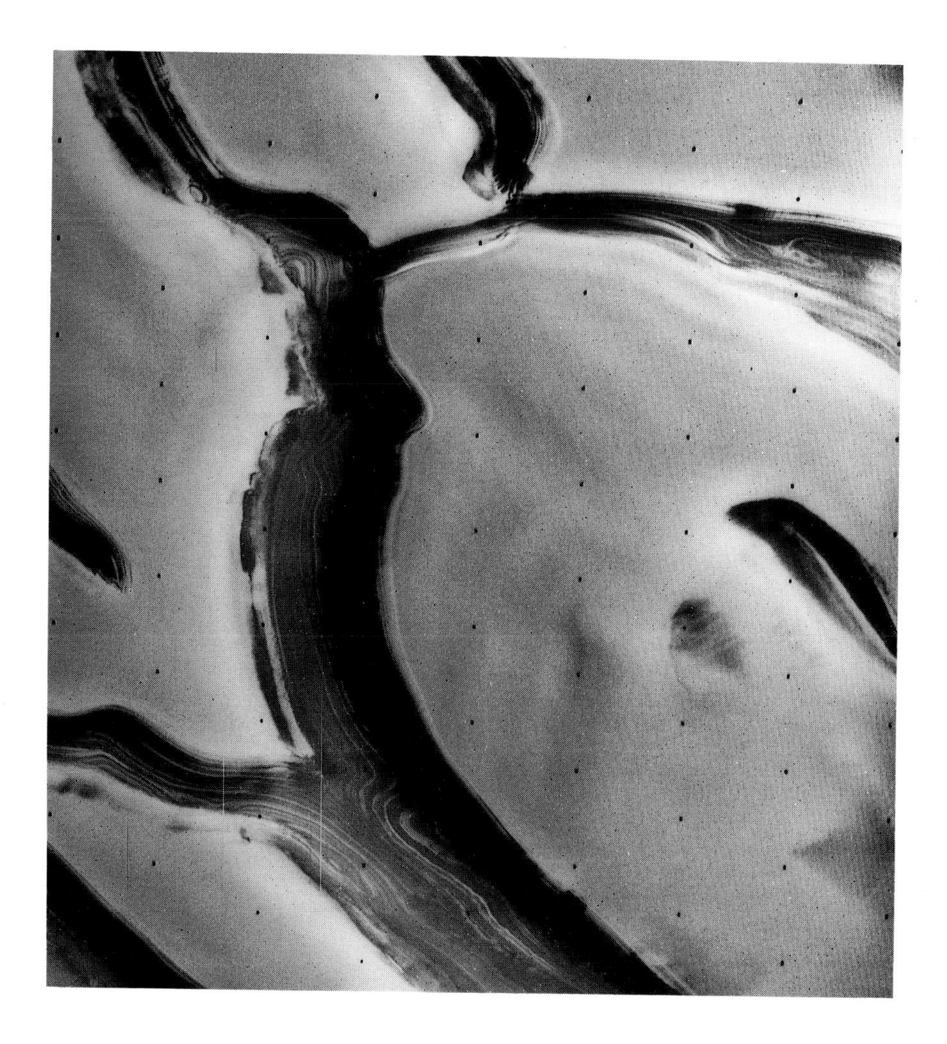

Fig. 11.3 Laminated polar deposits exposed in the walls of polar valleys. The frame is 65 km wide. Viking orbiter frame 065B74.

11.3 LAMINATED POLAR DEPOSITS

Close to the poles, a thick sequence of layered deposits obscures the older cratered plains. They extend to about the 80° latitude circle around both poles. These extensive blankets of sediment are each broken by a series of swirling valleys in whose walls are revealed a prominent light–dark stratification (Fig. 11.3). Within the valleys, whose width may be as much as 100 km, the stratification is horizontal. The surface of the layered units is everywhere smooth and lacks impact craters, and implies they must be very young. Estimates suggest this age may be around 10^8 years (Plaut *et al.*, 1988). It is likely, however, that only the visible deposits are this young, and that they obscure older yet similar deposits beneath. Interestingly, similar but older deposits have been located elsewhere on the planet, and although these now appear to be devoid of ice, Schultz and Lutz (1988) have suggested that they may be evidence for polar wandering on the planet, due to changes in Mars' moment of inertia.

Various estimates of the thickness of the sequence have been made. The general consensus is that they are between 1 and 2 km thick in the south and between 4 and 6 km thick in the north (Dzurisin and Blasius, 1975). Around the northern pole, the valleys spiral outwards in a counterclockwise direction; around the southern pole, the reverse happens. Cutting across this general trend are two very prominent and much larger valleys, one in the north (Chasma Boreale) and the other in the south (Chasma Australe).

11.3.1 Origin of the polar valleys

Considerable debate has centred on the origin of the polar valleys and escarpments. Cutts (1973) and Sharp (1973) suggested that wind was responsible, although they noted that in the north the valleys mostly swirl in the anti-Coriolis direction. In a later paper, Cutts *et al.* (1979) made the alternative suggestion that the valleys and scarps are not due to erosion at all, but to zones where there was non-deposition because of higher surface temperatures on the darker sloping terrain. Howard (1978), on the other hand, proposed that the present polar topography is in dynamical equilibrium with the climate, with volatile ablation occurring on the dark scarps and deposition on the icy flats. The ablated dust is believed to be removed by wind. The consequence of this process is a rough linearity and parallelism of escarpments, probably due to scarp retreat on a regional slope. The spiralling pattern of valleys is, according to Howard, due to the more rapid retreat of escarpments facing slightly west of the equatorward meridian, that is, in the direction of greatest solar and atmospheric warming. Should this latter idea prove to be true, it seems to imply that the layered deposits are mostly composed of water ice, probably with relatively small amounts of volcanic ash and dust.

11.3.2 Composition of the layered deposits

The polar deposits are believed to be accumulations of ice and dust, the layering possibly being due to to variations in the rate of accumulation of the two components. The preferential accumulation of such deposits at the north pole may reflect the fact that the atmosphere is dust laden when the northern seasonal cap forms, the carbon dioxide 'scavenging' dust from the atmosphere as it condenses. As the precessional cycle takes its course, presumably the southern pole eventually will become the site of preferred dust and ice deposition, and so on.

11.4 HIGH-LATITUDE PLAINS AND DUNE FIELDS

Surrounding the laminated deposits and passing beneath them above latitude 80°, are sparsely cratered plains. In the southern hemisphere these plains are etched by pits which are presumed to have been hollowed out by the wind; they also exhibit aeolian fretting. Dunes, too, are found but are far less extensive than those around the northern pole. The dunes which do occur tend to be confined to the interiors of impact craters or within the vicinity of escarpments (Fig. 11.4). That considerable movement of fine debris is currently ongoing, is revealed by accumulations of wind-etched material that may almost completely bury 40 km craters (Fig. 11.5).

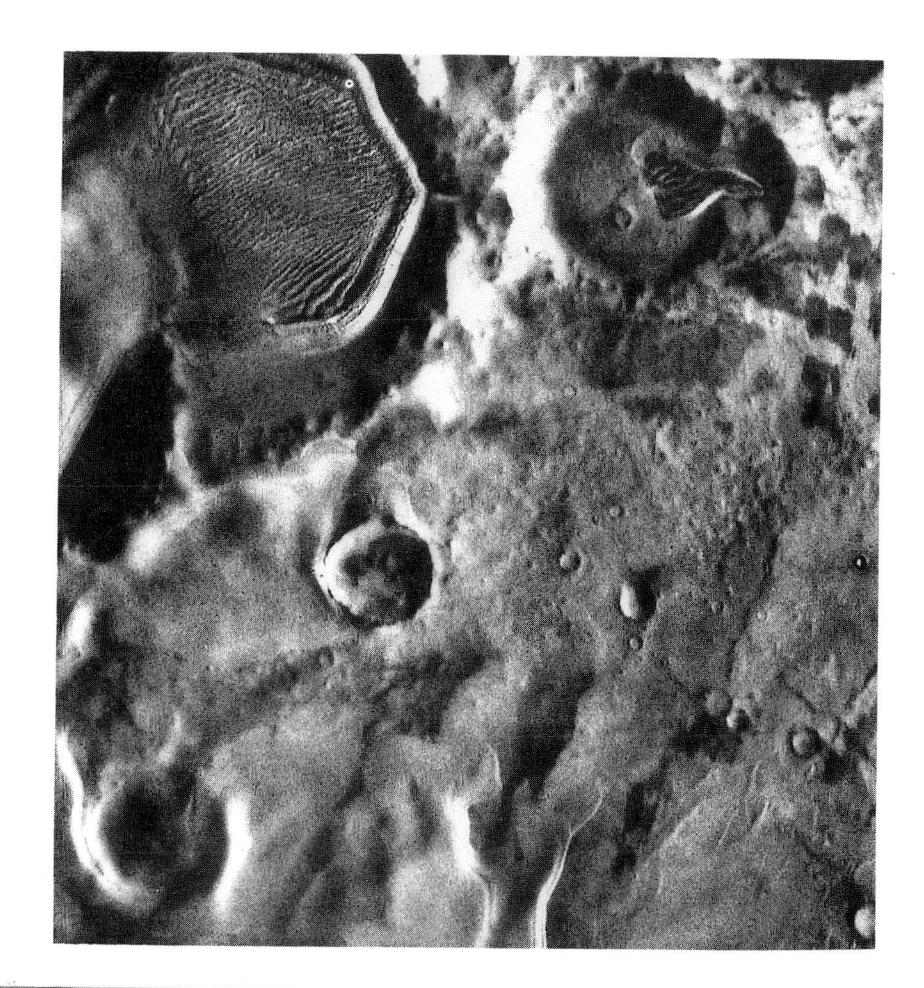

Fig. 11.4 High-latitude south polar plains, showing 50 km diameter impact crater with interior dune field. Adjacent plains show impact craters, windetching and albedo markings. Frame width 150 km. Viking orbiter frame 479B71.

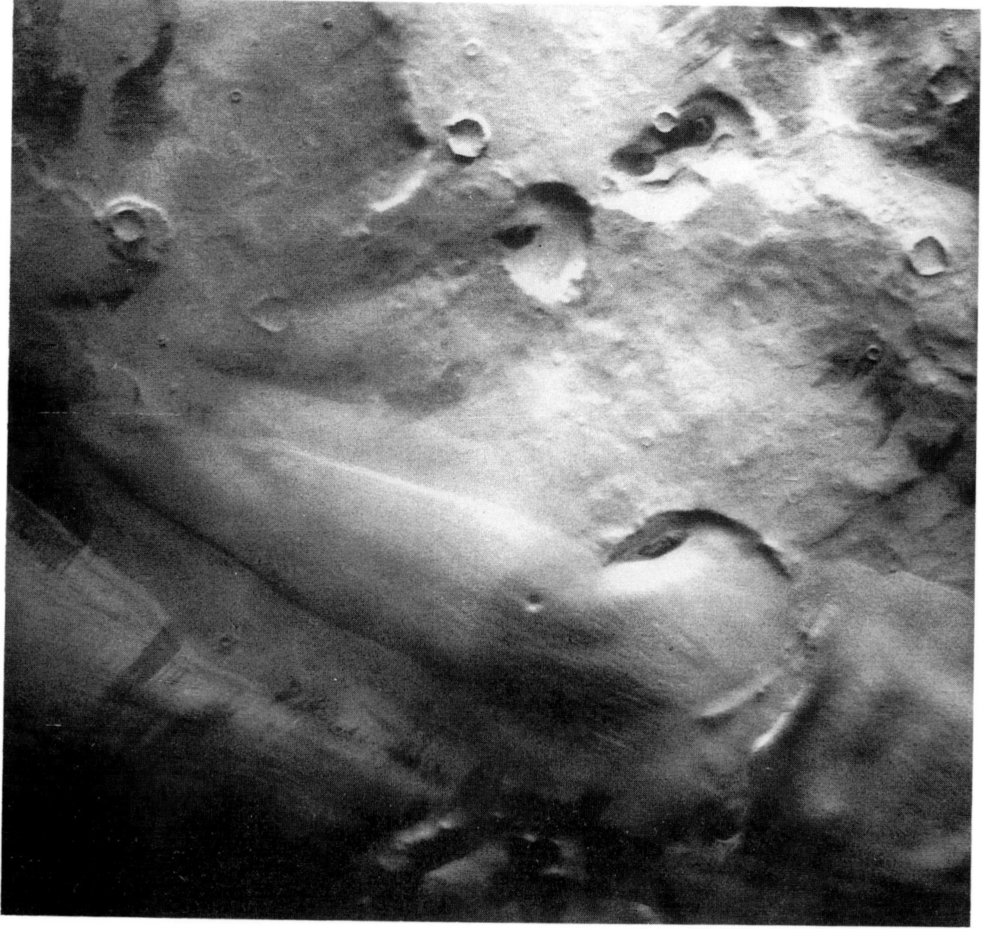

Fig. 11.5 Accumulation of fine-grained sediment in high-latitude south polar plains. The debris almost completely buries a 40 km impact crater. Frame width 175 km. Viking orbiter 479B57.

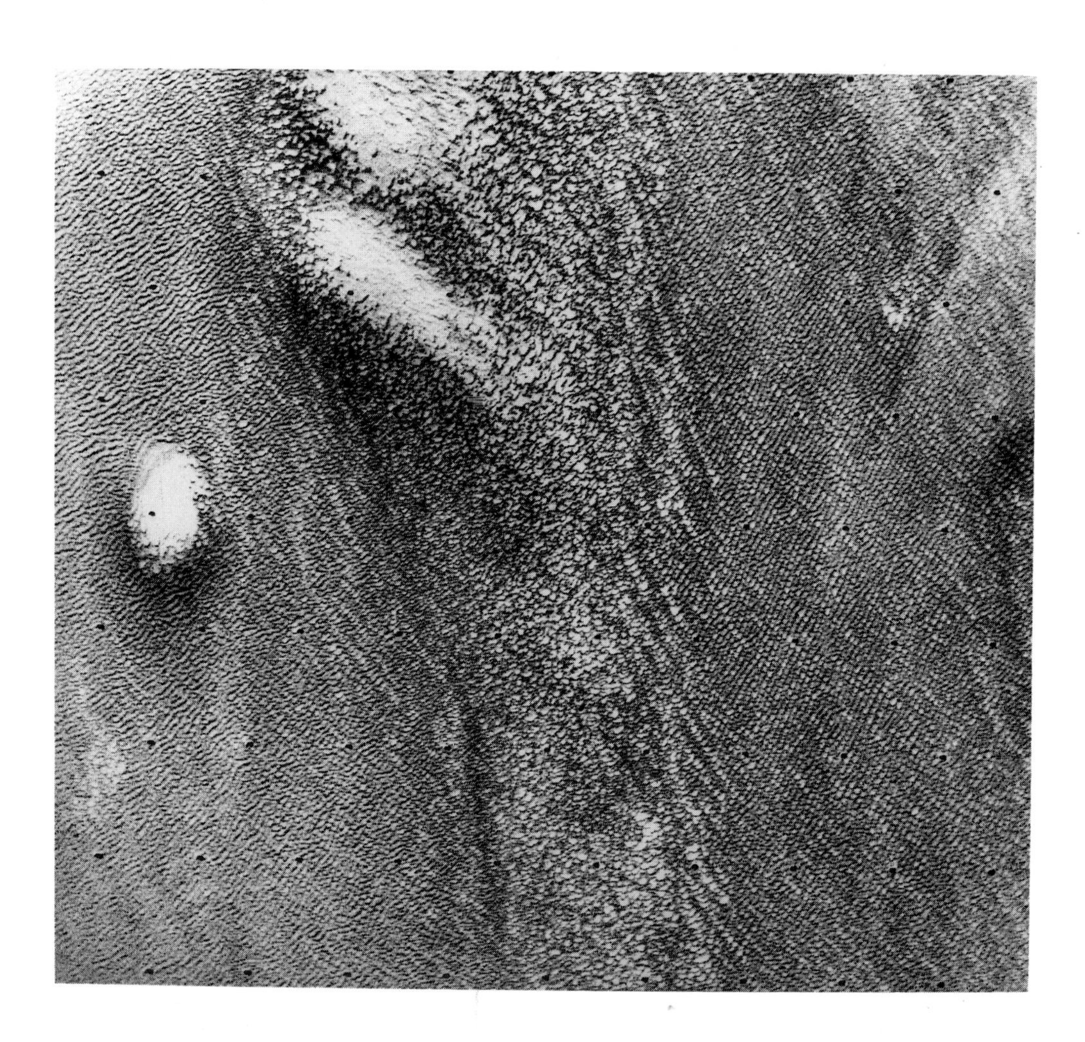

Fig. 11.6 *North polar dunefields at latitude 80°N. Note the predominance of longitudinal dunes. Frame width 114 km. Viking orbiter frame 060B03.*

Aeolian activity in the northern hemisphere has created vast circumpolar dune fields (Fig. 11.6). This northern dune collar is sometimes several hundred kilometres in width and covers an area of approximately 10^6 km^2 (Breed *et al.*, 1979). The predominant landforms within the northern collar are longitudinal dunes, although crescentic barchans tend to occur towards the edges of major dune fields or near escarpments. Transverse types are generally absent. On Earth the latter are typical of depositional sinks, while longitudinal dunes generally occur where sand is being actively moved through a region. Breed and colleagues (1979) measured the mean dune length as 0.34 km, mean width, 0.54 km and mean crest-to-crest distance, 0.55 km.

Since most terrestrial dunes are composed of quartz or, more rarely, calcite sand, it is interesting to speculate about where the material for the Martian polar dunes came from. Particulate material sampled at the Viking lander sites was in the micrometre (clay-sized) rather than the millimetre (sand-sized) grain size and did not appear to be composed of quartz. Indeed, it is difficult to envisage much quartz existing on the planet at all, since continental-type silicate rocks appear to be absent from Mars. On Earth, clay-sized particles do not normally form dunes; thus there is something of a problem here. There is, of course, the possibility that the considerable volume of potential fluvial material – from the large flood channels – which drained towards the northern

Fig. 11.7 Polygonally patterned ground at latitude 78°N, Vastitas Borealis. Frame width 65 km. Viking orbiter frame 070B08.

polar regions, may have been a major sand source, as are most terrestrial fluvial systems. However, at the present time, there is major uncertainty concerning this issue.

Another characteristic of plains at these high latitudes is a development of polygonal markings. The polygons, anything between 2 and 18 km in diameter, are picked out by low-albedo troughs separating rather rounded higher albedo mounds (Fig. 11.7). Helfenstein and Mouginis-Mark (1980) suggest that these are ice-wedge polygons. On Earth, these form where ice shrinks upon cooling in annual cycles; on Mars such cycles are significantly longer than on Earth (10^5–10^6 a), which may help to explain why they are so much larger than their potential terrestrial counterparts.

11.5 WIND ACTIVITY ON MARS

Wind erosion and transport is one of the few geological activities which currently can be observed on Mars. Meteorological instruments aboard the Viking landers on the plains of Chryse and Utopia measured wind speed and direction over more than one Martian year. Measurements were made at a height of 1.3 m above the ground surface and showed that nocturnal wind velocities were around $2\,m\,s^{-1}$, but that wind speed increased near sunrise, to reach speeds of around $7\,m\,s^{-1}$; occasional gusts registered speeds of $25\,m\,s^{-1}$. At sunset each day, wind direction changed from east to southwest then, at sunrise, swung towards the south-southwest.

Wind is very effective at moving fine-grained sedimentary material across the Martian surface, but there is scant evidence to suggest that it is an effective eroder of consolidated rocks. For instance, images of the ancient Hesperian-age rocks of the Chryse basin reveal perfectly preserved lava flows and crater details. On this basis, Arvidson (1979) extimates that erosion rates cannot be more than around $10^{-3}\,\mu m\,a^{-1}$.

11.5.1 Aeolian erosional landforms

Some of the best-developed wind-erosional features are developed in southern Amazonis, amongst the supposed pyroclastic plains deposits decribed in Section 8.4. Here, in a series of rather readily erodable strata, the wind has etched families of yardangs – elongate, streamlined ridges – which lie parallel to the dominant wind direction (Figs. 8.7 and 8.8). Similar landforms are widespread on the Earth and develop where abrasion and deflation combine together in major deserts (Ward, 1979).

In this area, too, are developed numerous deflation hollows (Fig. 11.8), fluted scarps and **pedestal craters**. The latter are common in latitudes higher than about 40° and represent a kind of crater modification produced by aeolian stripping (deflation). At these higher latitudes, impact craters often are surrounded by a relatively broad (2–6 crater radii wide) pedestal (Fig. 11.9). Each pedestal gives the impression of being a rather broad ejecta blanket, in which case it is considerably wider than blankets found around craters in the equatorial regions, where widths of 1–2 crater radii are the norm. Since there is no logical reason why such a discrepancy should occur, some other explanation for the pedestals is required.

Arvidson and his colleagues (1976) have suggested that in high latitudes, aeolian dust once blanketed the surface over very large regions. Subsequently, much of this has been removed by deflation, but around impact craters the surface was protected by blocky ejecta which retarded the rate of erosion locally, compared with that in the unprotected surrounding plains. The remnant pedestals are viewed, therefore, not as original ejecta blankets, but as residual landforms which may or may not share the extent of the original blanket of ejecta. Because there is a tendency for the pedestal edges to have a convex-upward profile, where the degree of surface stripping is great it gives the crater and its pedestal the misleading appearance of a volcanic dome. This had led in the past to considerable confusion with regard to the distribution of volcanic landforms on the planet, particulary when only low-resolution images were available for study.

11.5.2 Martian winds and wind streaks

Particulate material is moved either by being suspended in the atmosphere, by saltation (bouncing across the surface), or by creep (gradual slow movement induced by saltation). On Earth, the optimum particle size for movement by wind is 0.08 mm. Wind-tunnel experiments conducted at Arizona State University by Ron Greeley and his team (Greeley *et al.*, 1980), have shown that this figure is nearer 0.1 mm for Mars. Threshold wind velocities required to move particles are, however, somewhat higher than on Earth, where the figure is around $0.2\,m\,s^{-1}$, mainly because of the thinness of the Martian air. Thus, for an atmospheric pressure of 10 mbar the speed required to lift a 0.1 mm clast is $2.4\,m\,s^{-1}$, for a 5 mbar pressure, it is $4\,m\,s^{-1}$.

These figures are those that apply right on the surface; at the height of the Viking instrument boom, calculations suggest they should be approximately

Fig. 11.8 *Deflation hollows in plains units, Memnonia. Viking orbiter frame 764A14. Frame width 15 km. Centred at 7.12°S, 187.53°W.*

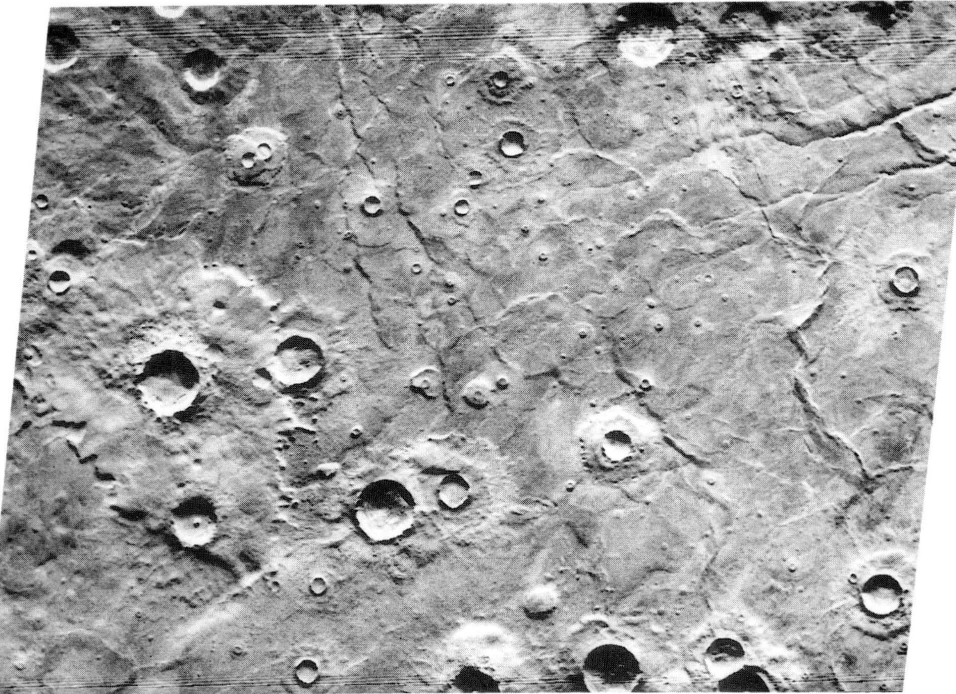

Fig. 11.9 *Pedestal craters developed in ridged plains south of Noachis. Frame width 365 km. Viking orbiter frame 361S16.*

Fig. 11.10 *Wind streaks on the surface of Syrtis Major Planum. Note the relationship with small impact craters. The graben at top right form Nili Fossae and the channel at top left, north of the prominent large crater Antoniadi, are Huo Hsing and Aquakuh Valles. Closely south of the centre of the mosaic may be seen the circular volcanic caldera described in Chapter 8. Viking orbiter mosaic 211–5601.*

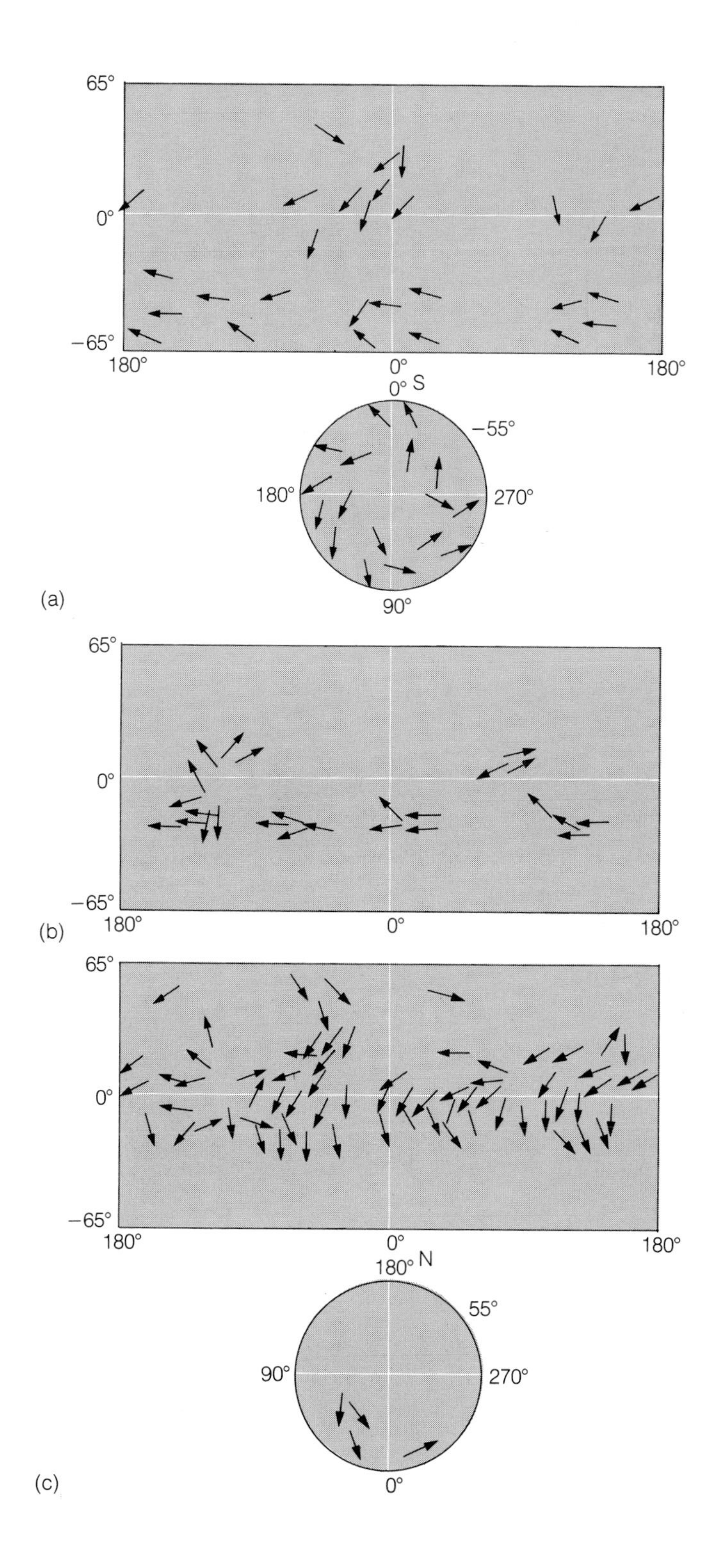

Fig. 11.11 Global distribution of wind streak directions from Viking post-dust storm data (after Thomas and Veverka, 1979): (a) dark splotch-associated streaks; (b) dark erosional streaks; (c) bright streaks.

one half of this. It is quite simple, therefore, to see why little particulate material was seen in the air around the lander sites, even during the stronger gusts. However, threshold velocities would be reached in the source areas of the major dust storms; indeed, it appears that only during such activity can large volumes of particulate matter be transported by the atmosphere. Martian dust storms currently appear to be the major redistributors of fine-grained sediment over the planet's surface.

Returning to the matter of aeolian landforms – there are widespread temporary albedo features that cover quite extensive parts of Mars. Particularly good examples of these outcrop in Syria Planum, where sharply defined albedo differences – which show little correlation with topography – can be traced over large distances. The most common of the albedo markings are **splotches** and **wind streaks**, however, and these do tend to have a relationship with local relief. Streaks are widespread on the surface of Syrtis Major (Fig. 11.10).

Three types have been noted (Thomas and Veverka, 1979; Sagan *et al.*, 1973; Greeley *et al.*, 1978, Veverka *et al.*, 1981) and each is believed to be the result of the movement of fine-grained particulate material across the surface. Bright streaks are depositional and show a preferential distribution in the lee of obstacles such as craters and knobs; dark streaks are erosional in origin and seem to represent areas where the thin veneer of light dust has been removed; dark splotches generally occur within impact craters and probably represent deposits which have been deflated from interior dunes. The dark streaks tend to be less stable than the light ones.

By mapping the planet-wide orientation and distribution of such features both during and after the 1971 and 1977 dust storms, it has been possible to gain some insights into the wind flow over the planet (Fig. 11.11). Thus, between the equator and 30°N, the wind flow is from the northeast, while south of the equator, winds tend to be more northerly and swing around between latitudes 20° and 30°S, to a more northwesterly direction. At higher latitudes, the winds tend mostly to blow from east to west.

Martian winds seem to be much better transporters of fine sediment than they do efficient eroders of coherent rocks. Few prominent erosional aeolian landforms are seen in low- or medium-resolution images. However, it should be noted that where higher-resolution imagery is available, many surfaces which appear unaffected by wind in lower-resolution frames, are seen to exhibit etched and pitted surfaces in the 10–20 m scale range. Our current inventory of aeolian landforms may, in consequence, be incomplete due to a lack of planet-wide high-resolution imagery.

THE GEOLOGICAL HISTORY OF MARS

12

12.1 INTRODUCTION

In the previous chapters I have attempted to describe most of the principal features of the Red Planet and some of the explanations which have been proffered by the planetary community to account for them. While it should be clear by now that there is still considerable room for improvement in our knowledge of Mars, and a need to continue Mars exploration, the surface mapping so far completed allows for an arrangement of the wide variety of Martian rock units into a temporal sequence, using the traditional methods of superposition/transection relations and impact crater age dating. There is still no absolute age data for Mars, therefore any attempts to attach an age handle must still be considered approximations. Despite this shortcoming, we can note confidently several major stages in the evolution of the planet, each being characterized by distinct events and processes which may or may not have overlapped in time.

1. Early Mars history includes events up to the Early/Late Hesperian boundary which share the common characteristic of being mainly global events. They include:
   (a) accretion and primary differentiation;
   (b) bombardment of the primordial crust by cosmic particles and production of major impact basins;
   (c) formation of the lowlands in the northern hemisphere.
   Subsequently there was:
   (d) uplift of the lithosphere in the Tharsis region, with the generation of fractures, particularly around Syria Planum;
   (e) production of extensive plains by flood-style volcanism (inter-crater plains formation) and widespread fluvial activity (runoff channel formation), together with phreatomagmatic and other eruptions at highland paterae.
   The final stage in this early phase saw:
   (f) continued uplift, fracturing and volcanism in the Tharsis region, particularly at Alba Patera and Tempe Terra;
   (g) instigation of faulting leading to formation of the Valles Marineris canyon system; and
   (h) formation of ridged plains with Lunae Planum ages and commencement of fretted channel development.
2. Late Mars history (Late Hesperian upwards) was marked by more localized geological activity, and commenced with:

(a) extensive volcanism in Tharsis, Elysium and Syria Planum, and catastrophic flooding to produce outflow channels which drained towards the northern lowlands; there was also a waning of tectonism in the Tharsis region.
(b) Later there were continued local volcanic eruptions in the Tharsis region and extensive volcanic plains formation in the lowland hemisphere; surface modification was accomplished mainly by aeolian processes.
(c) Later again, volcanism continued in Amazonis, Tharsis Montes and Olympus Mons, while mass wasting affected the uplands along the line of dichotomy.
(d) Most recently there was more central volcanism in the Tharsis region, while widespread flooding deposited sediments in the northern plains. Continued resurfacing of the northern lowlands and redistribution of polar sedimentary deposits has occurred up until the present time.

12.2 THE TECTONIC HISTORY OF MARS

Thus far, little has been said of the tectonic features of Mars. Now is the time to describe the extensive faulting which has been a feature of the planet throughout its evolution, and say more about the widespread plains ridges.

The faults which are such a characteristic feature of Mars are predominantly graben – extensional faults produced by fracturing of the relatively brittle crust. Individual graben vary between 1 and 5 km wide and may extend, usually as *en echelon* families, for thousands of kilometres. In ancient rock units, several sets of fractures, each of a different age and (sometimes) orientation, break up the surface into complex blocks (Fig. 12.1). The more spectacular of the fracture families are related to major arches in the lithosphere which, in the case of those associated with Tharsis, spread across one-third of the planet.

The extensive ridge systems, such as those which characterize the ridged plains of Noachian and Hesperian age, also are believed to have a tectonic (compressional) origin and to have formed in rocks whose ages range from the Late Noachian to Lower Amazonian. Many geologists consider it likely that they formed preferentially in volcanic flows, as they did on the Moon. They are most readily mapped on the smooth plains of regions like the Chryse Basin, Lunae Planum and Syrtis Major Planitia, but are widespread, albeit less readily observable, elsewhere, including the northern lowlands. Studies by Chicarro (1989) have shown that the first recorded period of major ridge formation was during the Upper Noachian and that at this time, ridges with over 1 km amplitude formed. The most prominent ridges are, however, younger, and are represented by those which outcrop on Lunae Planum; these are of Lower Hesperian age – indeed, they are of the same age as most of the better-developed ridged plains units in the upland hemisphere. This 'Lunae Planum age' is widely used as a basis for comparing crater ages between different rock units on Mars. The development of such ridges appears in some regions to have been related more to localized tectonics than to a global compressive regime; thus old impact basins and major bulges may have dictated ridge orientations at this time. However, mapping by Lucchitta and Klockenbrink (1981), has shown that many ridges are concentric about a focus at 10°N, 112°W, that is, near to the centre of Tharsis (Fig. 12.2). Ridges also formed in the Lower Amazonian but are less prominent than those which were emplaced

Fig. 12.1 Intersecting fractures in Tempe Fossae. Viking orbiter frame 627A36. Centred at 35.38°N, 83.17°W. Frame width 58.5 km.

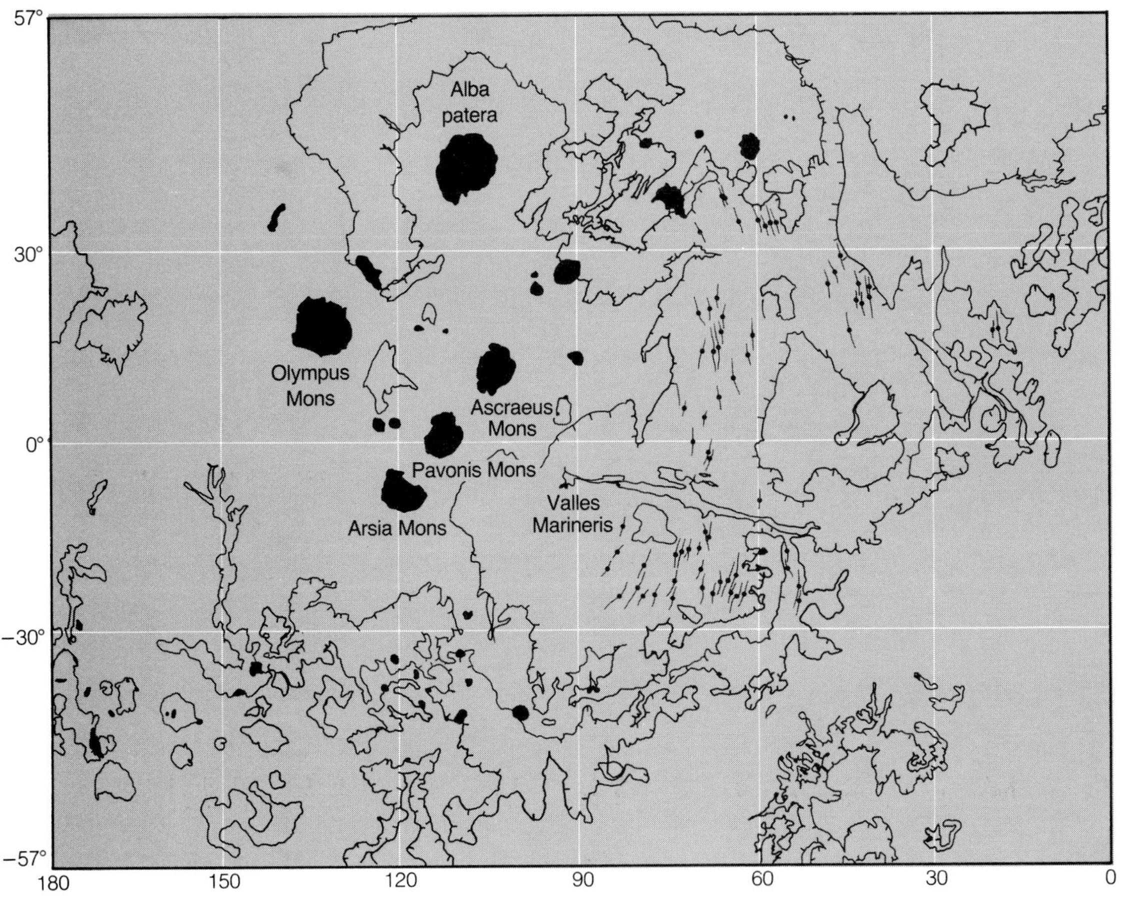

Fig. 12.2 Orientation of ridges in Hesperian-age units surrounding Tharsis. These appear to be roughly concentric about a focus at 10°N, 112°W (after Scott and Dohm, 1989).

in the preceding epoch. Crater counting suggests that ridge formation spans the period 3.85 Ga to about 1 Ga BP.

12.2.1 Faults and ridge systems in the Tharsis region

One way of seeking to comprehend the tectonic history is to map faults and ridges that outcrop within geological units of established age. One of the earliest structural analyses was that of Frey (1979). He noted that structures in the vicinity of Valles Marineris and Echus Chasma could not be explained by a single-stage uplift of the Tharsis region and, after studying fracturing in the south Tharsis region, he concluded that there had been at least two major upwarping events, the earlier having been centred on the ancient plateau of Thaumasia (Fig. 12.3a). The generally north-south structural pattern was later overprinted by regional Tharsis-related faults, particularly the west-east Valles Marineris system. Frey envisaged the Thaumasia region as an older Tharsis-type crustal dome.

In the early 1980s a further analysis was completed by Plescia and Saunders (1982), who defined four centres of tectonic activity (Fig. 12.3), the earliest being that defined earlier by Frey. The three subsequent foci were (i) northern Syria Planum (8°S, 100°W) and (ii) and (iii), two centres near Pavonis Mons.

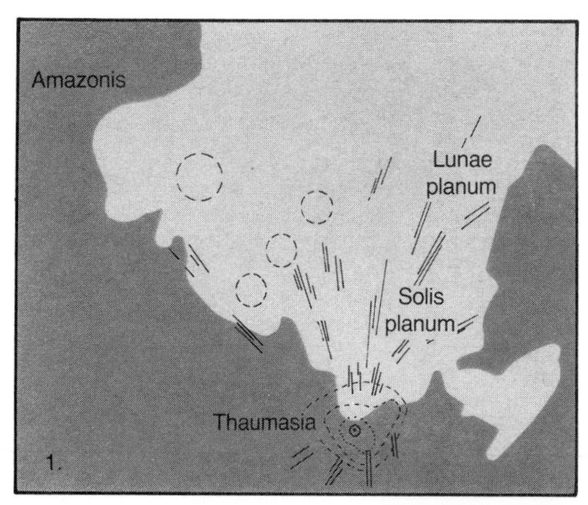

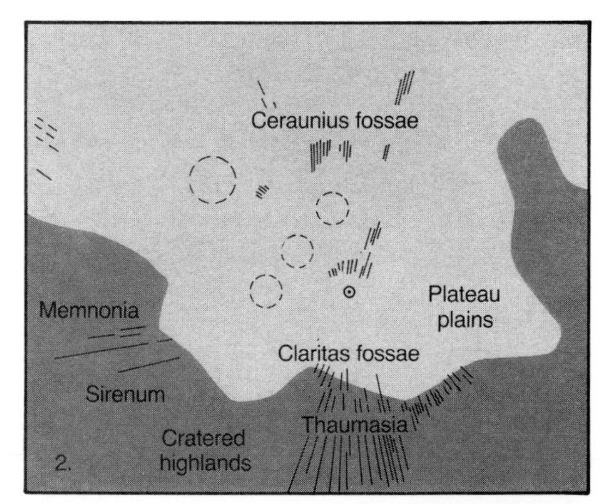

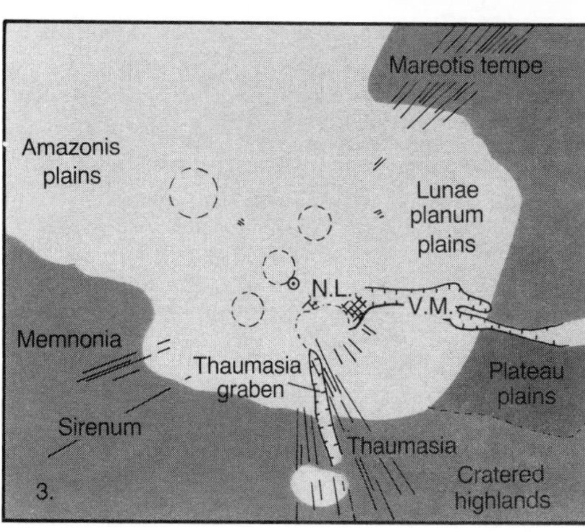

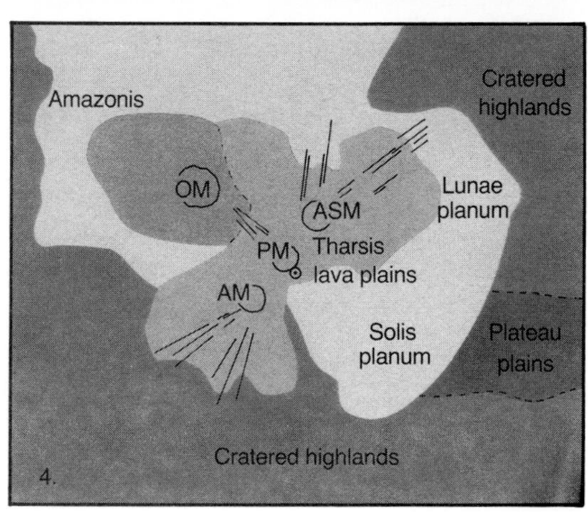

Fig. 12.3 *Centres of tectonic activity in the Tharsis region (after Frey, 1979 and Plescia and Saunders, 1982).*

The second of the Pavonis centres was characterized by fracturing which post-dated growth of the large Tharsis shield volcanoes. Their analysis clearly illustrated that each episode of tectonism was associated with a topographically elevated region and that it predated associated volcanism. This implies that fracturing was related to an early stage of the tectonism associated with the development of topographical highs, rather than to loading by extrusive volcanic deposits.

Since their work was published, considerable geological mapping has been undertaken and three new geological maps of Mars have been produced by the USGS. Structural analysis of the western hemisphere of Mars has very recently been undertaken by Scott and Dohm (1990), who produced three maps, one each for the Noachian, Hesperian and Amazonian periods. They conclude that in Noachian times (Fig. 12.4) the majority of faulting was associated with three foci: the Tharsis Montes with a prominent NE-SW axial trend, the Syria Planum rise (centred at 15°S, 105°W) and the Acheron Fossae structure which is located north of Olympus Mons. Ridges are widespread but it is not clear whether they have a Noachian age or were emplaced later, during Hesperian times. During the Hesperian period, widespread fracturing continued along the Tharsis Montes axis and in Syria Planum; it was also initiated with a radial pattern at the large volcanic centre of Alba Patera (Fig. 12.5). It did not

Fig. 12.4 *Tectonic features of Noachian times (after Scott and Dohm, 1989).*

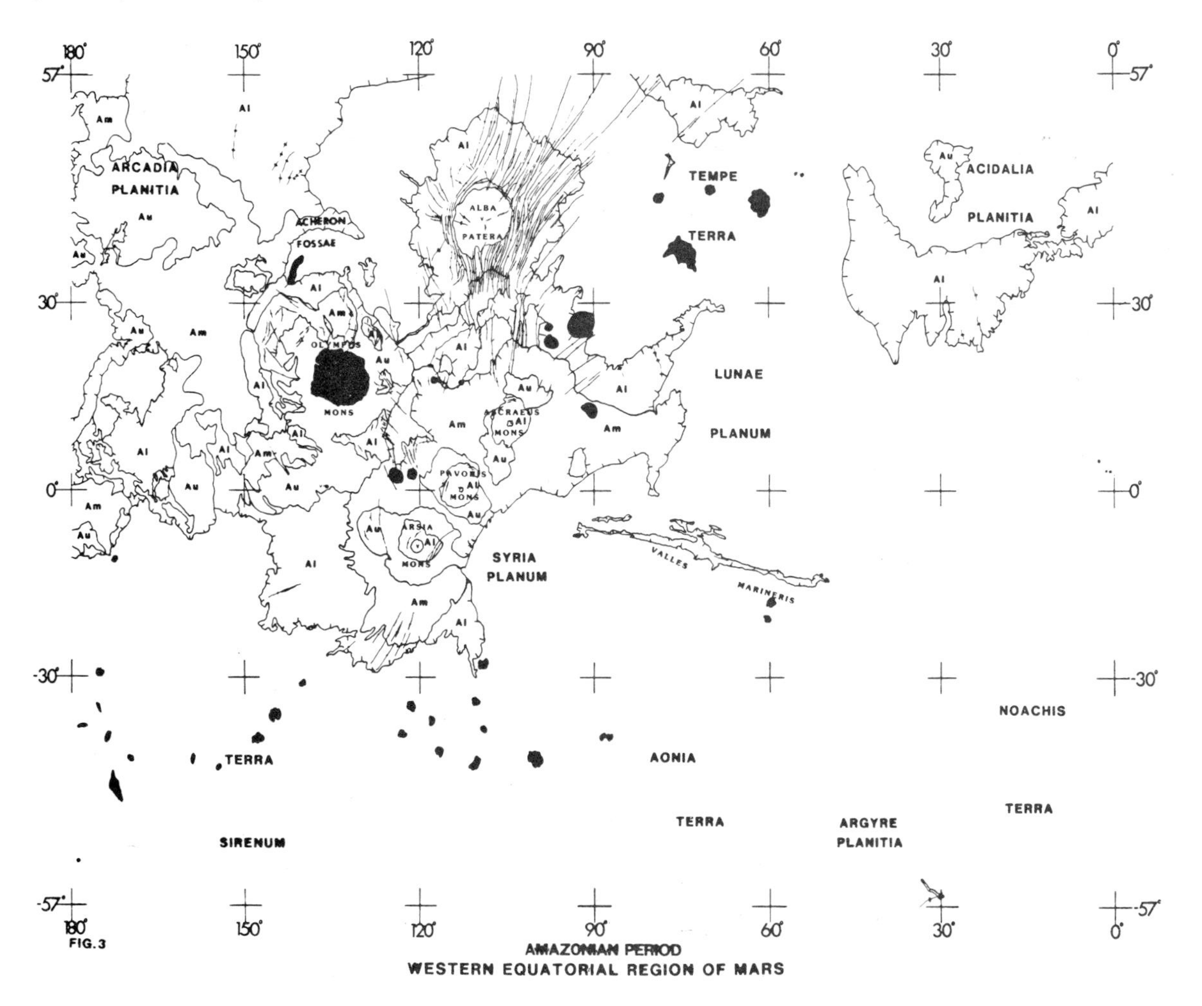

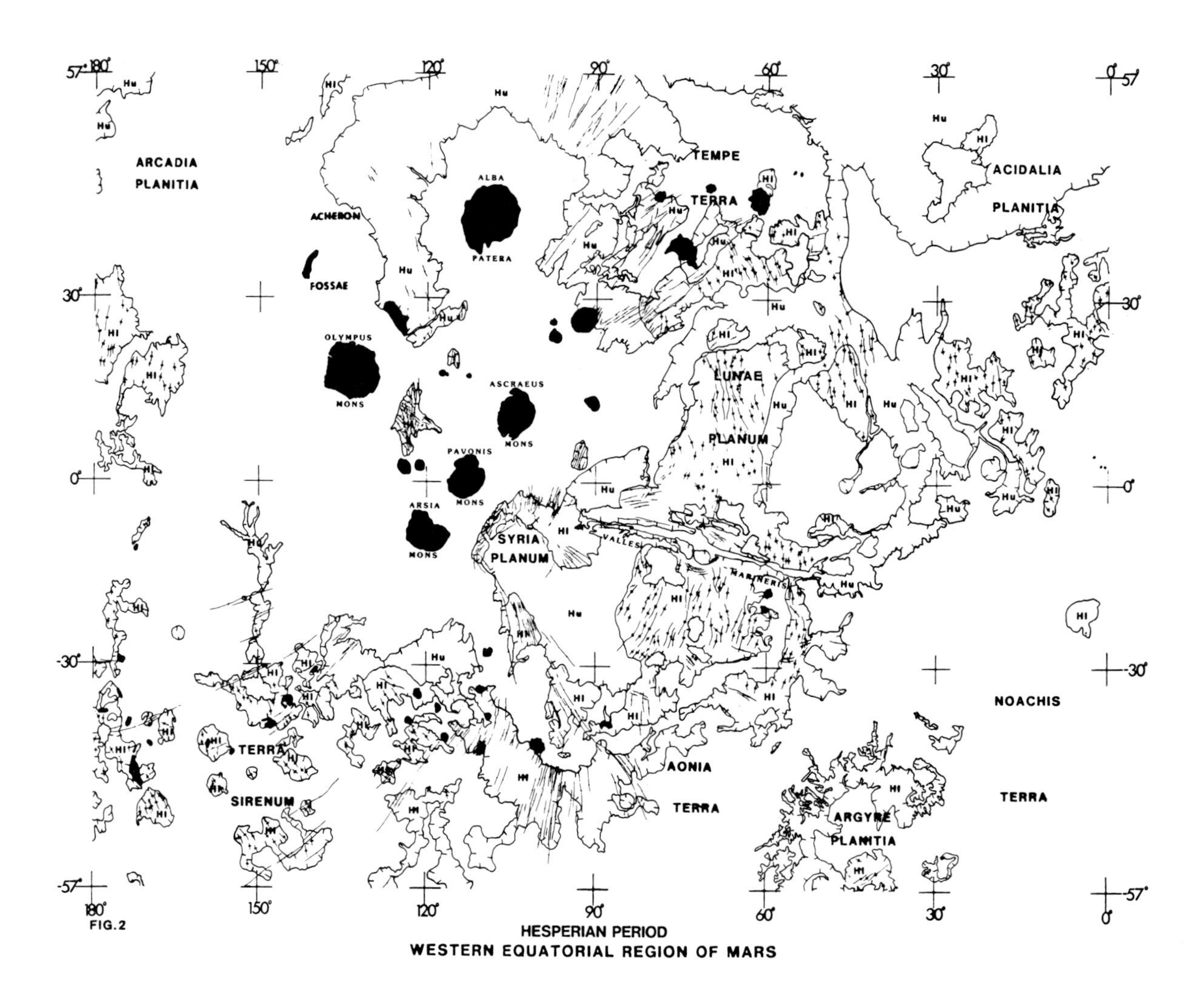

Fig. 12.5 *Tectonic features of Hesperian times (after Scott and Dohm, 1989).*

continue, however, at Acheron Fossae. Ridges, most of which predate the faults of this age, and which outcrop in an elliptical pattern surrounding major volcanic centres, ceased to form in the Hesperian. There was then a general decline in tectonism during the Amazonian period. However at Alba Patera some older radial fractures were rejuvenated, while concentric faulting was initiated around the summit (Fig. 12.6). Minor faulting also was associated with Olympus Mons and its aureole, some of which postdated several late Amazonian volcanic flows.

12.2.2 Faulting and ridge systems of Elysium

Faulting in the Elysium region was less intense than that experienced in Tharsis. The major episode of fracturing occurred during the Early Amazonian, when Elysium Fossae were produced (Mouginis-Mark *et al.*, 1984). The majority of graben have a pronounced WNW-ESE trend (Fig. 12.7). Subsequently, in the late Amazonian epoch, further fracturing occurred in southern Elysium Planitia, producing long, curving WNW-ESE fractures such as Cerberus Rupes. Curving ridges are most prominent in the east of the region, in southern Arcadia; the youngest are of early Amazonian age.

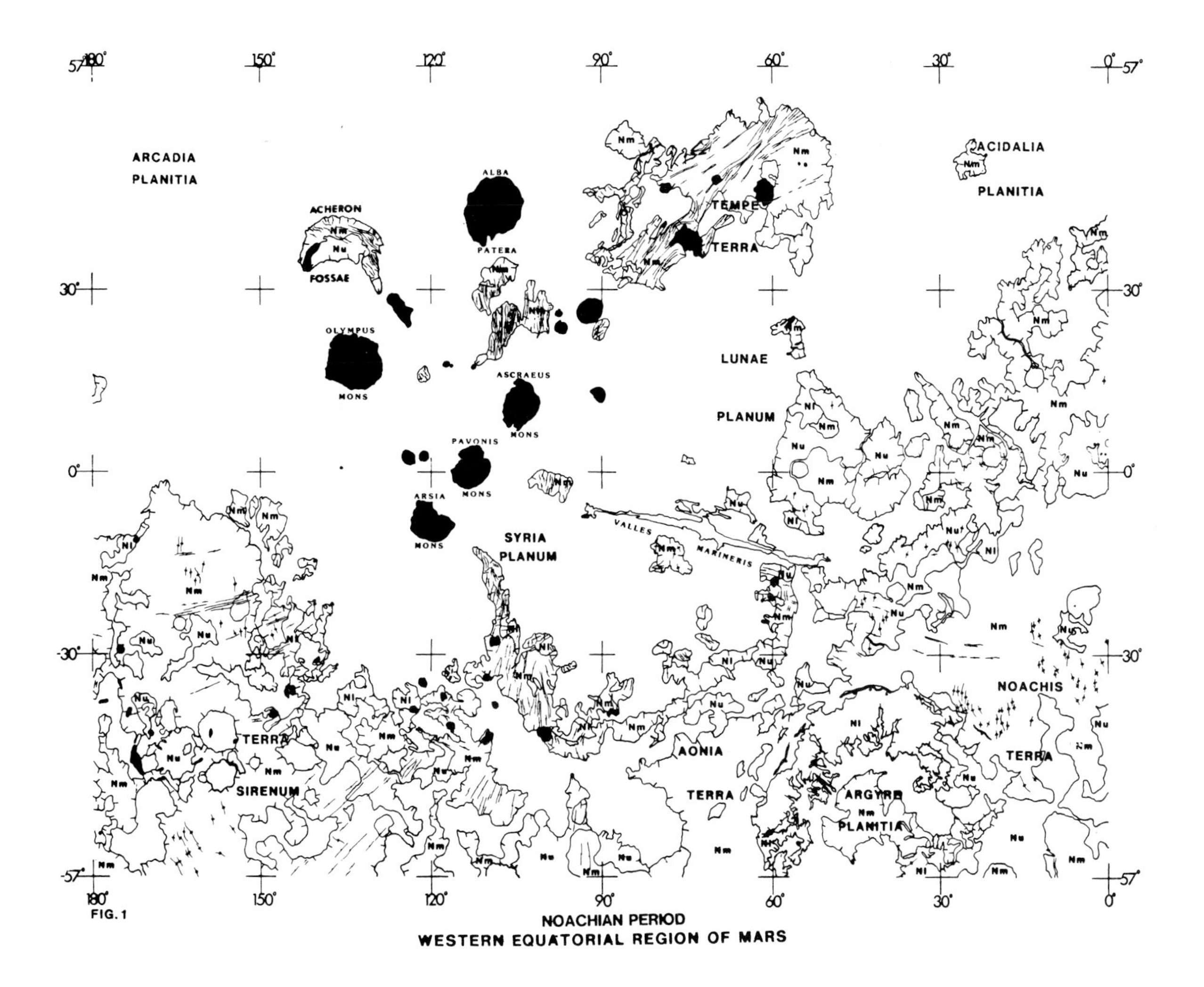

Fig. 12.6 Tectonic features of Amazonian times (after Scott and Dohm, 1989).

12.2.3 *Volcanotectonic provinces in the Tharsis region*

Tectonism and volcanism quite clearly have been connected intimately in space and time. Therefore, it should be instructive in the context of the planet's geological evolution, to consider the relationship between the two on a regional basis.

Photogeological studies have identified 13 fault sets in Syria Planum, nine sets in Tempe Terra, and 11 sets in Ulysses Fossae; all are related either to the Tharsis Rise or to Syria Planum. Detailed examination of the latest geological map of the western hemisphere of Mars (Scott and Tanaka, 1986) by Tanaka and Dohm (1989) has shown that of these 33 sets, less than half represent long regional faults whose origin is related to lithospheric doming. The rest tend to be much shorter and related to local volcanotectonic centres such as Alba Patera and Tempe Terra. This prediction had been made much earlier by Plescia and Saunders (1982). The volcanic and tectonic signatures exhibited by different areas allows for a subdivision of the region into seven broad **volcanotectonic provinces**, each of which contains faults and grabens which are radial with respect to Tharsis but which exhibit locally distinct volcanic and structural features (Fig. 12.8). The Elysium region constitutes an eighth province.

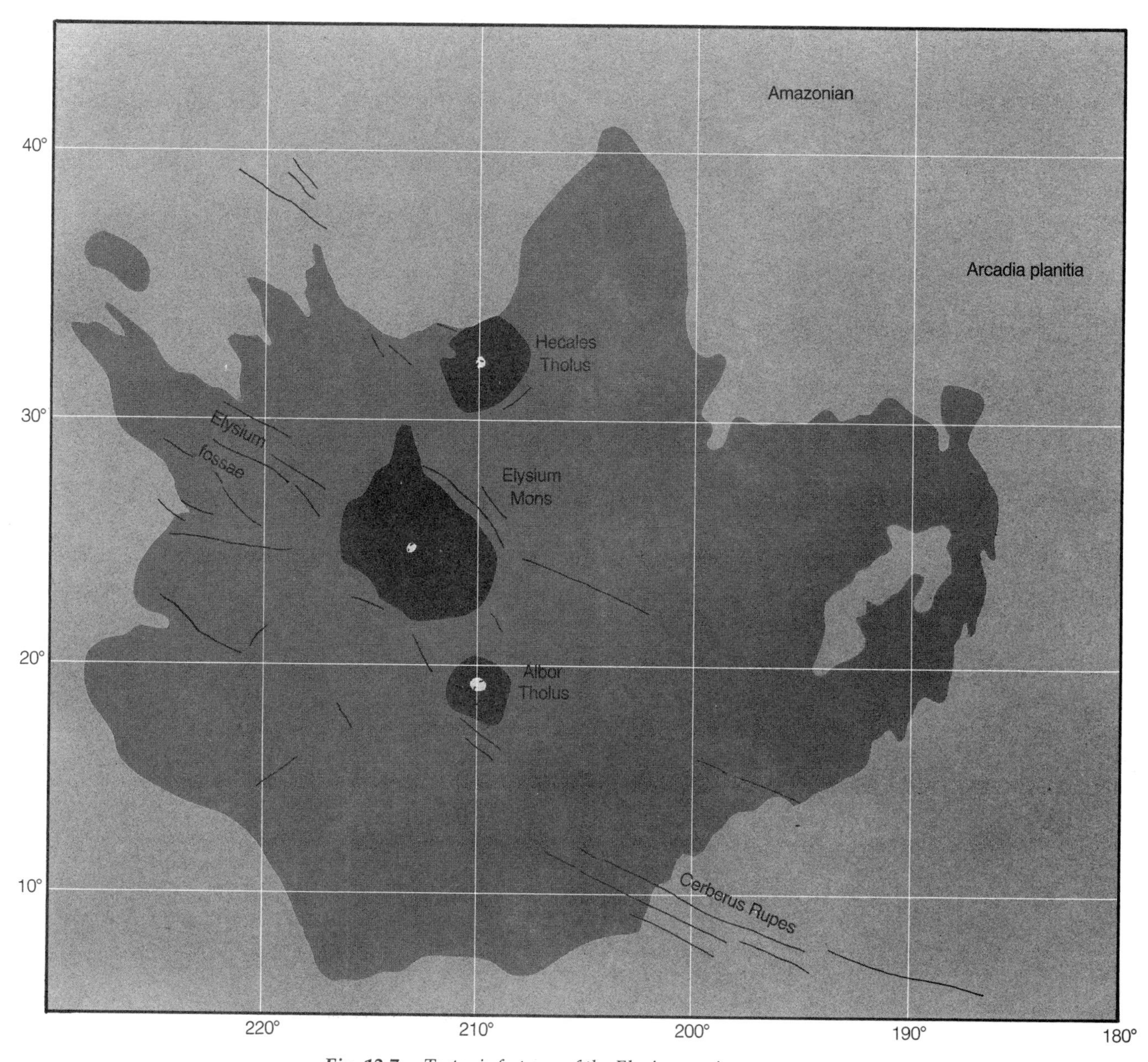

Fig. 12.7 *Tectonic features of the Elysium region.*

Each of the local volcanotectonic provinces developed along a major regional trend which was Tharsis-related. Substantial volumes of volcanic material were generated, and, at each focus, flows issued both from central vents and flank fissures, and from linear vents whose arrangement must have been controlled by the local stress regime; this is particularly clear in the case of Alba Patera (Cattermole, 1990). Table 12.1 illustrates the considerable diversity between them, which confirms that the development of the Tharsis region was a complex process.

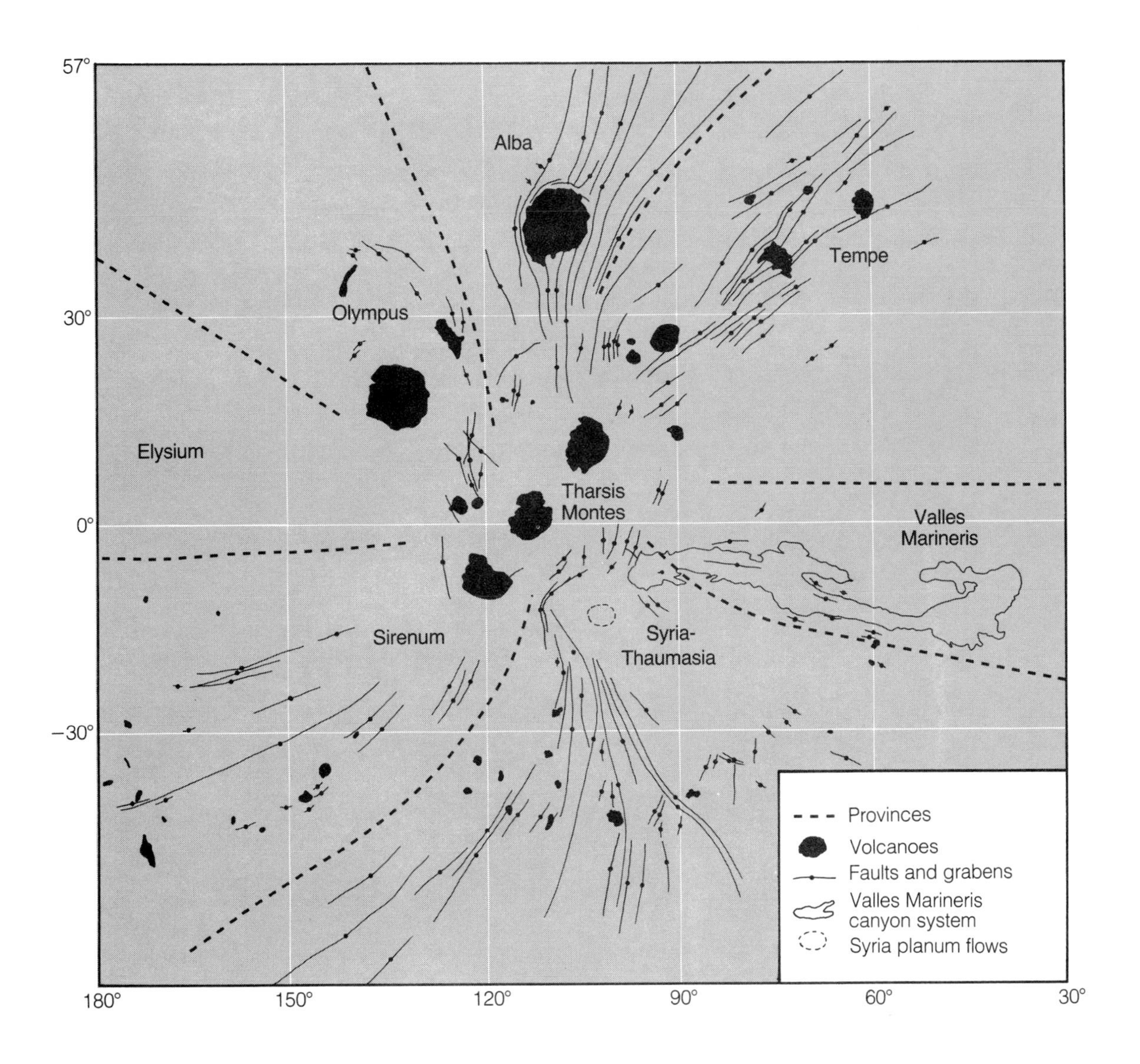

Fig. 12.8 *Volcanotectonic provinces of the Tharsis region (after Tanaka and Dohm, 1989).*

12.3 THE EVOLUTION OF THARSIS

The contentious issue of when, how and over what period the Tharsis lithospheric rise developed continues to provide much debate. However, in the past few years there has been much new work which has shed some light on each of these questions. One of the seemingly more plausible hypotheses has it as largely due to some form of mantle convection that, in actively uplifting the Martian lithosphere, also produced widespread fracturing. Convection could have been caused by radioactive heating, or have been due to the existence of lower-density mantle material beneath Tharsis, produced as a result of inhomogeneities inherent in the accretional process. However, such an explanation does not equate with the findings of Phillips and Ivins (1979) which suggest that the present distribution of ridges and fractures around Tharsis is best explained as a result of stresses generated in the lithosphere by the presence of the bulge and not in response to its uprise.

Table 12.1 The characteristics of Martian volcanotectonic provinces (after Tanaka and Dohm, 1989)

Province	Volcanic style	Tectonic style
Tharsis Montes	Large high shields with extensive flow fields	Few radial graben and some ring faults
Alba Patera	Broad low shield with very extensive flow fields	Many radial and some circumferential graben
Tempe	Small low shields and 'plains type' volcanism	Radial and concentric graben. Several foci
Valles Marineris	Pyroclastic rocks (?) Plateau flow field (?)	Broad uplift; deep rifting and collapse
Syria-Thaumasia	Widespread fissure activity	Local uplifts; radial and concentric grabens and normal faults
Sirenum	Dispersed volcanoes	Widely spaced graben radial to Tharsis Montes
Olympus	Large shield with extensive flow fields	Older radial graben; Acheron Fossae concentric graben
Elysium	Large shields and rilles	Widely spaced grabens radial to Tharsis Montes with local fractures.

12.3.1 Introduction

Various theories have been developed to account for the Tharsis Bulge. Global shrinking, tectonic effects antipodal to the Hellas basin, constructional volcanism and inhomogeneities in the Martian mantle are among those hypotheses to have been put forward (Phillips, 1978; Carr, 1984). Any plausible explanation for the Tharsis Bulge needs to satisfy both the gravity and topographic data, and also to be consistent with the nature and timing of tectonic and volcanic activity. This has proved to be a tall order and many hypotheses satisfy one or other of these criteria, but not necessarily all. It is for this reason that the problem of Tharsis remains in the forefront of Mars research.

The oldest unequivocal structures identifiable as having a link with the Tharsis Bulge are certain circumferential faults which outcrop in the Claritas Fossae region, on the southwest side of Syria Planum (Phillips *et al.* 1990). These have a Noachian age. The rise must, therefore, be as old as this. However, it should be noted that while flexural uplift may have been responsible for the Claritas Fossae structures, it does not necessarily imply that the subsequent massive elevation of the Tharsis region was largely a response to such a phenomenon. For instance, it could equally well be that major volcanic construction preceded fracturing. Uplift of Tharsis continued until at least the Late Hesperian and possibly the Early Amazonian period.

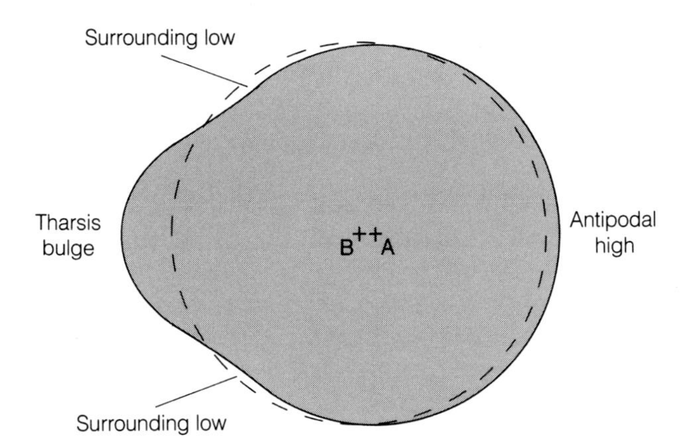

Fig. 12.9 *Equatorial cross-section of the Martian geoid.*

12.3.2 The Martian geoid

Gravity varies from place to place on the Martian surface, as it does on the Earth; this reflects the asymmetric distribution of mass within the planet's crust and mantle. One of the best ways of describing the shape is by the **geoid**; on the Earth this is a level surface that closely approximates the ocean surface.

Mariner 9 and Viking gravity data were analysed by Balmino *et al.* (1982) and a new model for the Martian gravity field produced. Because there is an unequal mass distribution within Mars, the geoid departs from the ideal fluid ellipsoid, in some places rising above it and in others falling beneath it (Plate 14). There is a marked geoid high associated with Tharsis. This indicates that isostatic compensation at shallow depths is not complete. The fact that the geoid high is shifted somewhat to the north of the equator, whereas the topographic high is to the south, is explicable by noting that the Tharsis rise straddles the upland-lowland boundary, and so the load is a few kilometres thicker to the north. Another high is antipodal to Tharsis, while a geoid low surrounds it. If we consider an equatorial cross-section of the geoid (the solid line in Fig. 12.9) the explanation for the surrounding low and antipodal high becomes clear: beyond Tharsis, the figure of the geoid approximates to a circle which is centred at point A. The dashed circle centred at point B, on the other hand, represents the equatorial cross-section of Mars without the Tharsis bulge. The displacement of the centre of the circle from point B to A results in a geoid low surrounding Tharsis (the part of the section where the dashed line is outside the solid one) and a geoid high opposite to it (where the solid line is exterior to the dashed one).

12.3.3 Lithospheric stresses beneath Tharsis

Willemann and Turcotte (1982), by 'thin shell' modelling, have been able to demonstrate that the lithospheric stresses needed partially to support the mass of Tharsis are reasonable and consistent with the observed radial fracture pattern; their model predicts that the thickness of the elastic lithosphere ranges from 110 to 260 km, while the thickness of the Tharsis load itself is between 40 and 70 km. (The latter is in broad agreement with the figures arrived at by Comer *et al.* (1985), who modelled the response in the elastic lithosphere to volcanic loading, by analysing the radial distance of circumferential graben surrounding Tharsis.) Their modelling also indicates that near the perimeter and beyond Tharsis, the stress at the top of the Martian lithosphere indicates a horizontal deviatoric stress near the surface. This reaches a maximum between 30° and 60° from the centre of loading which, therefore, is where fracturing is to be expected. If Tharsis were a dome in the elastic lithosphere, then linear deflection theory predicts that faults in this region should be thrust faults. They are not; they are graben. This precludes, therefore, the possibility that Tharsis is simply a lithospheric dome. The radial graben surrounding the Bulge are better explained by assuming that Tharsis is due to a crustal dome over a downward deflection of the lithosphere. Because the observed lava flows in the Tharsis region are relatively thin (2–3 km), they speculate that much of the dome was constructed from igneous intrusions.

Modelling by a slightly different approach – thick-shell tectonics – Banerdt *et al.* (1982) sought to establish the roles of isostatic support, flexural loading and buoyant support due to dynamical motions within the mantle. They were able to rule out the possibility that Tharsis was supported dynamically, since such a hypothesis would require that west-east trending grabens should characterize the highest parts of the Bulge whereas the observed faults are radial, as we have seen earlier. The same modelling procedure predicts that in an isostatic configuration, grabens will tend to form on and be confined to topographic highs, and that where associated with such highs there are preferred directions of maximum gradient, grabens will form normal to these directions. For Tharsis, this means they would trend either north-south or, in the north of the region, towards NNE-SSW. This is precisely what was noted in the earlier stages of the growth of Tharsis (Section 12.2.1), and implies that an isostatic regime operated at this time. The same group also confirmed that at a later stage, flexural loading of a laterally homogeneous elastic lithosphere had occurred. This gave rise to the radial graben found at radial distances greater than 40°.

On the basis described above, Banerdt and colleagues propose the following scenario for Tharsis (Fig. 12.10):

Stage 1 Tharsis began to be built by a combination of constructional volcanism and isostatic uplift. Most of the differentiated igneous materials remained above their source regions (Finnerty and Phillips, 1981; Finnerty *et al.*, 1988). Because mass was neither added nor removed from Tharsis, an isostatic regime prevailed; however because isostatic stresses are proportional to topography, at some stage they exceeded the finite strength of the near-surface rocks, with the result that grabens formed in the elevated regions.

Stage 2 By this time the lithosphere had become thicker and had reached the point where complete isostatic compensation of the growing volcanic pile was not achieved. Furthermore, volcanic materials may have been extruded from regions not beneath Tharsis itself, increasing the likelihood that non-isostatic conditions held. At this stage the volcanic load began to cause flexure of the lithosphere. Eventually, the load increased to such a degree that failure took place, and graben faults radial to Tharsis were generated.

Stage 3 By this time the lithosphere had become so thick that addition of increased volcanic loading had little if any effect on the stress levels; no further

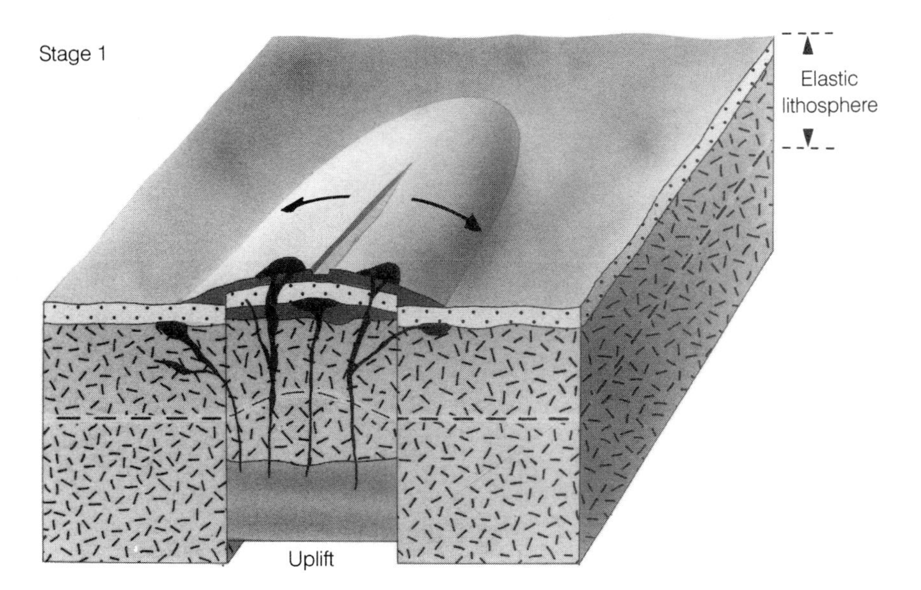

Fig. 12.10 *Cartoon showing hypothetical development of Tharsis. Stage 1: Elevation of Tharsis occurs due to both volcanic uplift and isostatic uprise. Stage 2: Lithosphere has thickened to stage where the load has to be supported flexurally, with formation of radial faults. Stage 3: Lithosphere is sufficiently thick that increasing load does not necessitate further graben formation (after Banerdt et al., 1982).*

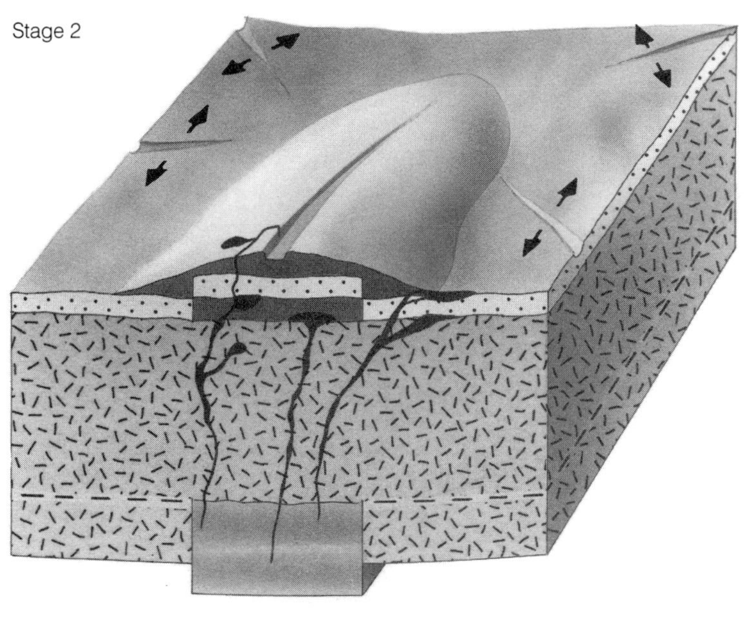

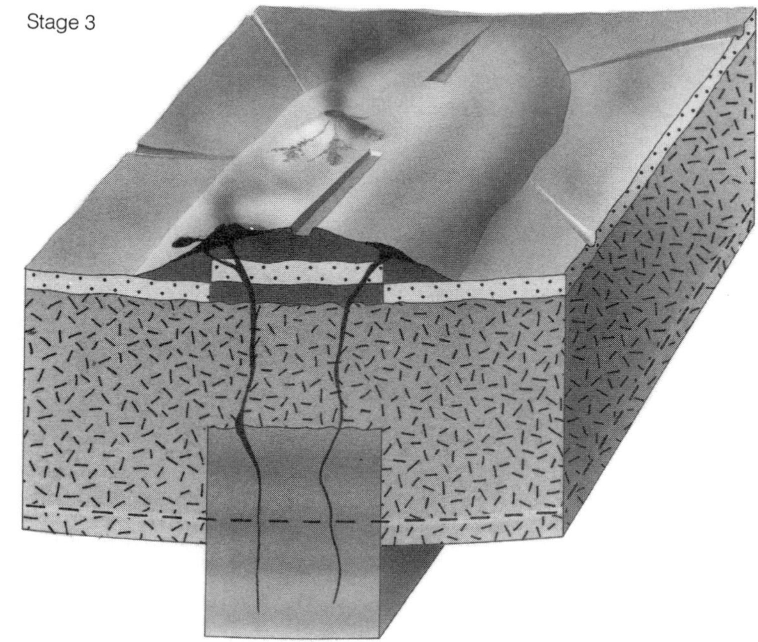

Mantle

Primordial crust

Basalt-like crust

Depleted source region

Basalt-like magma

regional failure thus occurred. This phase is represented by the youngest volcanic flow plains and shield volcanoes. As we have noted earlier, these developed their own local stress regimes.

12.3.4 *How was Tharsis uplifted?*

Many of the earlier explanations for the uplift of Tharsis invoked some long-lived thermal, chemical or dynamical anomaly in the Martian mantle or crust (Wise *et al.*, 1979a,b) which led to lithospheric fracturing and extended volcanism at a later time. As we have seen above, however, if the majority of the topographic elevation is due to dynamical uplift, the predicted stress field does not match that observed. This alone cannot be the true explanation. Furthermore, since it is now believed that the development of Tharsis spanned at least 3 Ga, there is the difficulty of maintaining internal dynamical motions over very long periods of time indeed.

However, a very plausible alternative has been proposed by Solomon and Head (1982). These authors see the bulge as a massive volcanic pile and not a tectonically produced rise. They suggest that because the primordial Martian lithosphere was laterally inhomogeneous, global and local stresses were caused prefentially where there was thin lithosphere, i.e. under Tharsis and probably Elysium. Once fractures had been propagated there, they would have provided channels of easy access for magma to reach the surface. In this way, once an elevated flow of thermal energy had been established, the high heat flow would have maintained relatively thin lithosphere beneath the sites of active volcanism and would have had the effect of concentrating fracturing there too (Fig. 12.11). Is there any evidence that the lithosphere is thin beneath Tharsis?

To constrain the thickness of the lithosphere beneath Tharsis, Comer *et al.* (1985) used elastic shell theory and analysed the effects of volcanic loading. They estimated the elastic lithosphere to be between 20 and 50 km thick beneath the regions surrounding Tharsis Montes, Alba Patera and Elysium Mons. This may be compared with a thickness of 150 km below Olympus Mons (a late-forming volcano with few radial graben) and >120 km beneath Isidis. Their results confirmed, therefore, the apparent thinning of the Martian crust beneath the crest of both the Tharsis and Elysium bulges. This tends to give credence to the above hypothesis.

According to Solomon and Head, during the Early Hesperian, with the onset of volcanism in the Tharsis region, the lithosphere probably remained relatively thin; in consequence the response of the lithosphere to volcanic loading would have been indistinguishable from local isostatic compensation by crustal subsidence. As Mars cooled, however, so the elastic lithosphere's thickness increased, and as this happened, so the amount of topographic relief and volcanic loading which could be supported by it became greater. Eventually, a stage was reached when the thickened lithosphere could support at least some of the loading by regional flexure and the increased strength of the lithosphere itself.

This model scores over several of the others since it requires no abnormal chemical or dynamical properties to be sustained for lengthy periods in the Martian mantle. It also predicts that the remnants of ancient cratered terrain that outcrop on the surface of the Tharsis Bulge must also be of volcanic origin, since they were an integral part of the growing volcanically generated dome.

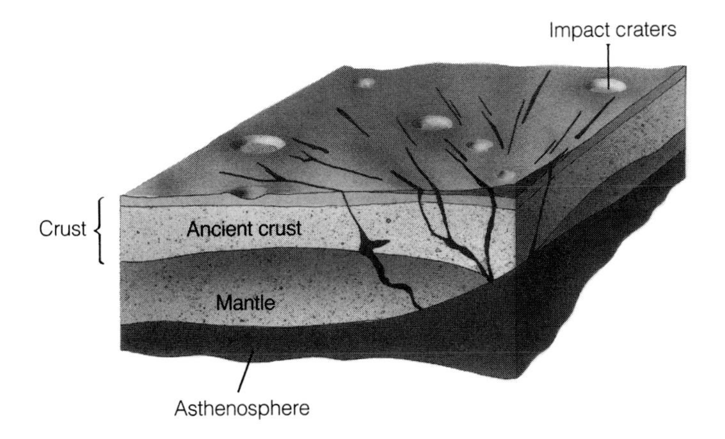

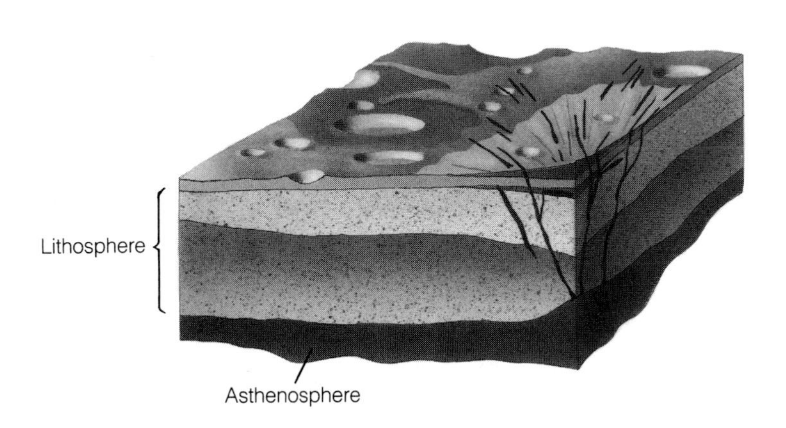

Fig. 12.11 *Cartoon of Tharsis evolution according to Solomon and Head (1982).*

Unfortunately, there is no way of testing this prediction; all that can be said is that if a Mars lander should sample such units, although impact breccias might be the predominant rock type, within these ought to be found abundant igneous clasts. Furthermore, it also implies that much of the volcanic construction that went into building Tharsis, occurred before the end of the heavy bombardment.

12.4 SOME FINAL THOUGHTS CONCERNING THARSIS

Discussions about the rise of Tharsis have tended to focus on the relative roles of volcanic construction and structural uplift. The literature about Tharsis is considerable and not a little confusing to the non-geophysicist; there are so many possibilities. For instance, the process of 'compensation uplift' whereby lateral loss of mass from a region leads to compensatory isostatic rise, can lead to regional elevation. This process is well known from the Earth and occurs where erosion of a newly uplifted mountain belt is followed by isostatic rebound. Another means of generating a rise is by 'intrusive uplift', where igneous bodies are intruded into the crustal regions from the mantle below. The contribution of the latter, a distinct possibility on Mars, unfortunately would be very difficult to measure. Then again, Phillips *et al.* (1990) recently have pursued a new line of thinking and have suggested an isostatic 'open-system magmatic differentiation' model. In simple terms, this assumes that all the components of a partial melting process, such as the melt products, the residuum and the lateral mass loss, must be equivalent to the mass of the original source region that experienced the original partial melting. Surpisingly, they discovered an isostatic effect which arises due to the difference between the isostatic and mass balance. This can have a profound effect on the degree of elevation which can be achieved in a magmatic system when the depth of compensation exceeds a few per cent of the planetary radius.

They have also shown that in order to achieve the requisite elevation of Tharsis, a depth of compensation of around 150 km is needed. Their calculations also show that partial melting in the Martian mantle began at quite shallow depths and never moved deeper than 250 km. This is in conflict with earlier gravity models of a purely isostatic nature (Sleep and Phillips, 1985), for these demand the lithosphere to be at least 400 km thick, which would push the depth of magma generation too deep. However, it is in agreement with the simplistic hydrostatic argument developed by Carr (1976). However, while a purely isostatic model cannot satisfy the requirements imposed by the observed total elevation and gravity data (Phillips *et al.*, 1973), it should be noted that by adding a small flexural load to an otherwise isostatic configuration, it is possible to accommodate both requirements. It is something of a delicate juggling act!

Other workers have proposed mantle convective models to explain Tharsis. Walter Kiefer and Bradford Hager (1989) recently suggested that the Tharsis and Elysium bulges were supported by internally heated convection which generated narrow plumes. On this model, there might be one or two major plumes beneath Tharsis, plus a small number of smaller ones. As we have seen also, Solomon and Head (1990), by utilizing variations in the thickness of Mars' elastic lithosphere estimated from volcanic loading to constrain near-surface temperature gradients and crustal thickness, have shown that lithosphere thickness variations must be largely due to mantle dynamic processes. These may be similar to the lithospheric reheating which occurs beneath hot-spot centres on the Earth.

12.5 THE CRUSTAL DICHOTOMY

The escarpment which marks the line of separation or dichotomy between the cratered uplands to the south and the lowlands to the north, is another global-scale feature of the planet (Plate 15). Yet, despite an ever-increasing awareness of its importance, manifested in intensive research into its nature, mode and age of formation, there is still no single, firmly held hypothesis to account for it.

One way to explain the north-south differences is to assume that the whole of the northern hemisphere has been lowered by erosion of the original cratered crust. However, this raises the seemingly insoluble problem of where the eroded material went. Another early suggestion was that of Wise *et al.* (1979a,b), who suggested that first-order convective overturn of the mantle led to a lowering of the entire northern hemisphere. However, there is no hard evidence that such a process ever occurred, indeed that it could have occurred, bearing in mind what we know of planetary development.

Today, the dichotomy zone appears to be a zone of active erosion, where scarp retreat is eating into the older cratered terrain. Such activity must have been going on for a considerable time, as is shown by the extensive occurrence on the plains side of the dichotomy of knobby terrain – consisting of isolated knobs and small mesas that appear to be relicts of the old plateau (Fig. 12.12). This knobby terrain outcrops in two main settings: along one half of the upland-lowland boundary, and in the lowlands north of it. Since a great deal of such terrain has the knobs arranged in roughly circular patterns, it has long been assumed that it developed from the upland cratered plateau. By studying impact crater densities within various regions of such modified terrain, Wilhelms and Baldwin (1989a) have been able to show that much of the knobby terrain has an Early Hesperian age. In other words, that there are surfaces as old as the upland cratered terrain, north of the line of dichotomy.

In an attempt to explain the situation, Wilhelms and Squyres (1984) proposed that one of the very earliest events to have affected Mars was a giant impact which gave rise to a vast 7700 km diameter basin. This supposed Borealis Basin was centred at 50°N, 190°W. They then argue that the northern lowlands occupy what was the interior of the multi-ring structure while the uplands outcrop over the exterior regions. It follows logically that, because this major cosmic event predated all others, Noachian-age heavily cratered units were emplaced on its floor and therefore must exist north of the lowland-upland boundary. They further make the suggestion that enhanced heat flow associated with the giant impact was responsible for melting ice locked into the super-basin materials, degrading them into the observed knobby landforms (in much the same way as removal of ground ice is cited in explaining the formation of Early Hesperian outflow channels).

The notion of a single huge impact early in Martian history has not received universal support; indeed, many are sceptical. The gravity data certainly does not strengthen the argument. By analogy with the Moon, it would be anticipated that mascons would be associated with large Martian basins. A brief look at the gravity data (Plate 14) shows quite clear anomalies over both Isidis and Utopia; however, no such anomaly is to be seen in the polar regions. Frey and Schultz (1989) have proposed that perhaps several large impacts were responsible for lowering the northern third of the planet, and they cite statistical evidence in support of this view. By invoking several overlapping large

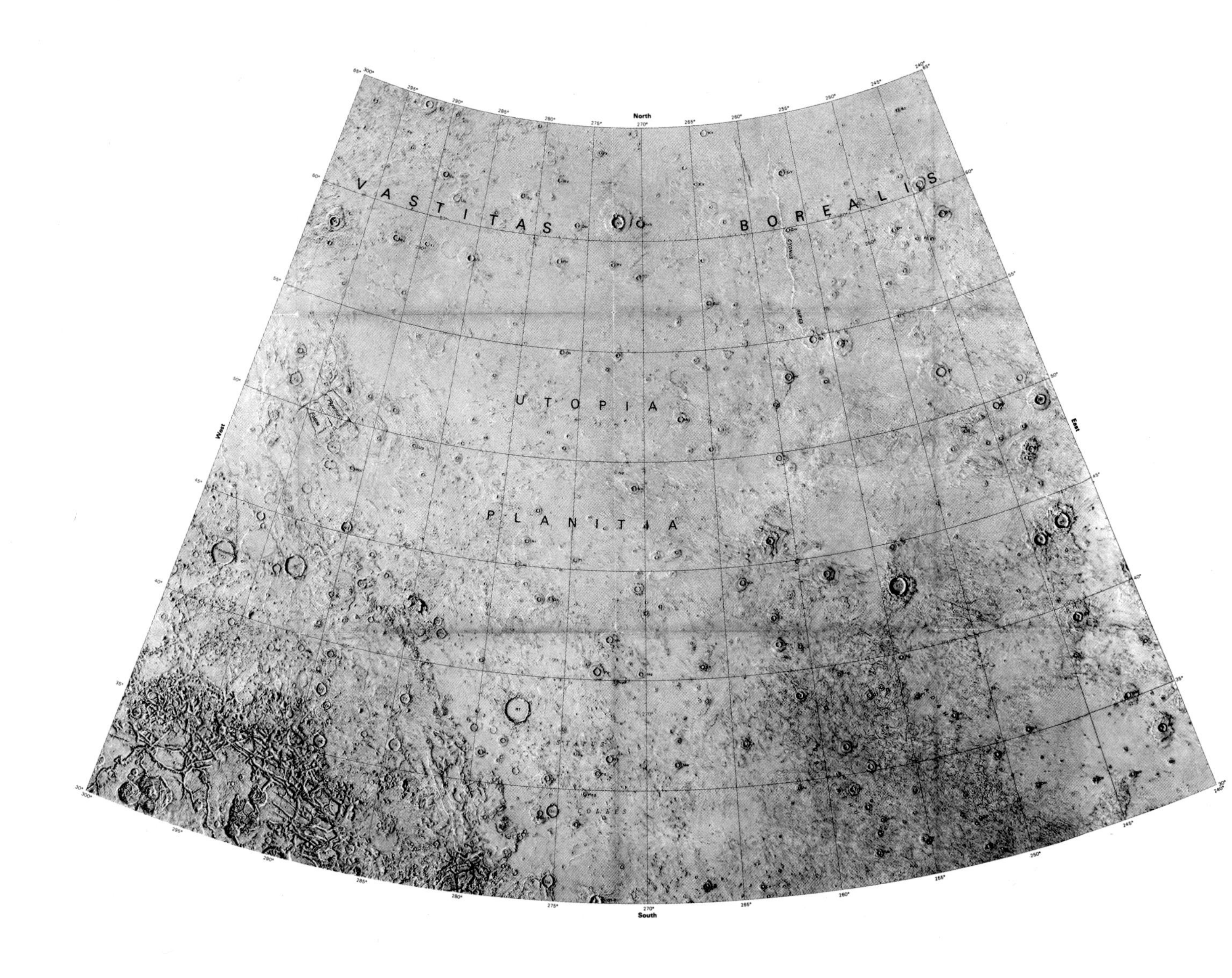

Fig. 12.12 *Shaded relief map showing development of knobby terrain in Utopia Planitia, north and east of the upland/plains boundary. USGS Map I-1646 (part of).*

basins, Frey and Schultz consider that the pattern of knobs within the northern lowlands can more reasonably be explained. The principal problem with this idea is that it is difficult to envisage why concentration of such large basins should be so effectively polarized into one third of the surface area. Furthermore, there seems no getting away from the fact that if several overlapping basins had existed, regions of topography higher than the original ancient surface must have been produced. There is no clear evidence for these.

Recent work by McGill (1989) has shown that relatively little thinning of the ancient crust by removal of surface materials can have occurred since Middle Noachian times. He bases his conclusions on the following observations: the oldest impact craters now buried beneath the northern plains were substantially fractured and eroded prior to burial, surviving now as isolated knobs and knob rings. By applying established crater dimensional equations, McGill shows that a maximum of around 200 m of lowlands-wide erosion of the pre-plains surface could have occurred. If this is so, then clearly it invalidates suggestions that 2–3 km of ancient crustal material must have been removed north of the line of dichotomy to produce the northern lowlands. Furthermore, it implies that the present position of the boundary cannot be not far removed from its pre-Noachian position. This appears to imply that the prominent scarp so characteristic of the upland-lowland boundary in the eastern hemisphere, far from being erosional, must be very largely structural in origin, albeit modified since its time of formation.

Structural mapping along that part of the boundary between Nilosyrtis Mensae and Aeolis (Maxwell, 1989) confirms this suspicion. Mapping indicates little correlation between the trends of ridges and scarps and the dichotomy itself; the exception to this is in a wide trough extending from Isidis to the large crater Herschel, where scarp orientations are axial-symmetric, suggestive of continued deformation by downfaulting through the period of plains formation. Maxwell concludes that this implies the downdropping of the terrain north of the present dichotomy as a discrete block 300 km wide, and postulates that this could also be the case for the partially buried ground – now represented as the knobby terrain – which outcrops elsewhere to the north of the dichotomy. While the faulting along this section of the dichotomy can be dated as Late Noachian-Early Hesperian, elsewhere along the boundary fracturing occurred at different times. This suggests a protracted episode of adjustment to the original dichotomy. If this were the case, an endogenic origin for the lowering of the northern one-third of Mars becomes the most plausible means of explaining it.

A number of other studies suggest that the dichotomy is not the ancient stable feature of the planet, as once was widely believed. Such a conclusion stems from assessing what geological activity was ongoing during Late Noachian-Early Hesperian times. There was, for instance, major erosion in the northern plains and demonstrable Early Hesperian faulting; similar activity has been less precisely dated as Late Noachian to Hesperian (McGill, 1987). It is probable, too, that fretted channels formed along the dichotomy at about this time. Then again, the widespread small valley networks characteristic of the cratered uplands also are believed to have developed at this stage. Recently, Wilhelms and Baldwin (1989) have suggested that igneous sills underlie much of the intercrater plains in the uplands and may have played a major role in releasing ground ice, both within the intercrater regions, and also along the dichotomy. Major volcanic resurfacing also took place at the Noachian-Hesperian boundary (Frey *et al.*, 1988). Finally, there is a narrow negative gravity anomaly associated with the upland-lowland boundary, suggestive of activity more recent than that implied if the dichotomy had its origin in an ancient impact (Phillips, 1988).

Because so many related geological events affected the northern hemisphere of Mars during Late Noachian-Early Hesperian times, the balance seems to tilt in the direction of some internal process as having been responsible for the lowering of the northern regions. Since it has already been acknowledged that first-order convective overturn seems unlikely to have occurred, some other (as yet unknown) internal mechanism must have thinned the lithosphere from below.

12.6 THE GEOLOGICAL DEVELOPMENT OF MARS

Mars accreted from the solar nebula like all the other planets. This took place around 4.6 Ga BP and may have taken less than half a million years for it to attain planetary dimensions. Mars, being only half the size of the Earth, had a heating capacity at least one order of magnitude lower, thus the temperature maximum attained after accretional heating and decay of short-lived radioactive nuclides would have been rather close to the surface. If a completely melted region ever developed, then this must have been located at about 0.9 of the planetary radius.

Mars, being less compressed than the Earth, is also of lesser density. Its moment of inertia indicates that it has a core but this is either smaller or significantly less dense than Earth's. Estimates of its diameter vary between 1400 and 2000 km, accounting for between 7% and 21% of its total volume (Fig. 12.13a). Geophysical and geochemical data imply that the Martian mantle has to be relatively enriched with respect to iron compared with the Earth, perhaps by as much as a factor of three (McGetchin *et al.*, 1981). The lengthy history of volcanism and tectonism which followed suggests that both core infall and mantle segregation may have taken longer to achieve for Mars than for the other terrestrial planets (Solomon and Chaiken, 1976). Furthermore, the widely developed extensional faulting suggests that significant cooling cannot have affected Mars during the last 1–2 Ga of its geological history.

During this early phase, the Martian lithosphere – which may have been of variable thickness – had become sufficiently thick, cool and elastic to preserve the cavities and ejecta associated with impacts. These are the first records of the planet's geological story.

12.6.1 Early Noachian geology

The earliest preserved geological materials are, not surprisingly, the rim materials of large impact basins like Isidis, Hellas and Argyre (which were formed in that order). These and other large impacts characterized the Lower Noachian epoch. Heavily cratered materials are also exposed peripheral to Tharsis and elsewhere north of the dichotomy, where the terrain is consistently 1–2 km lower than to the south. Knobby terrain also extends as far north as 55°N. This appears to indicate that the lowering of the northern hemisphere took place during Early Noachian times (Fig. 12.13b).

12.6.2 Middle Noachian geology

Continued bombardment of the crust occurred, with the formation of many smaller basins and large numbers of craters. The embayment of many of these craters by plains deposits indicates early resurfacing by volcanism. Also during this epoch, extensional faulting affected areas like Acheron, Claritas and Melas Fossae.

12.6.3 Late Noachian geology

Cratering continued, but the lack of 0.6–1.2 km diameter craters implies a phase of crater obliteration at this time. There was a widespread emplacement of intercrater plains and their formation may have buried older deposits over

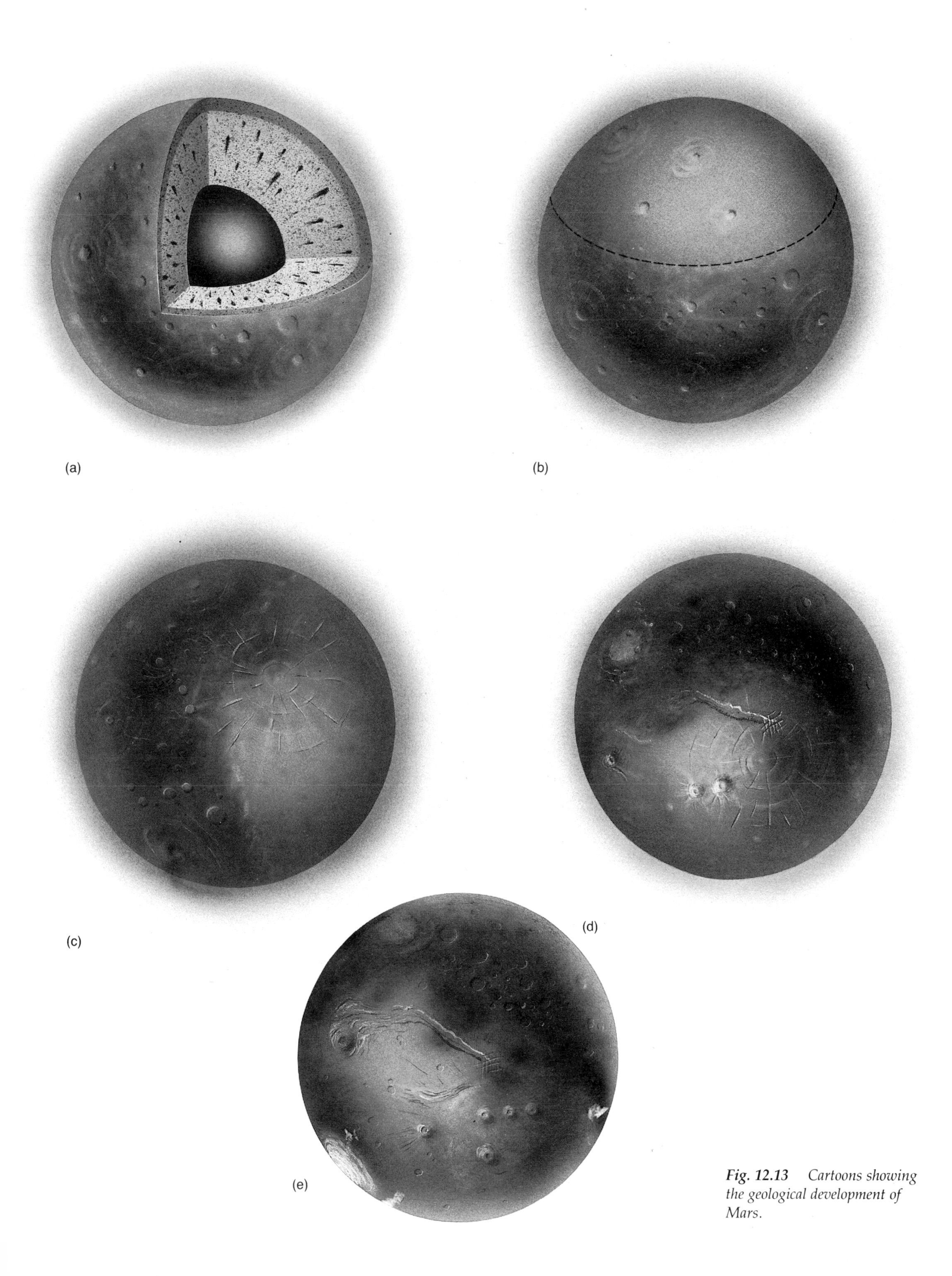

Fig. 12.13 Cartoons showing the geological development of Mars.

wide areas. Furthermore, there was also erosion, since the cratered plateau became incised by fluvial runoff channels. This has generally been taken to mean that at this time the climate was more temperate and the atmosphere denser than at present. Large regions of plains became ridged and there was also extensive fracturing in the rising Tharsis region, with most faults being radial about a locus in Syria Planum (Fig. 12.13c). Fissure volcanism occurred along Acheron Fossae, while centralized and often phreatomagmatic activity characterized foci like Apollinaris Patera and other highland paterae. It is possible that early faulting on the future site of Valles Marineris also occurred in the Late Noachian.

12.6.4 *Early Hesperian geology*

The cratering rate had by now fallen off, marking the end of the great bombardment period. Extensive ridged plains characterized the Early Hesperian. These are particularly extensive between Amazonis and Elysium Planitia, at Hellas, and in elevated regions such as Lunae, Sinai, Syrtis Major, Hesperia and Malea Plani. The development of wrinkle-type ridges and the presence of other volcanic landforms suggests that the plains were generated by flood volcanism and it is possible that centres active at this time – such as Amphitrites, Hadriaca, Tyrrhena and Peneus Paterae – may have supplied some of the plains materials. However, lavas undoubtedly were also extruded from widespread fissures. North of the dichotomy (in Vastitas Borealis) and along the line of the highland-lowland escarpment, there was intense erosion of earlier cratered terrain and Noachian age intercrater plains. Other units within Vastitas Borealis appear to represent deposits due to modification of the Early Hesperian ridged plains, erosion having been achieved by sapping or fluvial processes, and by aeolian activity. Significant here is the observation that in the region of Protonilus-Deuteronilus Mensae, where fretted channel development left extensive mesa remnants, relationships clearly indicate that fretting occurred after the formation of the Late Noachain plains but preceded the Early Hesperian ridged plains. The same relationships occur elsewhere along the upland-lowland boundary.

Volcanism became important from centres on Syrtis Major Planum, in Tempe Terra and at Alba Patera. Some flows emanated from calderae, but many appear to have had their origins in fissures and linear vents whose orientation may have been controlled by local tectonic regimes. Extensional faulting was intensive and widespread around Tharsis, which continued to rise, many fractures being radial about Syria Planum. The Noctis Labyrinthus-Valles Marineris rift faulting was in large part initiated during Lower Hesperian times (Fig. 12.13d).

12.6.5 *Late Hesperian geology*

By this time, intercrater plains formation had declined. However, volcanic activity began or continued at a number of major centres, notably on the northeast and southwest flanks of Tharsis, at Alba Patera, Ceraunius Fossae, Tempe Terra, Elysium Planitia and Syria Planum. Volcanism at this stage had become more centralized. Faulting was very intense surrounding Alba Patera; elsewhere faulting was rather weak, that which did occur being radial to a centre near Pavonis Mons. Faulting and collapse continued along Valles Marineris.

Erosion occurred in the south polar regions, cutting into the stratified deposits (? volcanic flows) of Lower Hesperian age. Outflow channel activity peaked during this epoch, the huge channel systems originating in collapsed regions of chaotic terrain mainly situated in the equatorial regions. Channel formation was accompanied by widespread fluvial deposition, particularly over Chryse Planitia and Amazonis, material being washed there from the northern uplands. Volatile release appears to have been very rapid and since it was synchronous with vigorous activity in Tharsis, may have been achieved by the increased thermal levels in the planet's crust.

12.6.6 *Early Amazonian geology*

Mars' youngest impact basin – the Lyot basin – was emplaced at this time, north of Deuteronilus Mensae. Its ejecta deposits are partially buried by Lower Hesperian sediments. Centralized volcanism was very important, the main centres being the huge shields of Tharsis Montes, Alba Patera and Elysium Mons. Flows were extensive and some entered Kasei Vallis to the east of the Tharsis Bulge. The aureole of Olympus Mons was also formed early in this epoch and it is probable that the shield had attained much of its final height by the close of it. Faulting was focussed around the southern edge of the Tharsis volcanic zone, at Alba Patera and west of Elysium Mons. There was localized fracturing, too, at Valles Marineris, alongside which deposition and erosion occurred, while ground-water (ice) sapping generated complex side-canyons whose pattern suggests control by existing faults. Several major landslides within Valles Marineris indicate continuing instability in this zone, and spread stratified wall materials across the floor. Fluvial activity at Mangala Valles cut into older channel materials at this time, while debris was spread out over Utopia Planitia from Elysium Fossae (Fig. 12.13e).

In the northern lowlands, some lava flows and extensive aeolian sediments were laid down. The latter mantled the pre-existing terrain over large areas, producing 'softened' topography. There was continued degradation of the highlands along the line of dichotomy and deposition of material north of the scarp.

12.6.7 *Middle Amazonian geology*

At this time centralized volcanic activity was largely confined to Tharsis Montes and Olympus Mons. However, there was extensive plains formation in Arcadia, associated with which appear to be small volcanic constructs; these may therefore be volcanic flows. Also in southern Amazonis (Medusae Fossae), along the line of dichotomy, wind-etched plains have been hypothesized to be ash-flow sheets.

Landslides continued to affect Valles Marineris at this time, while debris flows were active along the upland-lowland boundary at Protonilus Mensae. Faulting was minimal during the Middle Amazonian.

12.6.8 *Late Amazonian geology*

Limited volcanism took place on the flanks of Tharsis Montes and Olympus Mons and was accompanied by circumferential faulting. Limited channelling to the west of Olympus Mons and landslides may have been instigated by melting of ground ice by volcanic heating. It is also possible very young volcanic rocks were deposited in Valles Marineris.

Minor fracturing occurred along Valles Marineris, and landslides continued to occur here. Extensional faulting also affected southern Elsyium Planitia, opening fractures along Cerberus Rupes from which some fluvial channel material appears to have been generated and spread out over western Amazonis Planitia.

Considerable geological activity characterized the high latitudes, laminated deposits being laid down at both poles. These were overlain by permanent (residual) ice caps. The shape and extent of the residual caps changed with time, being largely governed by precessional phenomena over 50 000 and 2 Ma cycles. In the circumpolar zone there was (and still is) extensive movement of dust deposits, particularly at the northern pole, maintaining circumpolar dune fields of wide extent. This dust mantle may be as much as 200 m thick in northern high latitudes. Aeolian activity continued to modify the entire planet and does so at the present time. Along the dichotomy, there was (and is) movement of mass-wasted material along fretted channels and deposition of sediment north of the boundary escarpment.

12.7 CLIMATIC AND VOLATILE HISTORY

Naturally enough, the climatic history of Mars played an important part in the development of its surface. Since fluvial activity was extensive during Noachian times, implying that water could exist in a stable form on the planet's surface, it seems inevitable to conclude that the climate was more clement and the atmosphere considerably denser at this stage than it now is. Subsequently, in the Hesperian period, there was a change in the style of channel development, as outflow channels became the norm. These exhibit all the characteristics of having been formed by catastrophic floods which had their origins in extensive regions of collapsed and chaotic ground south of the dichotomy and largely along the equatorial zone. The only plausible explanation for their formation is that there was massive release of volatiles from huge aquifers within the brecciated cratered uplands. It has been estimated that the volume of water required to cut the channels would have been equivalent to a global ocean at least 50 metres deep.

It was thought for many years that these flood waters must have been extremely short-lived. Recently, however, the discovery of possible ancient shorelines in the low-lying northern plains, and of lakes within the equatorial canyon system, challenges this assumption. Consequently, there is considerable research activity in this area at present, aimed at understanding whether or not such oceans and other bodies of standing water could exist, and if so, for how long, and also whether they could exist under a cover of ice. Since some of the stratified (supposed lake) deposits within the canyons are 5 km thick this is a very stimulating area of enquiry!

In the Hesperian and the succeeding Amazonian period, melting of ground ice also seems to have initiated landslides and debris flows, with channel development being characterized largely by mass-wasting processes. Currently, the planet is locked into an icy grip by subfreezing temperatures. Geological activity is restricted largely to the polar regions, where dust storms and the seasonal cycle of cap advance and retreat constantly are modifying the surface. Elsewhere, surface modification is slow, and effected predominantly by aeolian activity. The planet is well past the peak of its geological and climatic activity.

Given the evidence of past water on Mars, it is clearly relevant to enquire how much may once have been present and where it has all gone. Measurements of the hydrogen escape flux from the atmosphere, and kinetic energy calculations, show that, over the course of its evolution, Mars must have lost the equivalent of a global layer of water 3 metres thick by processes like photodissociation and exospheric escape. This implies that most of the original volatile inventory must still reside somewhere; the potential reservoirs are 1. the polar caps, 2. the atmosphere, and 3. the regolith.

The MAWD experiment on the Viking spacecraft showed that if all the water locked into the atmosphere condensed on the planet's surface, it would produce a layer a mere 15 μm deep. In similar vein, measurements of the volumes of the perennial polar ice caps show quite clearly that they could not generate a planet-wide ocean more than a few tens of metres in depth. Both of these reservoirs thus fall considerably short of the inventory indicated by the geomorphological evidence. The only logical conclusion, therefore, is that most of this volatile material must remain locked into the subsurface.

Could this material – the regolith – hold so much fluid? Well, the answer appears to be, yes! Yes, at least, if we are right in assuming that the Martian regolith is similar to that of the Moon, which is known to be porous and brecciated to a depth of at least 20 km. By scaling to account for differences in the gravity between the two worlds, there is every likelihood that the Martian crust may be relatively porous to a depth of at least 10 km. On this basis, it could store a volume of volatiles equivalent to a global layer between 0.5 and 1 km deep. To test this theory, there is a need to land on Mars and probe beneath the surface either from a mobile roving vehicle or a manoeuvrable balloon. This is exactly what is planned for the Mars 94 and Mars 96 Russian missions. May they meet with consummate success!

13 THE NEXT STEPS

The ultimate goal of planetary exploration is to comprehend the nature and development of the planets from their origins to the present time. In order to do this we have to grasp the principles of those evolutionary processes – physical, chemical, geological and biological – that have shaped the planets in the past, and continue to modify them at the present time.

13.1 FUTURE MARS MISSIONS

At the present time various new Mars mission plans are under scrutiny. Specific studies have been made at NASA, in the Soviet Union and by the European Space Agency (ESA), and have identified, in the eyes of the planetary community, the most pressing needs for enhancing our understanding of Mars, and the best means of accomplishing this. In a report by the Solar System Exploration Committee of the NASA Advisory Council entitled *Planetary Exploration Through Year 2000* (1983), the following high-priority objectives were identified:

1. characterize the internal structure, dynamics and physical state of the planet;
2. characterize the chemical composition and mineralogy of surface and near-surface materials on a regional and global scale;
3. determine the chemical composition, mineralogy, and absolute ages of rocks and soils for the principal geologic provinces;
4. determine the interaction of the atmosphere with the regolith;
5. determine the chemical composition, distribution and transport of volatile compounds that relate to the formation and chemical evolution of the atmosphere and their incorporation in surface rocks;
6. determine the quantity of polar ice, and estimate the quantity of permafrost;
7. characterize the dynamics of the atmosphere on a global scale;
8. characterize the planetary magnetic field and its interaction with the upper atmosphere, solar radiation, and the solar wind;
9. characterize the processes that have produced the landforms of the planet;
10. determine the extent of organic chemical and biological evolution of Mars and explain how the history of the planet constrains these evolutionary processes; and
11. search for evidence of the signature of the early atmosphere in ancient sediments.

Recognizing that various models have been developed in each of these areas, NASA scientists rightly state that real progress can only be made by acquiring new data with state-of-the-art instrumentation. Ideally, a number of different missions would tackle specific problems; for instance a geoscience orbiter would be geared to obtain geochemical and geophysical data, while a climatology orbiter would collect appropriate atmospheric data over a period of at least one Martian year, 687 days. Unfortunately, the financial situation being as it is, 'ideal' plans can no longer be entertained, and most space-exploration funding bodies see the necessity to build all-purpose missions which will carry a variety of instrumentation which can obtain data over a wide spectrum. Thus NASA currently has advanced plans for a Mars Observer mission, which in essence is a geoscience-climatology orbiting spacecraft.

Most planning documents identify the urgent need to sample the Martian surface rocks; this means a lander probe. In ESA's publication *Mission to Mars* (Chicarro *et al.*, 1989), the committee for space exploration concludes that surveillance of the planet would profit by usage of microwave radar as well as high-resolution optical imaging systems, while a network of small surface stations would allow hard-landers (penetrators) and semi-hard landers (probes) to collect data at interesting sites, over a period of at least one Martian year. They see this as a precursor to a Mars rover sample-return mission. Certainly, the latter really marks the greatest need, and various groups currently are developing ideas for intelligent robotic-arm samplers. Only by returning samples to Earth can we obtain detailed geochemical data and, very importantly, radiometric dates.

Russian plans are well advanced, and two Mars missions are at an advanced stage of planning: Mars 94 and Mars 96. Since there is now a more open atmosphere between East and West, the Russians have invited scientists from the international scientific community to submit experiments that will fly to the planet at those times. Thus far plans are progressing well, and a variety of experiments will address many of the objectives cited above; for instance, there are definite plans to send a surface radar sounder on board either a balloon or surface rover, to utilize atmospheric balloons to collect climatological data, and a wide variety of other advanced instrumentation which is still under appraisal. If launches are made according to schedule, and budgetary constraints do not degrade the scope of the missions, our knowledge of the Red Planet should improve dramatically within the space of five years. However, at the present time it is unclear what effects the changing political situation in the Russian 'Commonwealth' will have on existing 'Soviet' space plans. Certainly there are severe problems.

13.2 CLIMATE AND THE SEARCH FOR LIFE

There is concrete evidence from geology and climatology that early in its history, Mars had a rather different atmosphere and climate than it has today. Mars may have had a relatively dense carbon dioxide atmosphere, with liquid water at the surface, much as we anticipate did the primaeval Earth. Modern estimates suggest that a pressure of 1 atmosphere of CO_2 would have been sufficient to warm Mars to above the freezing point.

Also, theoretical considerations of all of the planets suggest that during the initial few hundred million years of their evolution, the atmospheric composit-

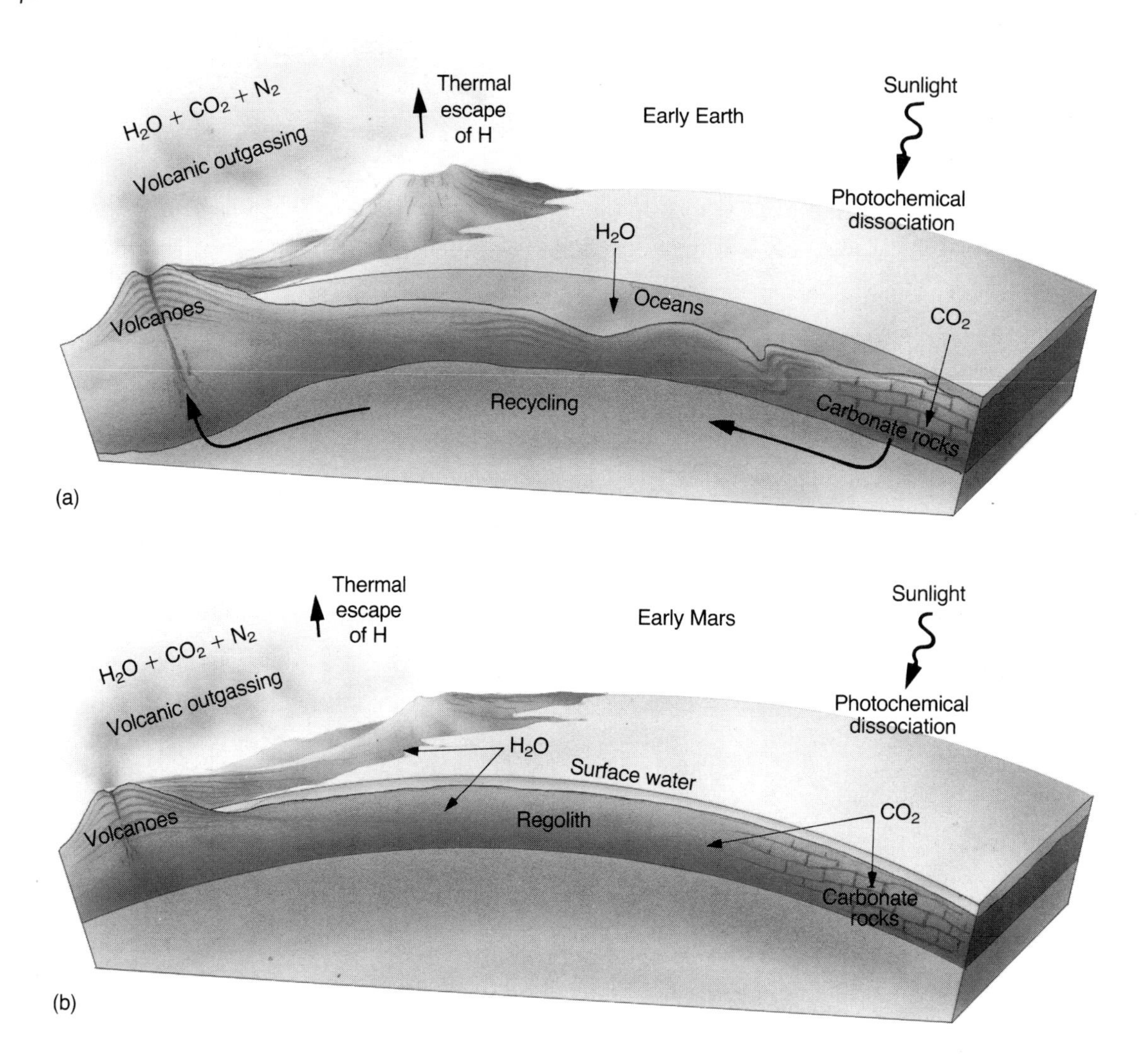

Fig. 13.1 *Cartoon showing CO_2-recycling system for (a) the Earth and (b) Mars.*

ions and pressures of the terrestrial planets were predominantly conditioned by volcanic outgassing of CO_2, N_2 and H_2O. Earth, Venus and Mars may at this time have experienced very similar surface conditions; however, thereafter developments diverged. On the Earth, atmospheric CO_2 was locked into carbonate rocks which eventually became recycled back into the atmosphere. In contrast, on Mars the carbonate rocks could have been similarly recycled by volcanism, but, eventually, as the latter became less intense and more localized, the CO_2 recycling process would decay, slowly depleting the Martian atmosphere (Fig. 13.1). On Venus, of course, things were different again, and currently the planet is suffering from a runaway greenhouse effect.

There is still a need to know exactly how much water Mars inherited from the primordial solar nebula, and what proportion of this was eventually outgassed. It would also be interesting to know if there was any later-stage input of volatiles from comets and meteorites. These and other vital questions have been under analysis in NASA's now-completed study project *Mars, Evolution of Climate and Atmosphere* (MECA) and in the continuing three-year project, *Mars: Evolution of Volcanism, Tectonism and Volatiles* (MEVTV). Some of the answers, it is hoped, will be obtained when NASA's Mars Observer spacecraft lifts off in September 1992.

Organic life was not found at either of the Viking lander sites; however, this does not rule out the possibility that it developed on Mars and that evidence may be found elsewhere. On Earth, primitive algae proliferated about 3.8 Ga ago; could such organisms have developed on Mars, and under what conditions? This is a potent question, and a fascinating one. Therefore, when new lander missions eventually do reach the Red Planet, there will be a serious search for fossilized remains in sedimentary rocks, perhaps nearer the poles than either of the Viking sites, or in different sedimentary environments. Certainly, if, as is suspected, seas and lakes did once exist one Mars, marine and lacustrine sediments naturally would be the prime targets for the search for a Martian biota.

13.3 EPILOGUE

Over a century ago, the Red Planet was held to be the world most likely to have developed life in some form. When the first spacecraft images were returned, revealing the barren cratered landscape of the upland hemisphere, hopes for such an exciting prospect faded sharply. Now, over two decades further on, there is a greater optimism that some form of organic life may be found. Science is an amazing discipline; ideas cycle and recycle as new evidence is forthcoming and new ideas are developed to account for it. This is its fascination. Mars – the Red Planet – may not currently be a very hospitable world for future astronauts, but once, several billions of years ago, there may have been primitive organisms struggling to survive, only to have their life snuffed out by some harsh change in the planet's climate, the details of which we have yet to understand.

Bon voyage to all future missions to Mars! Even if they do not find fossil remains, may they successfully send or bring back those vital scraps of information which will allow us to resolve at least some of the tantalizing questions raised by the last two decades of space exploration. Above all, let us hope the changes affecting life in what was the Soviet Union do not set back the welcome phase of international co-operation which had attended the planning of the Mars 94 and 96 missions.

APPENDIX A Astronomical data

Diameter	6787 km
Mass	6.4191×10^{21} kg
Volume	162.6×10^{12} km^3
Mean density	3930 kg m^{-3}
Surface gravity	3.71 m s^{-2}
Escape velocity	5.02 km s^{-1}
Axial inclination	23.98°
Visual geometric albedo	0.16
Rotation period	24.623 h
Orbital period	686.98 d
Maximum solar distance	249×10^6 km
Minimum solar distance	206×10^6 km
Ellipticity	0.0059
Surface temperature range	148–310 K
Mean surface pressure	6.1 mbar
Magnetic field strength	$<0.000\,03$ Earth

Phobos and Deimos

APPENDIX B

Fig. B.1 Phobos as imaged by Viking from 612 km in October 1978. The image shows the side of Phobos which always faces Mars, with linear grooves and the west wall of the crater Stickney at left, near the terminator. Viking orbiter frame P-20776.

Mars has two tiny moons, both of which are captured asteroids. Because they are composed of material stronger than their own gravitational force, they have irregular shapes, rather like cosmic potatoes! Of the two, Phobos is the larger, has a longest dimension of 27 km, and a mean density of 2200 kg m^{-3}. Deimos, the smaller moon, measures between 11 and 15 km across, and has a mean density of only 1700 kg m^{-3}.

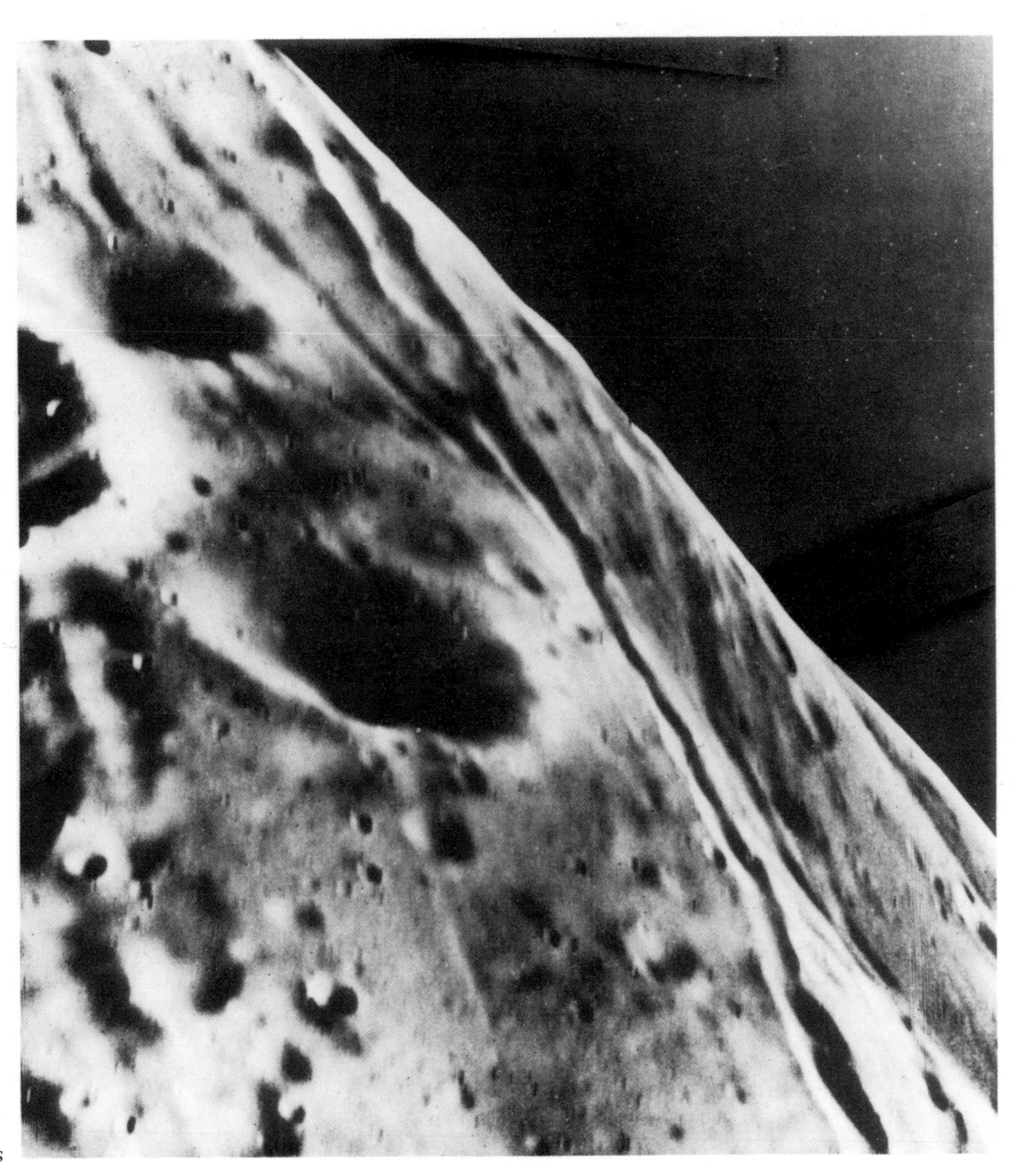

Fig. B.2 *Viking close-up of Phobos, showing bead-like grooves and small impact craters from a range of 120 km. Viking orbiter frame P-18634.*

The surface of Phobos is brecciated and heavily pockmarked with impact craters, the largest being Stickney, 10 km in diameter (Fig. B.1). This weird structure covers a major fraction of the satellite and has associated with it a series of 100–200 m wide grooves which are presumed to have formed in response to the Stickney-forming event. Some of these have a beaded structure (Fig. B.2). Like most asteroids, Phobos has a regolith which may be as much as 100 m thick.

Deimos is rather different from its companion. It, too, is densely cratered but is generally far smoother than Phobos (Fig. B.3). It has a lower albedo and seems to be mantled in a thick layer of dust that has buried many of the original craters. Both moons are widely believed to be composed of chondritic material.

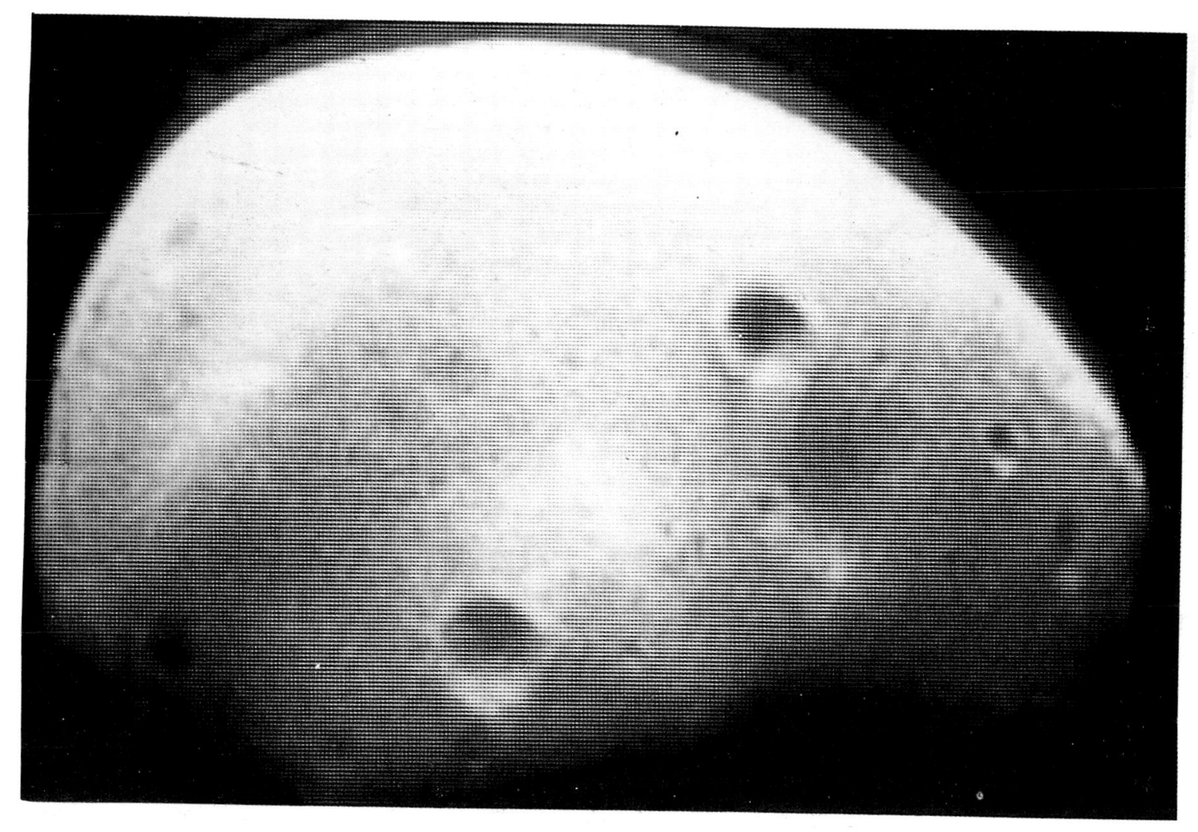

Fig. B.3 *Martian moon Deimos photographed by Viking orbiter.*

It was the intention of the Soviet Phobos mission to study the larger moon in detail and, indeed, to bounce a 'hopper' probe on its surface to conduct detailed chemical analysis. Unfortunately, the mission was prematurely terminated and detailed chemical results were not forthcoming.

APPENDIX C Model chronologies for Mars

There are no radiometric ages for any Martian rocks. Thus the only estimates of such ages have to be derived by predicting the relationship between crater densities and absolute age for Martian surfaces based on lunar experience. This necessitates certain assumptions, in particular that of the ratio between the cratering fluxes of Mars and the Moon. The flux for Mars is not well constrained, thus the results from the two most widely used chronologies, those of Neukum and Wise (1976) and Hartmann *et al.* (1981), differ considerably. In Table C.1 the preferred chronology of Hartmann *et al.* is presented.

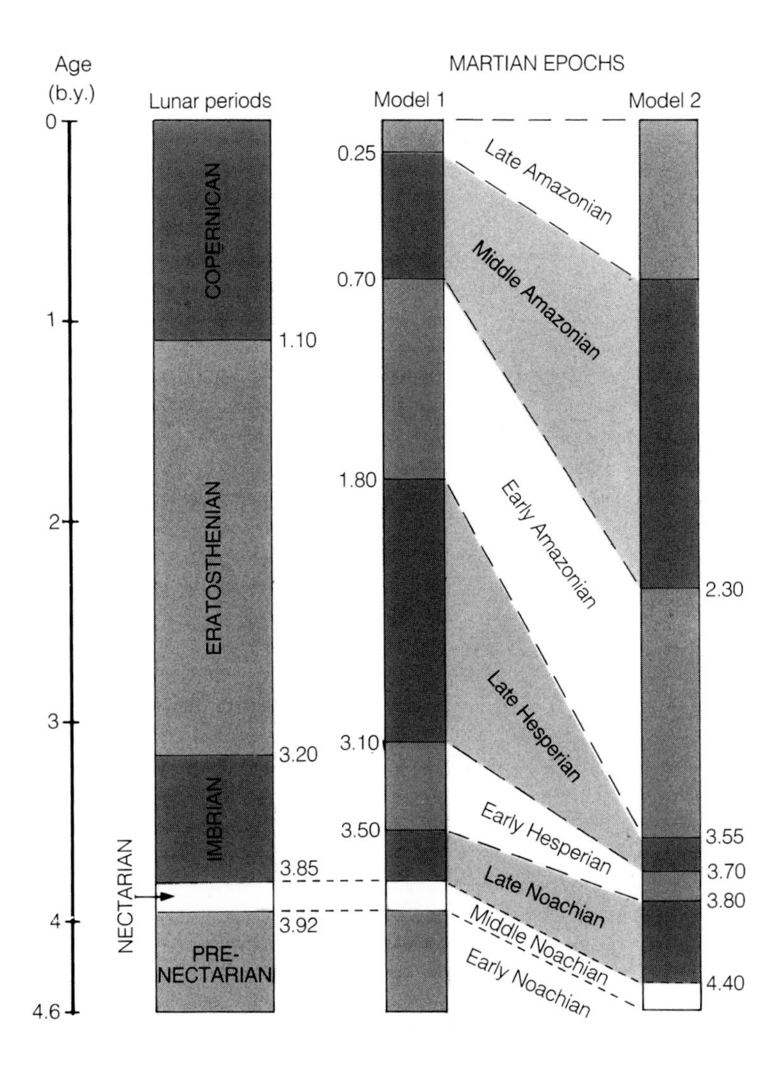

Table C.1 Chronology for Martian units (after Hartmann *et al.*, 1987)

Geologic province	Crater density relative to average lunar mare	Estimated crater retention age (billions of years)		
		Minimum likely	*Best estimate*	*Maximum likely*
Central Tharsis volcanic plains	0.1	0.06	0.3	1.0
Olympus Mons volcano	0.15	0.1	0.4	1.1
Extended Tharsis volcanic plains	0.49	0.5	1.6	3.3
Elysium volcanic rocks	0.68	0.7	2.6	3.5
Isidis Planitia	0.76	0.8	2.8	3.6
Solis Planum volcanic rocks	0.90	0.9	3.0	3.7
Chryse Planitia volcanic plains	1.1	1.2	3.2	3.8
Lunae Planum	1.2	1.3	3.2	3.8
Noachis ridged plains	1.3	1.7	3.3	3.8
Tyrrhenum Patera volcano	1.4	1.8	3.4	3.8
Tempe Fossae faulted plains	1.6	2.3	3.4	3.8
Plains on South Hellas rim	1.7	2.6	3.5	3.8
Alba Patera volcano	1.8	2.6	3.5	3.8
Hellas floor	1.8	2.6	3.5	3.8
Syrtis Major Planitia plains	2.0	2.0	3.6	3.9
Heavily-cratered plains				
– small D (<4 km)	1.4	1.8	3.4	3.8
– large D (>64 km)	13.0	3.8	4.0	4.2

APPENDIX D Stratigraphy of Martian geological features and units

Table D.1 has been adapted from that compiled by Kenneth Tanaka (1986) and is used with his permission.

Table D.1 Stratigraphy of Martian geological features and units

Unit or feature	Stratigraphic position
A. Plains and Highland Units	
Polar dunes, mantle, ice, and layered deposits	UA (and lower?)
Landslides	
Olympus Mons	UA
Valles Marineris	MA-UA
Debris aprons	
Tharsis Montes	UA
mensae terrain	MA-UA
Thick deposits, southern Amazonis Planitia	MA-UA
Smooth plains	
Malea Planum	UH-LA (or higher)
northern	LA to UA
Vastitas Borealis	UH
Layered deposits, Valles Marineris	UH
Argyre and Hellas Planitiae	UN-UH
Ridged plains material	LH
Intercrater plains material	UN-LH
Cratered plains material	
craters larger than 10 km in diameter	MN
craters smaller than 10 km in diameter	UN-LH
Basement material	LN
B. Volcanoes and Associated Lava Flows	
Olympus Mons (including aureoles)	LA (or lower)-UA
Ceraunius Fossae flows	UH-LA
Tharsis Montes (including flows)	UH (or lower)-UA
Small Tharsis volcanoes	UH-LA
Small volcanoes(?) in northern plains	UH-LA (or higher)
Elysium volcanoes	UH-LA
Syria Planum flows	UH
Alba Patera flows	LH-LA
Tempe Fossae flows	LH-UH
Dorsa Argentea flows	LH to UH

Table D.1 *continued*

Unit or feature	Stratigraphic position
Syrtis Major flows	LH
Highland paterae	UN to LH
Highland volcanoes	MN to UN
C. Channels	
Cerberus Rupes channel	UA
Outflow channels, flood plains and chaotic terrain	UH
Mangala Valles	UH-LA
Fretted channels	UN-UH
Small runoff channels	UN (or lower)
Long, sinuous channels (e.g. Ma'adim Vallis)	UN-LH
D. Fracture Systems	
Cerberus Rupes	UA
Northeast of Ascraeus Mons	MA-UA
South of Ceraunius Tholus	LA-MA
Elysium Fossae	UH-LA
Eastern Memnonia Fossae	UH-LA
East of Ceraunius Tholus	UH-LA
Alba and Tantalus Fossae	LH-LA
Noctis Labyrinthus	LH-UH
Tempe and Mareotis Fossae	LH-UH
Ulysses Fossae	LH
Uranius Fossae	LH
Western Memnonia Fossae	LH
Thaumasia Fossae	LH
Valles Marineris, main development	LH
Sirenum Fossae	UH-LH
Ceraunius Fossae	UN
Claritas Fossae (north)	UN
Noctis Fossae	UN
Faults east of Elysium Planitia	UN
Acheron Fossae	MN
Claritas Fossae (south)	MN
Basin faults	LN-MN
E. Impact Basins	
Lyot	LA
Lowell	LH
Most impact basins	LN-MN
South polar	LN
Argyre	LN
Hellas	LN
Isidis	LN

BIBLIOGRAPHY

Anders, E. and Owen T. (1977) Mars and Earth: Origin and abundance of volatiles. *Science*, **198**, 453–65.

Antoniadi, E. M. (1975) *The Planet Mars* (Trans. P. Moore), Reid, Devon.

Arvidson, R. E. (1979) A post-Viking view of Martian geologic evolution. *NASA TM-80339*, 80–1.

Arvidson, R. E., Carusi, A., Coradini, A. *et al.* (1976) Latitudinal variation of wind erosion of crater ejecta deposits on Mars. *Icarus*, **27**, 503–16.

Baker, V. R. (1982) *The Channels of Mars*, Texas, Austin.

Baker, V. R. and Kochel, R. C. (1978) Morphometry of streamlined forms in terrestrial and Martian channels. *Proc. 9th Lunar Sci. Conf.*, 3193–203.

Baker, V. R. and Milton, D. J. (1974) Erosion by catastrophic floods on Mars and Earth. *Icarus*, **23**, 27–41.

Baker, V. R. and Partridge, J. B. (1986) Small Martian valleys: Pristine and degraded morphology. *J. Geophys. Res.*, **91**, 3561–72.

Balmino, G., Moynot, B. and Vales, N. (1982) Gravity field model of Mars in spherical harmonics up to degree and order eighteen. *J. Geophys. Res.*, **87**, 9735–46.

Baloga, S. M. and Pieri, D. C. (1985) Estimates of lava eruption rates at Alba Patera, Mars. *Report Planet. Geol. Geophys. Program 1984, NASA TM-87563*, 245–7.

Banerdt, W. B., Phillips, R. J., Sleep, N. H. and Saunders, R. S. (1982) Thick shell tectonics on one-plate planets: Applications to Mars. *J. Geophys. Res.*, **87**, 9723–33.

Barlow, N. G. (1988) The history of martian volcanism determined from a revised relative chronology (abstract). MEVTV Workshop on Nature and Composition of Surface Units on Mars, 20–1. *LPI Tech. Rpt 88-05*, LPI, Houston.

Belton, M. J. S., Broadfoot, A. L. and Hunten, D. M. (1968) Abundance and temperature of CO_2 on Mars during the 1967 opposition. *J. Geophys. Res.*, **73**, 4795–806.

Binder, A. B. (1969) Internal structure of Mars. *J. Geophys. Res.*, **74**, 3110–18.

Blasius, K. R. (1976) Topical studies of the geology of the Tharsis region of Mars. PhD Thesis, California Institute of Technology.

Blumsack, S. L. (1971) On the effects of topography on planetary circulation. *J. Atmos. Sci.*, **28**, 1134–43.

Boyce, J. M. (1979) A method for measuring heat flow in the Martian crust using impact crater morphology. *NASA TM-80339*, 114–18.

Boyce, J. M. and Roddy, D. J. (1978) Martian rampart craters: Crater processes that may affect diameter-frequency distributions. *NASA TM-79729*, 162–5.

Brackenridge, G. R. (1987) Intercrater plains deposits and the origin of Martian valleys. MEVTV Workshop, Napa, California, 19–21.

Breed, C. S., Groller, M. J. and McCauley, J. F. (1979) Morphology and distribution of common 'sand' dunes on Mars: Comparison with Earth. *J. Geophys. Res.*, **84**, 8183–204.

Burns, R. G. and Fisher, D. S. (1989) Sulfide mineralization related to early crustal evolution of Mars. *LPI TM 89-04*, 20–2.

Carr, M. H. (1973) Volcanism on Mars. *J. Geophys. Res.*, **78**, 4049–62.

Carr, M. H. (1974) The role of lava erosion in the formation of lunar rilles and Martian channels. *Icarus*, **22**, 1–23.

Carr, M. H. (1976) Change in height of Martian volcanoes with time. *Geol. Romana*, **15**, 421–2.

Carr, M. H. (1979) Formation of Martian flood features by release of water from confined aquifers. *J. Geophys. Res.*, **84**, 2995–3007.

Carr, M. H. (1980) The morphology of the Martian surface. *Space Sci. Rev.*, **25**, 231–84.

Carr, M. H. (1981) *The Surface of Mars*, Yale, New Haven and London.

Carr, M. H. (1984) in *Geology of the Terrestrial Planets*, NASA SP-469.

Carr, M. H. and Clow, G. D. (1981) Martian channels and valleys: Their characteristics, distribution and age. *Icarus*, **48**, 91–117.

Carr, M. H., Crumpler, L. S., Cutts, J. A. *et al.* (1977) Martian impact craters and emplacement of ejecta by surface flow. *J. Geophys. Res.*, **82**, 4055–65.

Carr, M. H., Saunders, R. S., Strom, R. G. and Wilhelm, D. E. (1984) The geology of the terrestrial planets. *NASA SP-469.*

Cattermole, P. J. (1987) Sequence, rheological properties and effusion rates of volcanic flows at Alba Patera, Mars. *J. Geophys. Res.*, **92**, B553–60.

Cattermole, P. J. (1988) Mapping of volcanic units at Alba Patera, Mars (abstract). MEVTV Workshop on Nature and Origin of Surface Units on Mars, 37–9. *LPI Tech. Rpt 88-05*, LPI, Houston.

Cattermole, P. J. (1989) *Planetary Volcanism*, Ellis Horwood, Chichester.

Cattermole, P. J. (1990) Volcanic flow development at Alba Patera, Mars. *Icarus*, **83**, 453–93.

Chapman, C. R. and Jones, K. L. (1977) Cratering and obliteration history of Mars. *Ann. Rev. Earth Planet. Sci.*, **5**, 515–40.

Chicarro, A. F. (1989) Towards a chronology of compressive tectonics on Mars. MEVTV workshop on early tectonic and volcanic evolution of Mars. *LPI Tech. Report 89-04*, 23–5.

Chicarro, A. F., Scoon, G. E. N. and Coradini, M. (1989) Mission to Mars (Report of the Mars Exploration Study Team), *ESA SP-1117*, Paris.

Christensen, E. J. (1975) Martian topography derived from occultation, radar, spectral and optical measurements. *J. Geophys. Res.*, **80**, 2909–13.

Clifford, S. M. (1981) A model for the climatic behaviour of water on Mars. *Proc. 3rd Int. Colloq. on Mars*, Pasadena, LPI Contrib. 441.

Clifford, S. M., Greeley, R. and Haberle, R. M. (1988) Evolution of climate and atmosphere. *Eos*, **69**, 1585, 1595–6.

Cole, G. H. A. (1978) *The Structure of the Planets*, Wykeham, London.

Collins, S. A. (1971) The Mariner 6 and 7 Pictures of Mars, NASA SP-263, Washington DC.

Comer, R. P., Solomon, S. C. and Head, J. W. (1985) Mars: Thickness of the lithosphere from the tectonic response to volcanic loads. *Rev. Geophys. Space Phys.*, **23**, 61–92.

Cooper, H. S. (1980) *The Search for Life on Mars: The Evolution of an Idea*, Holt, Rinehart and Winston.

Crown, D. A. and Greeley, R. (1990) Styles of volcanism, tectonic association, and evidence for magma-water interactions in eastern Hellas, Mars. *J. Geophys. Res.* (In press).

Crown, D. A., Price, K. H. and Greeley, R. (1990) Evolution of the eastern rim of Hellas Basin, Mars. *Lunar Planet. Sci.*, **XXI**, 252–3.

Crumpler, L. S. and Aubele, J. C. (1978) Structural evolution of Arsia Mons, Pavonis Mons and Ascraeus Mons, Tharsis region of Mars. *Icarus*, **34**, 496–511.

Cutts, J. A. (1973) Nature and origin of the layered deposits in the Martian polar regions. *J. Geophys. Res.*, **78**, 4231–49.

Cutts, J. A., Blasius, K. R. and Roberts, W. J. (1979) Evolution of Martian polar landscape: Interplay of long-term variation in perennial ice caps and dust storm activity. *J. Geophys. Res.*, **84**, 2975–94.

de Hon, R. A. (1987) The Martian sedimentary record. Workshop on the nature and composition of surface units on Mars. *LPI NASA*.

De Vaucouleurs, G. (1950) *The Planet Mars* (Trans. P. Moore), Faber and Faber, London.

De Vaucouleurs, G. (1954) *Physics of the Planet Mars*, Faber and Faber, London.

Dohnyani, J. S. (1972) Interplanetary objects in review: statistics of their masses and dynamics. *Icarus*, **17**, 1–48.

Dzurisin, D. and Blasius, K. R. (1975) Topography of the polar layered deposits of Mars. *J. Geophys. Res.*, **82**, 4225–48.

Evans, J. E. and Maunder, E. W. (1903) Experiments on the actuality of the 'Canals' of Mars. *Monthly Notices Roy. Astr. Soc.*, **63**, 498.

Fanale, F. P. (1976) Martian volatiles: Their degassing history and geochemical fate. *Icarus*, **28**, 179–202.

Finnerty, A. A. and Phillips, R. J. (1981) A petrologic model for an isostatically-compensated Tharsis region of Mars. Presented at the 3rd Mars Colloquium, Pasadena.

Finnerty, A. A., Phillips, R. J. and Banerdt, W. B. (1988) Igneous processes and closed system evolution of the Tharsis region of Mars. *J. Geophys. Res.*, **93**, 10225–35.

Firsoff, V. A. (1980) *The New Face of Mars*, Ian Henry, Hornchurch, Essex.

Flammarion, C. (1892) *La Planète Mars*, Vols. 1 and 2, Gauthier-Villars, Paris.

Frey, H. (1979) Thaumasia: A fossilized early forming Tharsis uplift. *J. Geophys. Res.*, **84**, 1909–1023.

Frey, H. and Schultz, R. A. (1989) Origin of the Martian crustal dichotomy. *LPI Tech. Report 89-04*, 35–7.

Frey, H., Semeniuk, J. A. and Grant, T. (1989) Early resurfacing events on Mars. MEVTV Workshop on Early Tectonic and Volcanic Evolution on Mars. *LPI Tech. Report 89-04*, LPI, Houston, 38–40.

Frey, H., Semeniuk, A. M., Semeniuk, J. A. and Tokarcik, S. (1988) *Proc. Lunar Planet. Sci. Conf.*, **18**, 679–99.

Gault, D. E. and Baldwin, B. S. (1970) Impact cratering on Mars – some effects of the atmosphere. *Eos*, **51**, 342.

Gault, D. E. and Greeley, R. (1978) Exploratory experiments of impact craters formed in viscous-liquid targets: Analogs for Martian rampart craters? *Icarus*, **34**, 486–95.

Glasstone, S. (1968) The book of Mars, *NASA SP-179*.

Greeley, R. and Crown, D. A. (1990) Volcanic geology of Tyrrhena Patera, Mars. *J. Geophys. Res.*, **95**, 7133–49.

Greeley, R. and Guest, J. E. (1987) Geologic map of the eastern hemisphere of Mars. *USGS Map I-1802B*.

Greeley, R. and Spudis, P. D. (1981) Volcanism on Mars. *Rev. Geophys. Space Phys.*, **19**, 13–41.

Greeley, R., Papson, R. and Veverka, J. (1978) Crater streaks in the Chryse Planitia region of Mars: Early Viking results. *Icarus*, **34**, 556–67.

Greeley, R., Storm, D. and Wilbur, C. (1976) Frequency distribution of lava tubes and channels on Mauna Loa volcano, Hawaii. *Geol. Soc. Amer. Bull. Abstr.*, **8**, 192.

Greeley, R., Leach, R., White, J. *et al.* (1980) Threshold wind speeds for sands on Mars: Wind tunnel simulations. *Geophys. Res. Lett.*, **7**, 121–4.

Harris, S. A. (1977) The aureole of Olympus Mons, Mars. *J. Geophys. Res.*, **83**, 3099–107.

Hartmann, W. K. (1973) Martian cratering 4: Mariner 9 initial analysis of cratering chronology. *J. Geophys. Res.*, **78**, 4096–116.

Hartmann, W. K. and Raper, O. (1974) The new Mars: The discoveries of Mariner 9, *NASA SP-337*, Washington DC.

Hartmann, W. K., Strom, R. G., Weidenschilling, S. J. *et al.* (1981) Chronology of planetary volcanism by comparative studies of planetary cratering, in *Basaltic Volcanism on the Terrestrial Planets (BVSP)*, Pergamon, New York.

Head, J. W. and Wilson, L. (1981) Theoretical analysis of martian explosive eruption mechanisms (abstract). *Lunar Planet Sci.*, **12**, 427–9, LPI, Houston.

Helfenstein, P. and Mouginis-Mark, P. J. (1980) Morphology and distribution of fractured terrain on Mars. *Lunar Planet. Sci.*, **XI**, 429–31.

Hess, S. L., Henry, R. M. and Tillman, J. E. (1979) The seasonal variation of atmospheric pressure on Mars as affected by the south polar cap. *J. Geophys. Res.*, **84**, 2923–7.

Hess, S. L., Henry, R. M., Leovy, C. B. *et al.* (1977) Meteorological results from the surface of Mars: Viking 1 and 2. *J. Geophys. Res.*, **82**, 4559–74.

Hess, S. L., Ryan, J. A., Tillman, J. E. *et al.* (1980) The annual cycle of pressure measured on Mars measured by Viking 1 and 2. *Geophys. Res. Lett.*, **7**, 197–200.

Hodges, C. A. and Moore, H. J. (1979) The subglacial birth of Olympus Mons and its aureole. *J. Geophys. Res.*, **84**, 8061–74.

Holloway, J. R. (1990) Martian magmas and mantle source regions: Current experimental and petrochemical constraints. Lecture delivered at 21st Lunar and Planet Science Conference, Houston.

Holloway, J. R. and Bertka, C. M. (1989) Chemical and physical properties of primary Martian magmas. MEVTV workshop on early tectonic and volcanic evolution of Mars. *LPI TM-89-04*, 43–5.

Howard, A. D. (1978) Origin of the stepped topography of the Martian poles. *Icarus*, **34**, 581–99.

Jakosky, B. M. and Carr, M. H. (1985) Possible precipitation of ice at low altitudes on Mars during periods of high obliquity. *Nature*, **315**, 559–61.

Jakosky, B. M. and Martin, T. Z. (1987) Mars: North polar atmospheric warming during dust storms. *Icarus*, **72**, 528.

Jeffreys, H. (1970) *The Earth* (5th edn), Cambridge University Press, London.

Johnston, D. H. and Toksoz, M. N. (1977) Internal structure and properties of Mars. *Icarus*, **32**, 73–84.

Jones, K. L. (1974) Evidence for an episode of Martian crater obliteration intermediate in Martian history. *J. Geophys. Res.*, **79**, 3917–32.

Jordan, J. F. and Lorell, J. (1973) Mariner 9, an instrument of dynamical science. Paper presented to AAS/AIAA Astrodynamics Conference, Vail, Colorado, July.

Kaula, W. M. (1979) The moment of inertia of Mars. *Geophys. Res. Lett.*, **6**, 194–6.

Kerridge, J. F. and Matthews, M. S. (1988) *Meteorites and the Early Solar System*, University of Arizona Press, Tucson.

Kiefer, W. S. and Hager, B. H. (1989) The role of mantle convection in the origin of the Tharsis and Elysium provinces of Mars. MEVTV workshop on the early tectonic and volcanic evolution of Mars. *LPI Technical Report 89-04*, 48–50.

Kieffer, H. H. and Palluconi, F. D. (1979) The climate of the Martian polar caps. *NASA Conf. Publ. 2072*, 45–6.

Kieffer, H. H., Martin, T. Z., Peterfreund, A. R. and Jakosky, B. M. (1977) Thermal and albedo mapping of Mars during the primary Viking mission. *J. Geophys. Res.*, **82**, 4249–91.

King, E. A. (1978) Geologic map of the Tyrrhena Patera quadrangle of Mars. *USGS Map I-1073*.

King, E. S. and Riehle, J. R. (1974) A proposed origin for the Olympus Mons escarpment. *Icarus*, **23**, 300–17.

Kliore, A. J., Fjeldbo, G., Seidel, B. L. and Rasool, I. (1969) Mariners 6 and 7: Occultation measurements of the atmosphere of Mars. *Science*, **166**, 1393–7.

Kovach, R. L. and Anderson, D. L. (1965) The interiors of the terrestrial planets. *J. Geophys. Res.*, **70**, 2873–82.

Kuiper, G. P. (1952) *The Atmospheres of the Earth and Planets*, Chicago University Press.

Lee, S. W., Thomas, P. C. and Veverka, J. (1982) Wind streaks in Tharsis and Elysium: Implications for sediment transport by slope winds. *J. Geophys. Res.*, **87**, 10025–41.

Leighton, R. B. and Murray, B. C. (1966) Behaviour of carbon dioxide and other volatiles on Mars. *Science*, **153**, 136.

Leighton, R. B., Horowitz, N. H., Murray, B. C. *et al.* (1969) Mariner 6 and 7 television pictures: Preliminary analysis. *Science*, **166**, 49–67.

Leovy, C. B. and Mintz, Y. (1969) Numerical simulation of the weather and climate of Mars. *J. Atmos. Sci.*, **26**, 1167–90.

Lopes, R. M. C., Guest, J. E. and Wilson, C. J. (1980) Origin of the Olympus Mons aureole and perimeter scarp. *Moon and Planets*, **22**, 221–34.

Lopes, R., Guest, J. E., Hiller, K. and Neukum, G. (1982) Further evidence for a mass movement origin of the Olympus Mons aureole. *J. Geophys. Res.*, **87**, 9917–28.

Lorrel, J., Born, G. H., Christensen, E. J. *et al.* (1972) Mariner 9 celestial mechanics experiment: Gravity field and pole direction of Mars. *Science*, **175**, 317–20.

Lowell, P. (1895) *Mars*, Macmillan, Boston.

Lowell, P. (1906) *Mars and its Canals*, Macmillan, New York.

Lowell, P. (1909) *Mars as an Abode of Life*, Macmillan, New York.

Lowell, P. (1910) Schiaparelli. *Pop. Astr.*, **18**, 466.

Lucchitta, B. K. (1978) A large landslide on Mars. *Bull. Geol. Soc. Amer.*, **89**, 1601–9.

Lucchitta, B. K. (1979) Landslides in Valles Marineris, Mars. *J. Geophys. Res.*, **84**, 8097–113.

Lucchitta, B. K. (1981) Mars and Earth: Comparison of cold-climate features. *Icarus*, **45**, 264–303.
Lucchitta, B. K. (1982) Ice sculpture in the martian outflow channels. *J. Geophys. Res.*, **87**, 9951–73.
Lucchitta, B. K. (1987) Recent mafic volcanism of Mars. *Science*, **235**, 565–7.
Lucchitta, B. K. and Klockenbrink, J. L. (1981) Ridges and scarps in the equatorial belt of Mars. *Moon and Planets*, **24**, 415–29.
Lucchitta, B. K., Ferguson, H. M. and Summers, C. (1986) Sedimentary deposits in the northern lowland plains. *J. Geophys. Res.*, **91**, E166–74.
Lucchitta, B. K., Clow, G. D., Croft, S. K. *et al.* (1989) Canyon systems on Mars (abstract). *Fourth Int. Conf. Mars*, Tucson, 36–7.
McCauley, J. F. (1978) Geologic map of the Coprates quadrangle of Mars. *USGS Map I-897*.
McCauley, J. F. Carr, M. H., Cotts, J. A. *et al.* (1972) Preliminary Mariner 9 report on the geology of Mars. *Icarus*, **45**, 77–86.
McElroy, M. B., Kong, T. Y. and Yung, Y. L. (1977) Photochemistry and evolution of Mars's atmosphere: A Viking perspective. *J. Geophys. Res.*, **82**, 4379–88.
McGetchin, T. R., Pepin, R. O. and Phillips, R. J. (eds.) (1981) Basaltic Volcanism on the terrestrial Planets, *Basaltic Volcanism Study Project*, Pergamon, New York, 1246–54.
McGill, G. E. (1985a) Age and origin of large Martian polygons (abstract). *Lunar Planet. Sci.*, **XVI**, 535–7.
McGill, G. E. (1985b) Age of deposition and fracturing, Elysium/Utopia Region, Northern Martian Plains. *Geol. Soc. Amer., Abstr. Programs*, **17**, 659.
McGill, G. E. (1987) *Proc. 18th Lunar Planet. Sci. Conf.*, 620–1.
McGill, G. E. (1989) The Martian crustal dichotomy. *LPI Tech. Report 89-04*, 59–61.
Malin, M. C. (1977) Comparison of volcanic features of Elysium (Mars) and Tibesti (Earth). *Bull. Geol. Soc. Amer.*, **84**, 908–19.
Masursky, H., Boyce, J. M., Dial, A. L. *et al.* (1977) Classification and time of formation of Martian channels based on Viking data. *J. Geophys. Res.*, **82**, 4016–38.
Maxwell, T. E. (1989) Structural mapping along the cratered terrain boundary, eastern hemisphere of Mars. MEVTV workshop on early structural and volcanic evolution of Mars. *LPI Tech. Report 89-04*, 54–5.
Meyer, J. D. and Grollier, M. J. (1977) Geologic map of the Syrtis Major quadrangle of Mars. *USGS Map I-995*.
Michaux, C. M. and Newburn, R. L. (1972) Mars Scientific Model. *JPL 606-1*, Pasadena.
Moore, H. G., Arthur, D. W. G. and Schaber, G. G. (1978) Yield strengths of flows on Earth, Moon and Mars. *Proc. Lunar Planet. Sci. Conf. IXth*, 3351–78.
Moore, P. (1956) *Guide to Mars*, Muller, London.
Moore, P. (1977) *Guide to Mars*, 6th edn, Lutterworth, London.
Moore, P. and Cross, C. A. (1974) *Mars*, Mitchell Beasley, London.
Morris, E. C. (1979) A pyroclastic origin for the aureole deposits of Olympus Mons. *NASA TM-82385*, 252–4.
Morris, E. C. (1981) Structure of Olympus Mons and its basal scarp. *Abstr. 3rd Intl Colloquium on Mars*, 161–2.
Mouginis-Mark, P. H. (1981) Late-stage summit activity of Martian shield volcanoes. *Proc. 12th Lunar Planet. Sci. Conf.*, 1431–47.

Mouginis-Mark, P. H., Wilson, L. and Head, J. W. (1982) Explosive volcanism on Hecates Tholus, Mars: Investigation of eruption conditions. *J. Geophys. Res.*, **87**, 9890–904.

Mouginis-Mark, P. H., Wilson, L. and Head, J. W. (1984) Elysium Planitia Mars: Regional geology, volcanology, and evidence for volcano-ground ice interactions. *Earth, Moon and Planets*, **30**, 149–73.

Mouginis-Mark, P. H., Wilson, L. and Head, J. R. (1988) Polygenetic eruptions on Alba Patera, Mars. *Bull. Volcanol.*, **50**, 361–79.

Mutch, P. and Woronow, A. (1980) Martian rampart and pedestal craters' ejecta-emplacement: Coprates triangle. *Icarus*, **41**, 259–68.

Mutch, T. A., Arvidson, R. E., Head, J. W. and Saunders, R. S. (1976) *The Geology of Mars*, Princeton University Press.

Neukum, G. and Hiller, K. (1981) Martian ages. *J. Geophys. Res.*, **86**, 3097–121.

Neukum, G. and Wise, D. U. (1976) A standard crater curve and possible new time scale. *Science*, **194**, 1381–7.

Nier, A. O., McElroy, M. B. and Yung, Y. L. (1976) Isotopic composition of the Martian atmosphere. *Science*, **194**, 68–70.

Nummedal, D. and Prior, D. B. (1981) Generation of Martian chaos and channels by debris flows. *Icarus*, **45**, 77–86.

Öpik, E. J. (1965) Mariner IV and craters on Mars. *Irish Astron. J.*, **7**, 92–104.

Öpik, E. J. (1966) The Martian surface. *Science*, **153**, 255–65.

Owen, T. (1966) The composition and surface pressure of the Martian atmosphere: Results from the 1965 opposition. *Astrophys. J.*, **146**, 257–70.

Owen, T., Biemann, K., Rushneck, D. R. *et al.* (1977) The composition of the atmosphere at the surface of Mars. *J. Geophys. Res.*, **82**, 4635–9.

Palluconi, F. D. and Kieffer, H. H. (1981) Mars: The thermal inertia of the surface. *Icarus*, **45**, 415–26.

Parker, T. J., Schneeberger, D. M., Pieri, D. C. and Saunders, R. S. (1986) *LPI Tech. Report 87-01*, 96–8.

Pechmann, J. C. (1980) The origin of the polygonal troughs in the northern plains of Mars. *Icarus*, **42**, 185–210.

Peterson, J. E. (1977) Geologic map of the Noachis quadrangle of Mars. *USGS Map I-910*.

Phillips, R. J. (1978) Report on the Tharsis worshop. *NASA TM-79719*, 334–6.

Phillips, R. J. (1988) The geophysical signal of Martian global dichotomy. *Trans. Amer. Geophys. Union*, **69**, 389.

Phillips, R. J. (1990) Geophysics on Mars: Issues and answers. Lecture delivered at the 21st Lunar and Planet Science Conference, Houston.

Phillips, R. J. and Banerdt, W. B. (1990) Permanent uplift in magmatic systems with application to the Tharsis region of Mars. *J. Geophys. Res.*, **95**, 5089–100.

Phillips, R. J. and Ivins, E. R. (1979) Geophysical observations pertaining to solid state convection in the terrestrial planets. *Phys. Earth Planet Int.*, **19**, 107–48.

Phillips, R. J. and Saunders, R. S. (1975) The isostatic state of Martian topography. *J. Geophys. Res.*, **80**, 2893–8.

Phillips, R. J., Saunders, R. S. and Conel, J. E. (1973) Mars: Crustal structure as inferred from Bougeur gravity anomalies. *J. Geophys. Res.*, **78**, 4815–20.

Pieri, D. (1980) Geomorphology of Martian valleys, in *Advances in Planetary Geology* (ed. A. Woronow), NASA, Washington, DC.

Pike, R. J. (1978) Volcanoes on the inner planets: Some preliminary comparisons of gross topography. *Proc. 9th Lunar Planet. Sci. Conf.*, 3239–73.

Pike, R. J. (1979) Simple to complex craters: The transition on Mars. *NASA TM-80339*, 132–4.

Pike, R. J. (1980a) Formation of complex impact craters: Evidence from Mars and other planets. *Icarus*, **43**, 1–19.

Pike, R. J. (1980b) Terrain dependence of crater morphology on Mars: Yes and no. *Lunar Planet. Sci.*, **XI**, 885–7.

Plaut, J. J., Kahn, R., Guiness, E. A. and Arvidson, R. E. (1988) Accumulation of sedimentary deposits in the south polar region of Mars and implications for climatic history. *Icarus*, **75**, 361–83.

Plescia, J. B. (1979) Tectonism of the Tharsis region. *NASA Tech. Memo. 80339*, 47–9.

Plescia, J. B. (1980) Cinder cones of Isidis and Elysium. *NASA Tech. Memo. 82385*, 263–5.

Plescia, J. B. and Saunders, R. S. (1979) The chronology of Martian volcanoes. *Proc. 10th Lunar Planet. Sci. Conf.*, 2841–59.

Plescia, J. B. and Saunders, R. S. (1982) Tectonic history of the Tharsis region, Mars. *J. Geophys. Res.*, **87**, 9775–91.

Pollack, J. B. and Black, D. C. (1979) Implications of the gas compositional measurements of Pioneer Venus for the origin of planetary atmospheres. *Science*, **207**, 56–9.

Pollack, J. B., Leovy, C. B., Mintz, Y. H. and Van Camp, W. (1976) Winds on Mars during the Viking season: Predictions based on a general circulation model with topography. *Geophys. Res. Lett.*, **3**, 479–82.

Pollack, J. B., Colburn, D. S., Klaser, M. *et al.* (1979) Properties and effects of dust particles suspended in the Martian atmosphere. *J. Geophys. Res.*, **84**, 2929–45.

Potter, D. B. (1976) Geologic map of the Hellas quadrangle of Mars. *USGS Map I-941*.

Reasenberg, R. (1977) The moment of inertia and isostacy of Mars. *J. Geophys. Res.*, **82**, 369–75.

Reimers, P. E. and Komar, P. D. (1979) Evidence for explosive volcanic density currents on certain Martian volcanoes. *Icarus*, **39**, 88–110.

Ringwood, A. E. (1966) Chemical evolution of the terrestrial planets. *Geochim. Cosmochim. Acta*, **30**, 41–104.

Ringwood, A. E. and Clark, S. P. (1971) Internal constitution of Mars. *Nature*, **234**, 89–92.

Ryan, J. A., Henry, R. M. and Hess, S. L. (1978) Mars meteorology: Three seasons at the surface. *Geophys. Res. Lett.*, **5**, 715–18.

Sagan, C., Veverka, J. and Fox, P. (1973) Variable features on Mars. 2. Mariner 9 global results. *J. Geophys. Res.*, **78**, 4163–96.

Schaber, G. C., Horstman, K. C. and Dial, A. L. (1978) Lava flow materials in the Tharsis region of Mars. *Proc. 9th Lunar Planet. Sci. Conf.*, 3433–58.

Schiaparelli, G. V. (1894) The planet Mars (Trans. W. H. Pickering). *Astronomy and Astrophysics*, **13**, 635–40 and 714–23.

Schultz, P. H. (1977) Lunar and Martian floor-fractured craters, in *Basaltic Volcanism 2nd Inter-team Meeting*, LPI, Houston, 53–5.

Schultz, P. H. and Lutz, A. B. (1988) Polar wandering on Mars. *Icarus*, **73**, 91.

Schultz, P. H., Schultz, R. A. and Rogers, J. (1982) The structure and evolution of ancient impact basins on Mars. *J. Geophys. Res.*, **87**, 9803–20.

Scott, D. H. (1969) 'The geology of the southern Pancake Range and Lunar

Crater Volcanic Field, Nye County, Nevada'. Unpublished PhD thesis, UCLA.

Scott, D. H. (1982) Volcanoes and volcanic provinces: western hemisphere of Mars. *J. Geophys. Res.*, **87**, 9839–51.

Scott, D. H. and Carr, M. H. (1978) Geologic map of Mars. *USGS Map I-1083*.

Scott, D. H. and Dohm, J. M. (1990) Chronology and global distribution of fault and ridge systems on Mars. *Proc. 20th Lunar Planet. Sci. Conf.*, LPI, Houston, 487–501.

Scott, D. H. and Tanaka, K. L. (1980) Mars Tharsis region: Volcano-tectonic events in stratigraphic record. *Proc. 11th Lunar Planet. Sci. Conf.*, 2403–21.

Scott, D. H. and Tanaka, K. L. (1982) Ignimbrites of Amazonis Planitia region of Mars. *J. Geophys. Res.*, **87**, 1179–90.

Scott, D. H. and Tanaka, K. L. (1986) Geologic map of the western hemisphere of Mars. *USGS Map I-1802A*.

Sharp, R. P. (1973) Mars: South polar pits and etched terrain. *J. Geophys. Res.*, **78**, 4222–30.

Sharp, R. P. and Malin, M. C. (1975) Channels on Mars. *Geol. Soc. Amer. Bull.*, **86**, 593–609.

Sheehan, W. (1988) *Planets and Perception*. Tucson, Arizona.

Shreve, R. L. (1966) Sherman Landslide, Alaska. *Science*, **154**, 1639–43.

Simkin, T. and Howard, K. A. (1970) Caldera collapse in the Galapagos Islands, 1968. *Science*, **169**, 429–37.

Sjogren, W. L. (1979) Mars gravity: High-resolution results from Viking orbiter 2. *Science*, **203**, 1006–9.

Sjogren, W. L., Wong, L. and Downs, W. (1975) Mars gravity field based on a short arc technique. *J. Geophys. Res.*, **80**, 2899–908.

Sleep, N. H. and Phillips, R. J. (1979) An isostatic model for the Tharsis province. *Geophys Res. Lett.*, **6**, 803–6.

Sleep, N. H. and Phillips, R. J. (1985) Gravity and lithospheric stress on the terrestrial planets with reference to the Tharsis region of Mars. *J. Geophys. Res.*, **90**, 4469–89.

Slipher, E. C. (1963) *Mars*, Sky Publishing Corporation, Cambridge, Massachusetts.

Soderblom, L. A. (1977) Historical variations in the density and distribution of impacting debris in the inner solar system: Evidence from planetary imaging, in *Impact and Explosion Cratering* (ed. D. J. Roddy *et al.*), Pergamon, New York, 240–1.

Soderblom, L. A., Condit, C. D. and West, R. A. (1974) Martian planetwide crater distributions: Implications for geologic history and surface processes. *Icarus*, **22**, 239–63.

Solomon, S. C. (1979) Formation, history and energetics of cores in terrestrial planets. *Phys. Earth Planet. Int.*, **19**, 168–82.

Solomon, S. C. and Chaiken, J. (1976) Thermal expansion and thermal stress in the Moon and terrestrial planets: Clues to early thermal history. *Proc. 7th Lunar Sci. Conf.*, 3229–43.

Solomon, S. C. and Head, J. W. (1982) Evolution of the Tharsis province of Mars: The importance of heterogeneous lithosphere thickness and volcanic construction. *J. Geophys. Res.*, **87**, 9755–74.

Solomon, S. C. and Head, J. W. (1990) Heterogeneities in the thickness of the elastic lithosphere of Mars: Constraints on heat flow and internal dynamics. *J. Geophys. Res.*, **95**, 11073–83.

Sparks, R. S. J. and Wilson, L. (1976) A model for the formation of ignimbrite by gravitational column collapse. *J. Geol. Soc. Lond.*, **132**, 441–51.

Sparks, R. S. J., Wilson, L. and Hulme, G. (1978) Theoretical modelling of the generation, movement and emplacement of pyroclastic flows by column collapse. *J. Geophys. Res.*, **83**, 1727–39.

Spencer, J. R., Fanale, F. P. and Tribble, J. E. (1989) Karst on Mars? Origin of closed depressions in Valles Marineris by solution of carbonates in groundwater sulfuric acid (abstract). *Fourth Intl Mars Conf.*, 193–4, University of Tucson, Arizona.

Spitzer, C. R. (ed.) (1980) Viking orbiter views of Mars, *NASA SP-441.*

Spudis, P. D. and Greeley, R. (1978) Volcanism in the cratered uplands of Mars. *Eos*, **58**, 1182.

Squyres, S. W. (1978) Martian fretted terrain: Flow of erosional debris. *Icarus*, **34**, 600–13.

Stearns, H. T. and MacDonald, G. A. (1946) Geology of the Hawaiian Islands. *Bull. Hawaiian Inst. Hydrogeol.*, **8**, 1–203.

Strughold, H. (1954) *The Green and Red Planet*, Sidgwick and Jackson, London.

Swanson, D. A. (1973) Pahoehoe flows from the 1969–1971 Mauna Ulu eruptions, Kilauea Volcano, Hawaii. *Geol. Soc. Amer. Bull.*, **84**, 615–26.

Tanaka, K. L. (1986) The stratigraphy of Mars. *Proc. 17th Lunar Planet. Sci. Conf., J. Geophys. Res.*, **91**, E139–58.

Tanaka, K. L. and Dohm, J. M. (1989) Volcanotectonic provinces of the Tharsis region of Mars: Identification, variations, and implications. MEVTV workshop on early tectonic and volcanic evolution of Mars. *LPI Tech. Report 89-04*, 79–81.

Tanaka, K. L. and Golombek, M. P. (1989) Martian tension fractures and the formation of grabens and collapse features at Valles Marineris. *Proc. 19th Lunar Planet. Sci. Conf.*, 383–96.

Tanaka, K. L. and Scott, D. H. (1987) Geologic maps of the polar regions of Mars. *USGS Map I-1802C*, Reston, Virginia.

Theilig, E. and Greeley, R. (1979) Plains and channels in the Lunae Planum-Chryse Planitia region of Mars. *J. Geophys. Res.*, **84**, 7994–8010.

Thomas, P. (1982) Present wind activity on Mars: Relation to large latitudinally zoned sediment deposits. *J. Geophys. Res.*, **87**, 9999–10008.

Thomas, P. and Veverka, J. (1979) Seasonal and secular variations of wind streaks on Mars: An analysis of Mariner 9 and Viking data. *J. Geophys. Res.*, **84**, 8131–46.

Underwood, J. R. and Trask, N. J. (1978) Geologic map of the Mare Acidalium region of Mars. *USGS Map I-1048*, Reston, Virginia.

Urey, H. (1950) Structure and chemical composition of Mars. *Phys. Rev.*, **80**, 295.

Urey, H. (1952) *The Planets.* Yale, New Haven.

Veverka, J., Geirasch, P. and Thomas, P. (1981) Wind streaks on Mars: meteorological control of occurrences and mode of formation. *Icarus*, **45**, 154–66.

Wallace, D. and Sagan, C. (1979) Evaporation of ice in planetary atmospheres: Ice-covered rivers on Mars. *Icarus*, **39**, 385–400.

Ward, A. W. (1974) Climatic variations on Mars. 1. Astronomical theory of insolation. *J. Geophys. Res.*, **84**, 7934–9.

Ward, A. W. (1979) Yardangs on Mars: Evidence of recent wind erosion. *J. Geophys. Res.*, **84**, 8147–66.

Webster, P. J. (1977) The low-altitude circulation of Mars. *Icarus*, **30**, 626–49.

Wells, R. A. (1979) *Geophysics of Mars*, Elsevier.

Wilhelms, D. E. and Baldwin, R. B. (1989a) The relevance of knobby terrain to the Martian dichotomy. *LPI Tech. Report 89-04*.

Wilhelms, D. E. and Baldwin, R. B. (1989b) The origin of igneous sills in shaping the Martian uplands. *Proc. 19th Lunar Planet. Sci. Conf.*, 355–65.

Wilhelms, D. E. and Squyres, S. W. (1984) The martian hemispheric dichotomy may be due to a giant impact. *Nature*, **309**, 138–40.

Willemann, R. J. and Turcotte, D. L. (1982) The role of lithospheric stress in the support of the Tharsis Rise. *J. Geophys. Res.*, **87**, 9793–801.

Williams, R. S. (1978) Geomorphic processes in Iceland and on Mars: A comparative appraisal from orbital images. *Geol. Soc. Amer. 91st Ann. Mtg, Abtracts with Programs*, 517.

Wise, D. U. (1979) Geologic map of the Arcadia quadrangle of Mars. *USGS Map I-1154*, Reston, Virginia.

Wise, D. U., Golombek, M. P. and McGill, G. E. (1979a) Tharsis province of Mars: Geologic sequence, geometry and a deformation mechanism. *Icarus*, **21**, 1–11.

Wise, D. U., Golombek, M. P. and McGill, G. E. (1979b) Tharsis province of Mars: Geologic sequence, geometry and a deformation mechanism. *Icarus*, **38**, 456–72.

Wood, C. A. (1980) New observations of the Martian basins. *Lunar Planet. Sci.*, **X**, 1271–2.

Wood, C. A. and Head, J. W. (1976) Comparison of impact basins on Mercury, Mars and the Moon. *Proc. 7th Lunar Sci. Conf.*, 3629–51.

Wood, C. A., Head, J. W. and Cintala, M. J. (1978) Interior morphology of fresh Martian craters: The effects of target characteristics. *Proc. 9th Lunar Planet. Sci. Conf.*, 3691–709.

Woronow, A. (1981) Preflow stresses in Martian rampart ejecta blankets: A means of estimating the water content. *Icarus*, **45**, 320–30.

Zimbelman, J. R. (1985) Estimates of rheologic properties for flows on the martian volcano Ascraeus Mons. *Proc. Lunar Planet. Sci. Conf. 16th*, D157–62.

INDEX

Figure numbers are in bold, table numbers are in italic.

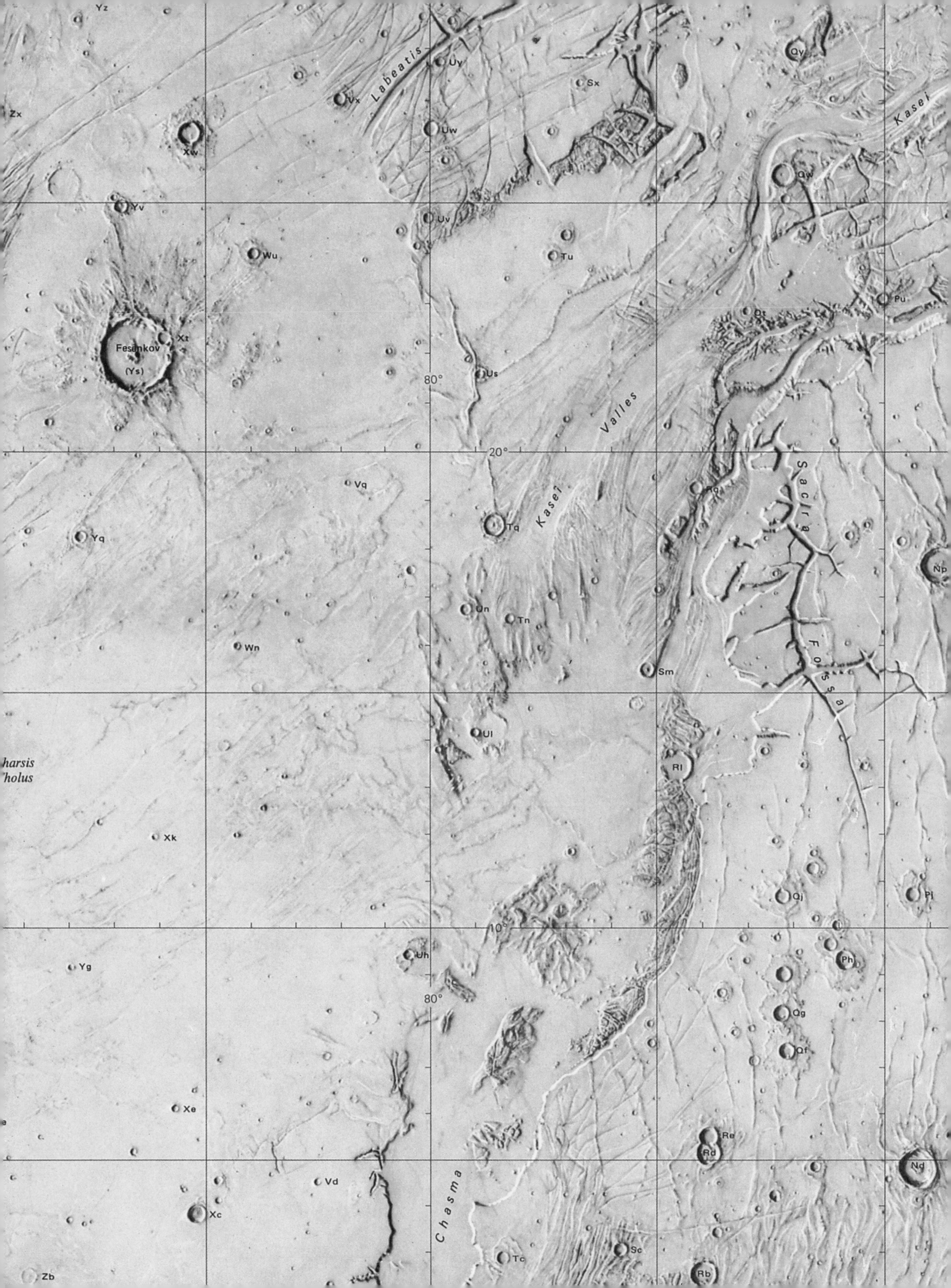

Yz
Zx
Xw
Labeatis
Vx
Uy
Uw
Sx
Qy
Kasei
Qw
Yv
Uv
Wu
Tu
Pu
Qt
Xt
Fesenkov
(Ys)
Us
80°
Valles
20°
Vq
Sacra
Rq
Yq
Tq
Kasei
Np
Un
Tn
Wn
Fossae
Sm
Ul
Rl
harsis
'holus
Xk
Qj
Pj
10°
Uh
Yg
Ph
80°
Qg
Qf
Xe
Re
Rd
Nd
Vd
Chasma
Xc
Tc
Sc
Zb
Rb